T0299720

CounterExamples

From Elementary Calculus to the Beginnings of Analysis

TEXTBOOKS in MATHEMATICS

Series Editors: Al Boggess and Ken Rosen

PUBLISHED TITLES

RISK ANALYSIS IN ENGINEERING AND ECONOMICS, SECOND EDITION
Bilal M. Ayyub

COUNTEREXAMPLES: FROM ELEMENTARY CALCULUS TO THE BEGINNINGS OF ANALYSIS
Andrei Bourchtein and Ludmila Bourchtein

INTRODUCTION TO THE CALCULUS OF VARIATIONS AND CONTROL WITH MODERN APPLICATIONS
John T. Burns

MIMETIC DISCRETIZATION METHODS
Jose E. Castillo

AN INTRODUCTION TO PARTIAL DIFFERENTIAL EQUATIONS WITH MATLAB®, SECOND EDITION
Mathew Coleman

RISK MANAGEMENT AND SIMULATION
Aparna Gupta

ABSTRACT ALGEBRA: AN INQUIRY-BASED APPROACH
Jonathan K. Hodge, Steven Schlicker, and Ted Sundstrom

QUADRACTIC IRRATIONALS: AN INTRODUCTION TO CLASSICAL NUMBER THEORY
Franz Holter-Koch

GROUP INVERSES OF M-MATRICES AND THEIR APPLICATIONS
Stephen J. Kirkland

AN INTRODUCTION TO NUMBER THEORY WITH CRYPTOGRAPHY
James Kraft and Larry Washington

REAL ANALYSIS AND FOUNDATIONS, THIRD EDITION
Steven G. Krantz

ELEMENTS OF ADVANCED MATHEMATICS, THIRD EDITION
Steven G. Krantz

APPLYING ANALYTICS: A PRACTICAL APPROACH
Evan S. Levine

ADVANCED LINEAR ALGEBRA
Nicholas Loehr

PUBLISHED TITLES CONTINUED

DIFFERENTIAL EQUATIONS WITH MATLAB®: EXPLORATION, APPLICATIONS, AND THEORY
Mark A. McKibben and Micah D. Webster

APPLICATIONS OF COMBINATORIAL MATRIX THEORY TO LAPLACIAN MATRICES OF GRAPHS
Jason J. Molitierno

ABSTRACT ALGEBRA: AN INTERACTIVE APPROACH
William Paulsen

ADVANCED CALCULUS: THEORY AND PRACTICE
John Srdjan Petrovic

COMPUTATIONS OF IMPROPER REIMANN INTEGRALS
Ioannis Roussos

TEXTBOOKS in MATHEMATICS

CounterExamples

From Elementary Calculus to the Beginnings of Analysis

Andrei Bourchtein
Ludmila Bourchtein
Pelotas State University, Brazil

CRC Press is an imprint of the
Taylor & Francis Group, an **informa** business
A CHAPMAN & HALL BOOK

CRC Press
Taylor & Francis Group
6000 Broken Sound Parkway NW, Suite 300
Boca Raton, FL 33487-2742

ISBN-13: 978-1-4822-4667-4 (hbk)

Visit the Taylor & Francis Web site at
http://www.taylorandfrancis.com

and the CRC Press Web site at
http://www.crcpress.com

To Haim and Maria with love and gratitude;
to Maxim with love and pride

Contents

Preface

In this manuscript we present counterexamples to different false statements, which frequently arise in the calculus and fundamentals of real analysis, and which may appear to be true at first glance. A counterexample is understood here in a broad sense as any example that is counter to some statement. The topics covered concern functions of real variables. The first part (chapters 1-6) is related to single-variable functions, starting with elementary properties of functions (partially studied even in college), passing through limits and continuity to differentiation and integration, and ending with numerical sequences and series. The second part (chapters 7-9) deals with function of two variables, involving limits and continuity, differentiation and integration.

One of the goals of this book is to provide an outlook of important concepts and theorems in calculus and analysis by using counterexamples. We restricted our exposition to the main definitions and theorems of calculus in order to explore different versions (wrong and correct) of the fundamental concepts and to see what happens a few steps outside of the traditional formulations. Hence, many interesting (but more specific and applied) problems not related directly to the basic notions and results are left out of the scope of this manuscript.

The selection and exposition of the material are directed, in the first place, to those calculus students who are interested in a deeper understanding and broader knowledge of the topics of calculus. We think the presented material may also be used by instructors that wish to go through the examples (or their variations) in class or assign them as homework or extra-curricular projects. In order to make the majority of the examples and solutions accessible to calculus students, we tried to avoid more advanced notions and results of real analysis. At the same time, we believe that some examples, despite having a simple formulation, are not so simple to solve, and can be of interest to students of real analysis as well.

It is assumed that a reader has knowledge of college algebra. For reading of the first two chapters, a familiarity with the university calculus courses is not required. The chapters 3-6 are accessible for students who have learned basic concepts of the first two semesters of calculus, and the chapters of the second part require some knowledge of functions of several variables. The logical sequence of the material just follows the chapter sequence, that is, for reading of a specific chapter one does not need any knowledge of the next chapter subjects.

This book is not appropriate as the main textbook for a course, but rather, it can serve as a supplement that can help students to master important concepts and theorems. So we think the best way to use this book is to read its parts while taking a respective calculus course: the chapters 2-4 correspond roughly to the topics of the first semester calculus, chapters 5-6 - to the second semester, and chapters 7-9 - to the third semester. On the other hand, the students already familiarized with the subjects of calculus sequence can find here deeper interpretation of the results and finer relation between concepts than in standard presentations. Also, more experienced students will better understand provided examples and ideas behind their construction.

To facilitate the reading of the main text (containing counterexamples) and make the text self-contained, and also to fix terminology, notation and concepts, we gather the relevant definitions and results in the introductory sections of each chapter. Additionally some basic notions of the theory of sets and functions used everywhere in the book are presented in the Introduction chapter. For many examples, we make explicit references to the concepts/theorems to which they are related.

A representative (but by no means exhaustive) bibliography can be found at the end of the book, including both collections of problems and textbooks in calculus/analysis. On the one hand, these references are the sources of different examples collected here, although it was out of our scope to trace all the original sources. On the other hand, they may be used for finding further information (examples and theory) on various topics. Some of these references are classic collections of the problems, such as that by Demidovich [3] and by Gelbaum and Olmstead [5]. Our preparation of this text was inspired in the first place by the latter book. We tried to extend its approach to a broader and less specialized range of topics in calculus and analysis by providing an alternative set of counterexamples, which require less training in mathematics and are accessible not only for mathematics majors, but also for natural science and engineering students. We hope that both mathematics and non-mathematics students will find our book as stimulating, challenging and useful as we found the above two books during our student days and as we continue to find it until now.

List of Figures

Introduction

0.1 Comments

0.1.1 On the structure of this book

This book consists of the Introductory chapter and nine chapters of counterexamples divided into two parts. In the first part (chapters 1-6), the subjects of single-variable functions, corresponding approximately to the first two terms of the standard university calculus sequence, are considered:
1) in chapter 1 - elementary properties of functions, such as elements of definition, periodicity, even/odd functions, monotonicity and extrema;
2) in chapter 2 - limits, with analysis of limit definition and properties;
3) in chapter 3 - continuity, with an emphasis on global properties;
4) in chapter 4 - differentiability, considering elementary properties, main theorems and applications;
5) in chapter 5 - different kinds of integrals (indefinite, definite and improper) with applications;
6) in chapter 6 - infinite sequences and series, including both basic and finer convergence/divergence tests.

The second part (chapters 7-9) contains the topics of the functions of two variables, which are quite representative for the most topics of the multivariable functions. All the chapters of the second part include a few examples similar to those presented in the first part in order to show how the ideas applied in one-dimensional case can be generalized/extended to many variables. However, the main part of the chapters 7-9 is devoted to examples that highlight a specificity of concepts and results for multi-variable functions. The subjects of the second part correspond to the third term of the university calculus courses:

1) in chapter 7 - limits and continuity, with details on relation between general, partial and iterated limits;
2) in chapter 8 - differentiation, including partial derivatives, differentiability, directional derivative and their applications;
3) in chapter 9 - double, iterated and line integrals with applications.

Each chapter is divided in sections corresponding to the (conditional) division of the material in main topics. Examples in each section usually start at a very basic level and attain certain level of complexity during the developing of exposition. At the end of each chapter supplementary exercises of different levels of complexity are provided, the most difficult of them with a hint to the solution.

The logical sequence of the material just follows the chapter sequence, which means that the reading of a specific chapter does not require any knowledge of the next chapter subjects. At the same time, different concepts, results and examples considered and analyzed in earlier chapters/sections are used in the subsequent sections.

Since the use of counterexamples in this book is aimed to analyze important concepts and theorems in calculus and analysis, the selection of the material is restricted by those examples that have direct connection with main concepts. Therefore, more specific and applied problems are left out of the scope of this manuscript. The interested reader can find these kinds of problems in different collections of problems on real analysis. Some well-known books of this type, containing problems of different levels of difficulty, are indicated in the bibliography: [2], [3], [7], [8], [9], [10].

The following structure is chosen for the presentation of each example:
1) False statement.
2) Solution (counterexample(s)). It provides a complete analytic solution to the posed problem.
3) Remark(s) (optional). It provides additional explanations, extensions of the counterexamples, comparisons with other similar situations, and also it gives links to a general theory and makes comparisons with the correct statements.
4) Figure(s) (optional, for some examples involving more geometric and complex constructions). It provides geometric representation of the analytic solutions and also it serves to develop a geometric intuition.

This form of presentation offers first a prompt for the reader to find out what is wrong with the proposed statement, substantiating a guess with analytic proof and looking for the corresponding correct results. If the reader wants to receive hints to the solution or corresponding results of a theory, he/she may proceed directly to items 2)-4) of each example.

In the choice of figures for illustration of the analytic solutions, we don't try to provide as many pictures as possible or to keep an equal distribution of pictures among chapters, sections, subsections, etc. Instead we follow the principle of choosing the figures that can be most helpful for understanding the presented counterexamples (at least from our point of view). In the first place, it is related to the nature of each counterexample (if it is more analytic or geometric) and to the complexity and singularity of the functions employed in solution.

All the false statements in this text are consistently placed within quotation marks.

0.1.2 On mathematical language and notation

The language in this manuscript, just as in the majority of mathematical books, is a mixture of a natural language with mathematical terminology and symbolism used traditionally for a concise expression of numerous concepts, results and logical reasoning. Depending on the topic, the use of mathematical symbols and terms can be more or less concentrated. In order to facilitate access to these symbols/terms inside the text, the list of notations and subject index are provided. Nevertheless, we find it useful to give below a few preliminary explanations of some common symbols used most frequently:
1) The symbol of equality $=$ has multiple (albeit close) meanings: it can mean equality between two objects (numbers, sets, functions, etc.), for example, $2 \cdot 9 = 3 \cdot 6$; it can be used to define a new object by means of other known ones, for example, $a^3 = a \cdot a \cdot a$; or it can be used for notation of an object, for instance, $\mathbb{R} =$ (the set of all real positive numbers)
2) The symbol $\to$ (or $\Rightarrow$) also has multiple meanings: it can mean a logical implication, as in "the number a is a multiple of 4" $\to$ "the number a is a multiple of 2"; or it can mean a tendency of a variable quantity, for example, $x \to a$ means that the variable x approaches the value a; it is also used to denote a transformation, for example, $f: X \to Y$ indicates that a function f maps a set X into Y.
3) The symbol $\{\ldots\}$ means collection of elements that form a set, for example, $\mathbb{N} = \{1, 2, 3, \ldots\}$ is the set of natural numbers, and $\mathbb{Z} = \{0, \pm 1, \pm 2, \ldots\}$ is the set of integers.
4) The symbol $\in$ ($\notin$) denotes that an element belongs (does not belong) to a set, for example, $-1 \in \mathbb{Z}$ ($-1 \notin \mathbb{N}$). Similarly, the symbol $\subset$ ($\not\subset$) denotes that one set is (is not) subset of another, for example, $\mathbb{Z} \subset \mathbb{R}$ ($\mathbb{Z} \not\subset \mathbb{N}$).
5) The symbol $\forall$ means, depending on context, "any", "every", "each", "for any", "for all", for example, $\forall x \in \mathbb{Z}$ can mean every element (or for all elements) of the set $\mathbb{Z}$, that is, every (or for all) integer(s).
6) The symbol $\exists$ ($\nexists$) means existence (non-existence), for example, $\exists x \in \mathbb{N}$ such that x is even or $\nexists x \in \mathbb{N}$ such that x is negative.

Throughout this work we try to follow the standard terminology and notation used in calculus/analysis books. To avoid any misunderstanding and ambiguity, the main concepts and results, along with the corresponding terminology and notation, are set out in the supplementary parts of the text containing background material.

0.2 Background (elements of theory)

The background material is split in several parts around the manuscript. The first part, the general background section presented below, contains some basic concepts and results from the set theory and function theory common for real-valued functions of one and several real variables and used in different sections of the manuscript. Other parts appear at the beginning of each chapter of counterexamples and bring the material relevant for the specific topics treated there. All the concepts and results presented can be found in almost every calculus and analysis book. In particular, one can consult the classical books on calculus and analysis indicated in the bibliography list, which treat the subjects with different levels of strictness, abstraction and generalization: [1], [4], [6], [11], [12], [13], [14].

0.2.1 Sets

General sets

Set. We consider a *set* to be a primary non-definable concept, which can be described as a collection of objects (elements) gathered by using some criterion. To distinguish between sets and their elements when they appear together we will use upper case for the former and lower case for the latter. If element a is in (belongs to) A we write $a \in A$; otherwise we write $a \notin A$.

The set which contains no element is called the *empty set* (usual notation $\varnothing$).

Subset. If every element of a set A is an element of a set B, we say that A is a *subset* of B (or A is contained in B) and write $A \subset B$.

If $A \subset B$ and $B \subset A$, then the two *sets coincide* and we write $A = B$. Otherwise $A \neq B$.

Operations on sets.

The *union* of the sets A and B (denoted $A \cup B$) is defined to be the set C such that $c \in C$ if $c \in A$ or $c \in B$.

The *intersection* of the sets A and B (denoted $A \cap B$) is defined to be the set C such that $c \in C$ if $c \in A$ and $c \in B$.

The *difference* of the sets A and B (denoted $A \backslash B$) is the set of all elements $a \in A$ such that $a \notin B$.

The *complement* of $B \subset A$ with respect to A is the set of all elements $a \in A$ such that $a \notin B$.

The *Cartesian product* $A \times B$ is the set of all the ordered pairs (a, b) where $a \in A$ and $b \in B$.

Basic properties : $A \cup B = B \cup A$, $A \cap B = B \cap A$, $(A \cup B) \cup C = A \cup (B \cup C)$, $(A \cap B) \cap C = A \cap (B \cap C)$, $A \cup (B \cap C) = (A \cup B) \cap (A \cup C)$, $A \cap (B \cup C) = (A \cap B) \cup (A \cap C)$, $A \cup \varnothing = A$, $A \cap \varnothing = \varnothing$.

Numerical sets

Natural numbers. The set $\mathbb{N}$ of the *natural numbers* is the set of all numbers used for counting: $\mathbb{N} = \{1, 2, 3, ...\}$.

Integers. The set $\mathbb{Z}$ of the *integers* is the set including all the natural numbers, zero and the negatives of the natural numbers: $\mathbb{Z} = \{0, \pm 1, \pm 2, \ldots\}$.

Rational numbers. The set $\mathbb{Q}$ of the *rational numbers* is composed of all the fractions $\frac{p}{q}$ such that $p \in \mathbb{Z}$, $q \in \mathbb{N}$.

Decimal fractions. A decimal fraction is the number represented in the form $(\pm)\, a_0, a_1 a_2 a_3 \ldots$, where $a_0 \in \mathbb{N} \cup \{0\}$ and the digits $a_i \in \mathbb{N} \cup \{0\}$, $0 \leq a_i \leq 9$, $i = 1, 2, 3, \ldots$. This is the decimal representation of every point on a coordinate line.

A decimal fraction is called finite if $a_i = 0$ for all $i \geq m$, where m is some natural number. A decimal fraction is called periodic if $a_{k+i} = a_{k+i+p}$ for some $k \in \mathbb{N} \cup \{0\}$, $p \in \mathbb{N}$, and all $i \in \mathbb{N}$, that is, there exists a finite string of digits (of length p), that is repeated infinitely many times starting from the $(k+1)$-th digit. Otherwise, a decimal fraction is called aperiodic.

Remark 1. A finite fraction can be considered as a special case of a periodic fraction.

Remark 2. The set of the rational numbers can be equivalently defined as the set of all decimal periodic fractions.

Irrational numbers. The set $\mathbb{I}$ of the *irrational numbers* is composed of all decimal aperiodic fractions.

Real numbers. The set $\mathbb{R}$ of the *real numbers* is the set of all decimal fractions.

Remark. Although $\mathbb{I}$ is not a common notation for the set of irrational numbers, we will use this symbol for brevity of notation (another, more common option is $\mathbb{R} \backslash \mathbb{Q}$).

Basic properties: $\mathbb{N} \subset \mathbb{Z} \subset \mathbb{Q} \subset \mathbb{R}$, $\mathbb{I} \subset \mathbb{R}$, $\mathbb{Q} \cup \mathbb{I} = \mathbb{R}$, $\mathbb{Q} \cap \mathbb{I} = \varnothing$.

Remark 1. Real numbers are also called real points (or simply points) due to existence of a one-to-one correspondence between the set $\mathbb{R}$ and the set of all points on a coordinate line called the real axis (see more about one-to-one correspondence in section 1.2)

Remark 2. Frequently it is convenient to consider the extended real axis, including positive $(+\infty)$ and negative $(-\infty)$ infinities. In this case all real (geometric) points are called the finite points, and infinities are called the infinite points.

Remark 3. The Cartesian product $\mathbb{R} \times \mathbb{R} = \mathbb{R}^2$ is the set of all the ordered pairs of real numbers. Due to existence of a one-to-one correspondence between the set $\mathbb{R}^2$ and the set of all points on a plane, with the above pairs being the Cartesian coordinates of the points, these elements of $\mathbb{R}^2$ are also called the points.

Sets in $\mathbb{R}^n$

The space $\mathbb{R}^n$. The *n-dimensional space* $\mathbb{R}^n$ can be considered as the set of all the ordered n-tuples $(x_1, \ldots, x_n)$, where $x_i \in \mathbb{R}$, $i = 1, \ldots, n$. It can be shown that there exists a one-to-one correspondence between $\mathbb{R}^n$ and the set of all the points x with the Cartesian coordinates $(x_1, \ldots, x_n)$ in the "geometric" n-dimensional space. In the case $n = 2$ and $n = 3$ it means one-to-one correspondence between all the ordered pairs and all the points in a plane, and between all the ordered triples and all the points in (three-dimensional) space, respectively. For this reason, n-tuples can be conveniently thought of as the points and the i-th element in n-tuple is called the i-th coordinate of the point $x = (x_1, \ldots, x_n)$.

The space $\mathbb{R}^n$ can be considered as the Cartesian product of n one-dimensional real spaces, $\mathbb{R} \times \mathbb{R} \times \ldots \times \mathbb{R} = \mathbb{R}^n$ that explains the used notation.

In the case $n = 1$, the used notation of the space is simply $\mathbb{R}$.

In the case $n = 1$, $n = 2$ and $n = 3$, for the sake of simplicity, the notations of the coordinates can be changed to (x) (or simply x), (x, y) and (x, y, z), respectively.

Distance between two points. The *distance* between two points $x = (x_1, \ldots, x_n)$ and $y = (y_1, \ldots, y_n)$ is defined by the formula $d(x, y) = \sqrt{\sum_{i=1}^{n} (x_i - y_i)^2}$. Another usual notation is $d(x, y) = |x - y|$. In the case $n = 1$, $n = 2$ and $n = 3$, it corresponds to the well-known Euclidean distance between two points on a line $d(x, y) = |x - y|$, where the vertical lines are the symbols of the absolute value, on a plane $d(x, y) = \sqrt{(x_1 - y_1)^2 + (x_2 - y_2)^2}$, and in a space $d(x, y) = \sqrt{(x_1 - y_1)^2 + (x_2 - y_2)^2 + (x_3 - y_3)^2}$.

The following inequality holds:

$$|x_i - y_i| \leq d(x, y) \leq \sqrt{n} \max_{1 \leq i \leq n} |x_i - y_i|,$$

which means, in particular, that the distance between x and y is small if, and only if, all the coordinates get close.

Sphere. The set of all points equidistant from a given point c is called a *sphere* , that is, for a given centerpoint c and radius $r > 0$, the sphere $S_{c,r}$ is defined by the analytical relation $S_{c,r} = \{ x \in \mathbb{R}^n : d(x, c) = r \}$.

Open ball. The set of all points located inside the sphere $S_{c,r}$ is called an *open ball* . The usual notation is $B_{c,r}$ and the analytical description is $B_{c,r} = \{ x \in \mathbb{R}^n : d(x, c) < r \}$. In the case $n = 1$, $n = 2$ and $n = 3$, the open ball is the open interval $(c - r, c + r)$, the open disk $(x_1 - c_1)^2 + (x_2 - c_2)^2 < r^2$, and the open ball (in the common geometrical sense of the word) $(x_1 - c_1)^2 + (x_2 - c_2)^2 + (x_3 - c_3)^2 < r^2$, respectively.

Closed ball. The set of all points located both inside and on the sphere $S_{C,r}$ is called a *closed ball* . The notation and analytical description is $\bar{B}_{c,r} = \{x \in \mathbb{R}^n : d(x, c) \le r\}$.

Neighborhood. The open ball $B_{c,\delta}$, $\delta > 0$ is called a *δ-neighborhood* of the point c. Frequently the term is shortened to neighborhood of c. The number δ is called the *radius* of the neighborhood.

Deleted neighborhood. The δ-neighborhood with the deleted central point, that is, the set $B_{c,\delta} \setminus \{c\}$, $\delta > 0$ is called the *deleted neighborhood* of c.

Limit point. A point c is a *limit point* of the set S if in any deleted neighborhood of c there exists at least one point of S. (Of course it implies that in any neighborhood of c there are infinitely many points of S.) Limit points are also often referred to as accumulation points.

Isolated point. If $c \in S$ and c is not a limit point of S, then c is called an *isolated point* of S.

Interior point. A point c is an *interior point* of S if there is a neighborhood of c contained in S.

Exterior point. A point c is an *exterior point* of S if there is a neighborhood of c which does not contain any point of S. (Equivalently, an exterior point of S is an interior point of the complement of S in $\mathbb{R}^n$.)

Boundary point. A point c is a *boundary point* of S if any neighborhood of c contains both points of S and out of S. (In other words, c is a boundary point of S if it is neither an interior nor an exterior point of S.)

Open set. A set S is *open* if every point of S is its interior point.

Closed set. A set S is *closed* if every limit point of S is a point of S.

Bounded set. A set S is *bounded* if there exists a ball that contains S. Otherwise, a set is *unbounded* .

Compact set. A set S is called *compact* if it is closed and bounded.

Closure of set. If L is the set of all limits points of a given set S, then $\bar{S} = S \cup L$ is called the *closure* of S.

Dense set. A set $S \subset D$ is dense in D if $\bar{S} = D$.

Separated sets. Two sets A and B are *separated* if $\bar{A} \cap B = \varnothing$ and $A \cap \bar{B} = \varnothing$.

Disconnected/connected set. A set S is *disconnected* if it can be represented as $S = A \cup B$, where A and B are nonempty separated sets. Otherwise a set is called *connected* .

Curves. The set of points $x = x(t)$, $t \in [a, b]$ is called a *curve* with endpoints $x(a)$ and $x(b)$. A curve is *continuous* if $x(t)$ is continuous on $[a, b]$. A continuous curve is *simple* if it does not intersect itself anywhere between its endpoints. A continuous curve is *closed* if its endpoints coincide.

Linearly connected set. A set S is *linearly connected* if any two of its points can be joined by a (continuous) curve which lies in S. If S is linearly connected then it is connected. For open sets the converse is also true.

Simply connected set. A connected set S is *simply connected* if every simple closed curve in S encloses only points that are in S.

General Remark. Any characterization of a set (such as connectedness, closeness, etc.) is related to the space where this set is considered. For example, an open disk is an open set in $\mathbb{R}^2$, while it is neither open nor closed set in $\mathbb{R}^3$.

Special sets in $\mathbb{R}$

Interval. The *interval* is a set of points x in $\mathbb{R}$, which satisfy one of the following inequality: $a < x < b$ (*open interval*), or $a \le x \le b$ (*closed interval*), or $a \le x < b$, or $a < x \le b$ (the last two intervals are called *half-open or half-closed*). The common notations are (a, b), $[a, b]$, $[a, b)$ and $(a, b]$, respectively.

Remark. In the case of a strong inequality, the point a can be $-\infty$ and the point b can be $+\infty$. In the former case, the formal inequality $-\infty < x < b$ (or $-\infty < x \le b$) just means $x < b$ (or $x \le b$), and similar interpretation is used for the latter case.

Neighborhood. The interval $(a - \delta, a + \delta)$, $\delta > 0$ is called a *δ-neighborhood* of the point a. Frequently the term is shortened to neighborhood of a. The number δ is called the *radius* of the neighborhood.

One-sided neighborhood. The *right-hand (left-hand)* neighborhood of a is the interval $(a, a + \delta)$ $((a - \delta, a))$.

Deleted neighborhood. The δ-neighborhood with the deleted central point, that is, the set $(a - \delta, a) \cup (a, a + \delta)$, $\delta > 0$ is called the *deleted neighborhood* of a .

Theorem. Any open set is the union of open intervals.

Bounded set. A set S is *bounded above (below)* if there exists a real number M (m) such that $a \le M$ $(a \ge m)$ for all $a \in S$. The number M (m) is called an upper (lower) bound of S. A set is *bounded* if it is bounded above and below. Otherwise, a set is *unbounded* .

Supremum/infimum. If a set S is bounded above, then there exists the least upper bound of S, which is called *supremum* and denoted $\sup S$. Similarly, if a set S is bounded below, then there exists the greatest lower bound of S, which is called *infimum* and denoted $\inf S$.

Bounded interval. An open interval (a, b) is *bounded* when a and b are finite points, that is a and b are some points on the real axis. If a or b are infinite points, the interval is *unbounded* . A closed interval $[a, b]$ is supposed to be always bounded.

Remark. Sometimes a bounded interval is called a finite interval and unbounded - infinite. Of course, the number of the points in any non-singular $(a < b)$ interval is always infinite. Usually, the specific meaning of the term "finite/infinite" as applied to intervals is clear in the used context.

Theorem. Any connected set in $\mathbb{R}$ is an interval.

0.2.2 Functions

Concepts

Function, domain, image, pre-image. A correspondence f between two sets X and Y such that with each element $x \in X$ is associated exactly one element $y \in Y$ is called a *function* from X to Y and is denoted by $y = f(x) : X \to Y$, or in short notation $f(x)$ or even f . The set X is called the *domain* of f (we also say f is defined on X) and the set Y is called the *codomain.* The part of Y that contains all values $f(x)$ for $x \in X$ is called the *image (or range)* of f and the usual notation is $f(X)$. Respectively, the *image of any set* $S \subset X$ is denoted by $f(S)$.

The set of all $x \in X$ such that $f(x) \in P$, $P \subset Y$ is called the *inverse image (or pre-image)* of P and the usual notation is $f^{-1}(P)$.

Remark 1. Here we consider only such functions that $X \subset \mathbb{R}^n$ and $Y \subset \mathbb{R}$.

Remark 2. Since a general codomain is of a little interest (it can always be chosen to be $\mathbb{R}$), when it is possible we will consider that in the definition of a function $y = f(x) : X \to Y$ the set Y is the image $f(X)$, that is, the smallest codomain.

Graph of a function. If f is defined on X, then the *graph* of f is the set of the points $(x, f(x)) \in \mathbb{R}^{n+1}$ such that $x \in X$.

Level surface. The surface $f(x) = k$, where the constant $k \in f(X)$, is the *level surface.*

Mapping onto. If a function $f(x)$ defined on X has image Y, that is $f(X) = Y$, then $f(x)$ maps X *onto* Y. Another common name is *surjective mapping (or surjective function).*

Mapping one-to-one. If for any pair $x_1, x_2 \in X$, $x_1 \neq x_2$ it follows that $f(x_1) \neq f(x_2)$, then $f(x)$ is called a *one-to-one mapping* on X. Another common name is *injective mapping (or injective function).*

One-to-one correspondence. If $f(x)$ is a one-to-one mapping of X

onto Y, then $f(x)$ is said to be a *one-to-one correspondence* between X and Y. Another common name is *bijective mapping (or bijective function).*

Remark. Frequently the term *one-to-one function* is used for one-to-one correspondence.

Composition of functions. Let $f(x)$ maps X into Y, and $g(y)$ maps Y in to Z. Then the *composition* of f and g is the function h with domain X and codomain Z defined by the formula $h(x) = g(f(x))$, $\forall x \in X$. The standard notation is $h(x) = g(f(x))$ or $h(x) = g \circ f(x)$.

Inverse function. Let $f(x)$ be a one-to-one function with domain X and image Y. In this case, it is possible to define the *inverse function* $f^{-1}(y)$, with domain Y and image X, that assigns to each $y \in Y$ the only element $x \in X$ such that $f(x) = y$.

Restriction of function. If $f(x)$ is defined on X and $S \subset X$, the *restriction* of $f(x)$ to S is the function $g(x)$ defined on S such that $g(x) = f(x)$, $\forall x \in S$.

Extension of function. If $f(x)$ is defined on X and $S \supset X$, the *extension* of $f(x)$ onto S is the function $g(x)$ defined on S such that $g(x) = f(x)$, $\forall x \in X$.

Equivalent sets. Two sets are called *equivalent* if there exists a one-to-one correspondence between them.

Finite/infinite sets. A set is called *finite* if it has a finite number of elements. Otherwise a set is *infinite.* The empty set is considered to be finite.

Countable/uncountable set. A set is *countable* if it is equivalent to $\mathbb{N}$. A set is *uncountable* if it is neither finite nor countable.

Elementary properties

Bounded function. A function $f(x)$ is *bounded above (below)* on a set S if there exists a real number M (m) such that $f(x) \leq M$ $(f(x) \geq m)$ for all $x \in S$. A function is *bounded* if it is bounded above and below. Otherwise, a function is *unbounded.*

Global maximum. A point x_0 is a *global maximum (strict global maximum)* point of $f(x)$ on a set S if for any $x \in S$, $x \neq x_0$ it follows that $f(x) \leq f(x_0)$ $(f(x) < f(x_0))$.

Global minimum. A point x_0 is a *global minimum (strict global minimum)* point of $f(x)$ on a set S if for any $x \in S$, $x \neq x_0$ it follows that $f(x) \geq f(x_0)$ $(f(x) > f(x_0))$.

Global extremum. A point x_0 is a *global extremum (strict global extremum)* if it is global maximum or global minimum (strict global maximum or strict global minimum).

Local maximum. A point x_0 is a *local maximum (strict local maximum)* point of $f(x)$ if there exists a neighborhood of x_0 such that for any $x \neq x_0$, which belongs to this neighborhood and to the function domain, it follows that $f(x) \leq f(x_0)$ ($f(x) < f(x_0)$).

Local minimum. A point x_0 is a *local minimum (strict local minimum)* point of $f(x)$ if there exists a neighborhood of x_0 such that for any $x \neq x_0$, which belongs to this neighborhood and to the function domain, it follows that $f(x) \geq f(x_0)$ ($f(x) > f(x_0)$).

Local extremum. A point x_0 is a *local extremum (strict local extremum)* if it is local maximum or local minimum (strict local maximum or strict local minimum).

Remark 1. There is an ambiguity in the terminology of extremum points, because the extremum point x_0, the extremum value $f(x_0)$ and the extremum pair $(x_0, f(x_0))$ all of them can be called the extremum point. However, usually it does not generate a misunderstanding regarding the use of this term within a specific context.

Remark 2. Sometimes a global extremum (minimum/maximum) is also called an absolute extremum (minimum/maximum), and a local extremum (minimum/maximum) - a relative extremum (minimum/maximum).

Part I

Functions of one real variable

Chapter 1

Elementary properties of functions

1.1 Elements of theory

Concepts

Function, domain, image, pre-image. A correspondence f between two sets X and Y such that with each element $x \in X$ is associated exactly one element $y \in Y$ is called a *function* from X to Y and is denoted by $y = f(x) : X \to Y$, or in short notation $f(x)$ or even f . The set X is called the *domain* of f (we also say f is defined on X) and the set Y is called the *codomain.* The part of Y that contains all values $f(x)$ for $x \in X$ is called the *image (or range)* of f and the usual notation is $f(X)$. Respectively, the *image of any set* $S \subset X$ is denoted by $f(S)$.

The set of all $x \in X$ such that $f(x) \in P$, $P \subset Y$ is called the *inverse image (or pre-image)* of P and the usual notation is $f^{-1}(P)$.

Remark. In chapters 1-6 we consider only such functions that $X \subset \mathbb{R}$ and $Y \subset \mathbb{R}$.

Remark. Since a general codomain is of a little interest (it can always be chosen to be $\mathbb{R}$), when it is possible we will consider that in the definition of a function $y = f(x) : X \to Y$ the set Y is the image $f(X)$, that is, the smallest codomain.

Graph of a function. If f is defined on X, then the *graph* of f is the set of all the ordered pairs $(x, f(x)) \in \mathbb{R}^2$ such that $x \in X$.

Mapping onto. If a function $f(x)$ defined on X has image Y, that is $f(X) = Y$, then $f(x)$ maps X *onto* Y. Another common name is *surjective mapping (or surjective function).*

Mapping one-to-one. If for any pair $x_1, x_2 \in X$, $x_1 \neq x_2$ it follows that $f(x_1) \neq f(x_2)$, then $f(x)$ is called a *one-to-one mapping* on X. Another common name is *injective mapping (or injective function).*

One-to-one correspondence. If $f(x)$ is a one-to-one mapping of X

onto Y, then $f(x)$ is said to be a *one-to-one correspondence* between X and Y. Another common name is *bijective mapping (or bijective function).*

Remark. Frequently the term *one-to-one function* is used for one-to-one correspondence.

Composition of functions. Let $f(x)$ maps X into Y, and $g(y)$ maps Y into Z. Then the *composition* of f and g is the function h with domain X and codomain Z defined by the formula $h(x) = g(f(x))$, $\forall x \in X$. The standard notation is $h(x) = g(f(x))$ or $h(x) = g \circ f(x)$.

Inverse function. Let $f(x)$ be a one-to-one function with domain X and image Y. In this case, it is possible to define the *inverse function* $f^{-1}(y)$, with domain Y and image X, that assigns to each $y \in Y$ the only element $x \in X$ such that $f(x) = y$.

Equivalent sets. Two sets are called *equivalent* if there exists a one-to-one correspondence between them.

Finite/infinite sets. A set is called *finite* if it has a finite number of elements. Otherwise a set is *infinite*. The empty set is considered to be finite.

Countable/uncountable set. A set is *countable* if it is equivalent to $\mathbb{N}$. A set is *uncountable* if it is neither finite nor countable .

Elementary properties

Bounded function. A function $f(x)$ is *bounded above (below)* on a set S if there exists a real number M (m) such that $f(x) \leq M$ $(f(x) \geq m)$ for all $x \in S$. A function is *bounded* if it is bounded above and below. Otherwise, a function is *unbounded*.

Even function. A function $f(x)$ defined on X is called *even* if for any $x \in X$ it holds that $f(-x) = f(x)$.

Odd function. A function $f(x)$ defined on X is called *odd* if for any $x \in X$ it holds that $f(-x) = -f(x)$.

Periodic function. A function $f(x)$ defined on X is called *periodic* with *period* $T \neq 0$ if for any $x \in X$ it holds that $f(x+T) = f(x)$. The smallest positive number T (if it exists) for which this property holds is called the fundamental period.

The primarily (geometric) properties of even functions (follow directly from the definition):
1) the domain of an even function is symmetric with respect to the origin;
2) the graph of an even function is symmetric with respect to the x-axis.

The primarily properties of odd functions (follow directly from the definition):
1) the domain of an odd function is symmetric with respect to the origin;
2) the graph of an odd function is symmetric with respect to the origin.

The primarily properties of periodic functions (follow directly from the definition):
1) the domain of a periodic function is not bounded either at the left or at the right;
2) the graph of a periodic function can be obtained by an infinite many translations of its part defined on an interval of length T.

Increasing function. A function $f(x)$ is called *increasing (strictly increasing)* on a set S if for any $x_1, x_2 \in S$, $x_1 < x_2$ it follows that $f(x_1) \leq f(x_2)$ $(f(x_1) < f(x_2))$.

Decreasing function. A function $f(x)$ is called *decreasing (strictly decreasing)* on a set S if for any $x_1, x_2 \in S$, $x_1 < x_2$ it follows that $f(x_1) \geq f(x_2)$ $(f(x_1) > f(x_2))$.

Monotone function. A function $f(x)$ is called *monotone (strictly monotone)* on a set S if it is increasing or decreasing (strictly increasing or strictly decreasing) on this set.

Global maximum. A point x_0 is a *global maximum (strict global maximum)* point of $f(x)$ on a set S if for any $x \in S$, $x \neq x_0$ it follows that $f(x) \leq f(x_0)$ $(f(x) < f(x_0))$.

Global minimum. A point x_0 is a *global minimum (strict global minimum)* point of $f(x)$ on a set S if for any $x \in S$, $x \neq x_0$ it follows that $f(x) \geq f(x_0)$ $(f(x) > f(x_0))$.

Global extremum. A point x_0 is a *global extremum (strict global extremum)* if it is global maximum or global minimum (strict global maximum or strict global minimum).

Local maximum. A point x_0 is a *local maximum (strict local maximum)* point of $f(x)$ if there exists a neighborhood of x_0 such that for any $x \neq x_0$, which belongs to this neighborhood and to the function domain, it follows that $f(x) \leq f(x_0)$ $(f(x) < f(x_0))$.

Local minimum. A point x_0 is a *local minimum (strict local minimum)* point of $f(x)$ if there exists a neighborhood of x_0 such that for any $x \neq x_0$, which belongs to this neighborhood and to the function domain, it follows that $f(x) \geq f(x_0)$ $(f(x) > f(x_0))$.

Local extremum. A point x_0 is a *local extremum (strict local extremum)* if it is local maximum or local minimum (strict local maximum or strict local minimum).

Remark. There is an ambiguity in the terminology of extremum points, because the extremum point x_0, the extremum value $f(x_0)$ and the extremum pair $(x_0, f(x_0))$ all of them can be called the extremum point. However, usually it does not generate a misunderstanding regarding the use of this term within a specific context.

1.2 Function definition

Example 1. "If A and B are two subsets of the domain X of $f(x)$, then $f(A \cap B) = f(A) \cap f(B)$."

Solution.

Consider the function $f(x) = \sin x$ and choose $A = [0, 2\pi]$ e $B = [\pi, 3\pi]$. Then $f(A \cap B) = f([\pi, 2\pi]) = [-1, 0]$, while $f(A) \cap f(B) = [-1, 1] \cap [-1, 1] = [-1, 1]$.

Remark 1. The following relation is true: $f(A \cap B) \subset f(A) \cap f(B)$.

Remark 2. For the union the statement is true, that is, $f(A \cup B) = f(A) \cup f(B)$.

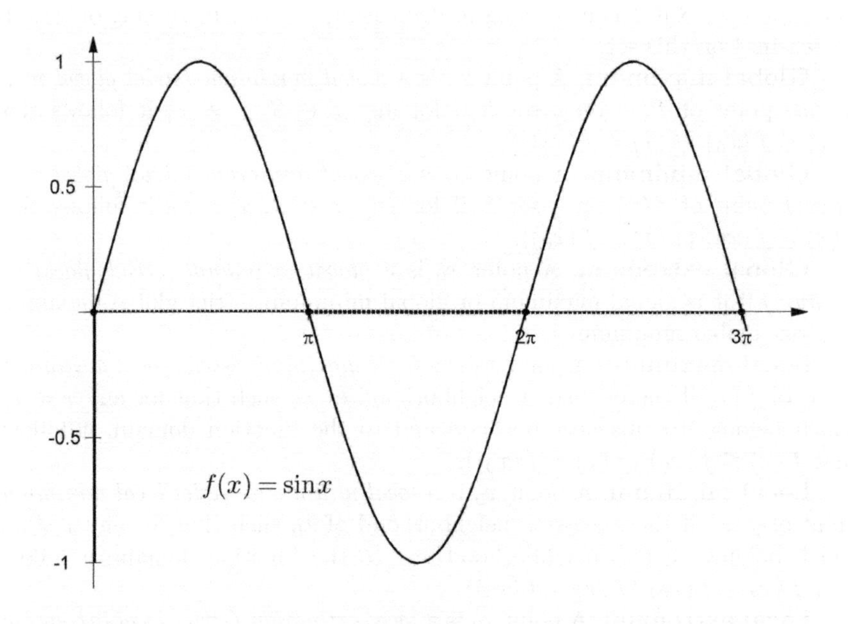

FIGURE 1.2.1: Example 1

Example 2. "If A is a subset of the domain X of $f(x)$, then $f(X \backslash A) = f(X) \backslash f(A)$."

Solution.

Consider the function $f(x) = x^2$ and choose $A = [0, +\infty)$. Then $f(\mathbb{R} \backslash A) = f((-\infty, 0)) = (0, +\infty)$, while $f(\mathbb{R}) \backslash f(A) = [0, +\infty) \backslash [0, +\infty) = \varnothing$.

Remark 1. The following relation is true: $f(X \backslash A) \supset f(X) \backslash f(A)$.

Remark 2. The statement is true for one-to-one functions.

Remark to Examples 1 and 2. The first two examples of this section show that some properties of domains cannot be extended to their images.

Example 3. "If $f(x)$ maps X onto X, then $f(x)$ is one-to-one on X."
Solution.
Consider the function $f(x) = \cos 2\pi x$ and choose $X = [-1, 1]$. Then $f(X) = [-1, 1] = X$, but $f(x)$ is not one-to-one (it gets the same values in the points x and $-x$ for $\forall x \in [0, 1]$, since $f(x)$ is even).

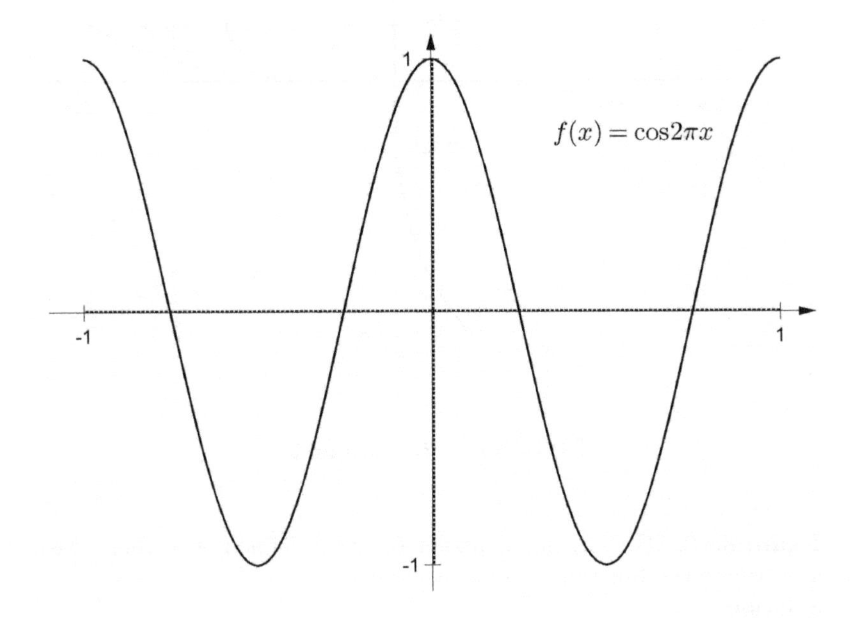

FIGURE 1.2.2: Example 3

Example 4. "If $f(x)$ maps X onto X and is one-to-one, then $f(x) = x$ or $f(x) = -x$."
Solution.
The function $f(x) = x^3$ transforms $X = [-1, 1]$ onto $[-1, 1]$ and is one-to-one correspondence.

Example 5. "If $f(x)$ maps X onto $Y \subset X$, $Y \neq X$, then there is no inverse function to $f(x)$ on Y."
Solution.
The function $f(x) = x^3$ transforms $X = \left[-\frac{1}{2}, \frac{1}{2}\right]$ onto $Y = \left[-\frac{1}{8}, \frac{1}{8}\right] \subset X$, but the inverse function $f^{-1}(y) = \sqrt[3]{y}$ is well defined on Y.

Remark. This example shows that the existence of an inverse does not related to the "sizes" of the domain and image.

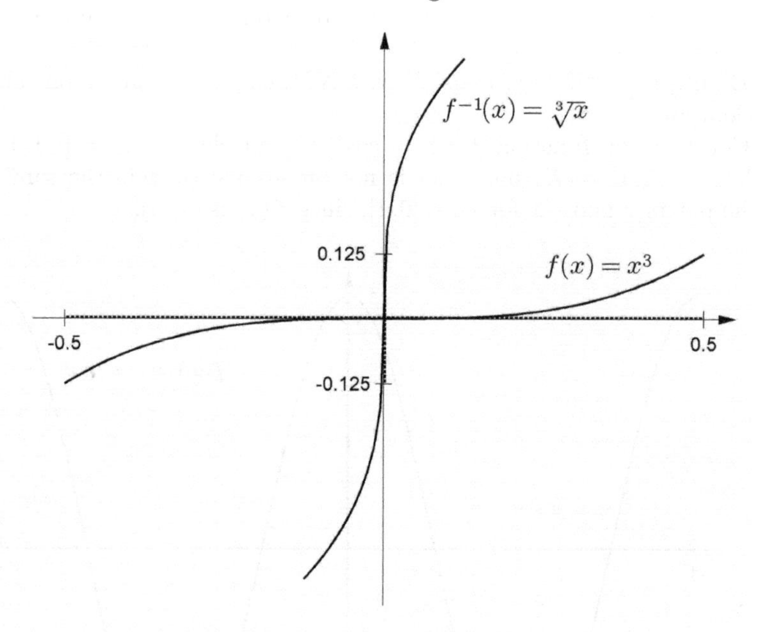

FIGURE 1.2.3: Example 5

Example 6. "For any non-constant function defined on $\mathbb{R}$ there exists an interval where the function admits the inverse."

Solution.

Dirichlet's function $D(x) = \begin{cases} 1, & x \in \mathbb{Q} \\ 0, & x \in \mathbb{I} \end{cases}$ is defined on $\mathbb{R}$, but it has repeated values in any interval.

Remark. It is seen that the existence of an inverse is not guaranteed even on a small (but still representative) part of domain.

Example 7. "If $f(x)$ is injective on A and B, then $f(x)$ is injective on $A \cup B$."

Solution.

Let $f(x) = x$ on $A = [0, 1]$ and $f(x) = x - 2$ on $B = [2, 3]$. Then $f(x)$ is injective separately on A and B, but it is not injective on $A \cup B$, because $f(A) = f(B)$.

Another kind of counterexample is $f(x) = x - [x]$ on $A = [0, 1)$ and $B = [1, 2)$. In this case A and B are not separated and $A \cup B = [0, 2)$ is connected set. However, the result is the same: $f(A) = f(B)$.

One more example is $f(x) = x^2$ on $A = [-1, 0]$ and $B = [0, 1]$. In this case

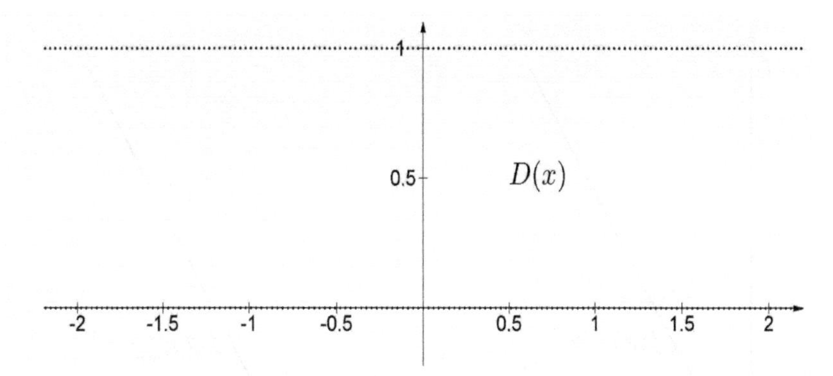

FIGURE 1.2.4: Example 6

neither the union $A \cup B$ nor the function $f(x)$ have "jumps": more strictly, the set $A \cup B$ is connected and $f(x)$ is continuous (continuous functions will be considered in chapter 3). However, again $f(A) = f(B)$, which means that $f(x)$ is not injective on $A \cup B$.

Remark 1. The following strengthened version of the statement is still false: "if $f(x)$ is injective on a neighborhood of each point of its domain X, then $f(x)$ is injective on X". The corresponding counterexample can be $f(x) = |x|$ or $f(x) = x^2$ considered on $X = \mathbb{R} \setminus \{0\}$.

Remark 2. The statement in Remark 1 (but not the original one) becomes true if $f(x)$ is continuous on a connected domain X (continuous functions will be considered in chapter 3).

Remark 3. Similar statement for the intersection of sets is true: if $f(x)$ is injective on A and B (and $A \cap B \neq \varnothing$), then $f(x)$ is injective on $A \cap B$.

Example 8. "If $f(x)$ is surjective on a pair $A \to C$ and on $B \to D$, then $f(x)$ is surjective on the pair $A \cap B \to C \cap D$."

Solution.

Let $f(x) = |x|$: $A = [0,1] \to C = [0,1]$ and $B = [-1,0] \to D = [0,1]$. Evidently, the conditions of the statement hold, but not the conclusion: $f(x) = |x|$: $A \cap B = \{0\} \to C \cap D = [0,1]$ is not surjective.

Remark. Similar statement for the union of sets is true: if $f(x)$ is surjective on a pair $A \to C$ and on $B \to D$, then $f(x)$ is surjective on the pair $A \cup B \to C \cup D$.

Remark to Examples 7 and 8. These two examples show that the injective property does not extend on the union of domains (although it does extend on the intersection) and the surjective property does not extend on the intersection of domains (while it does extend to the union).

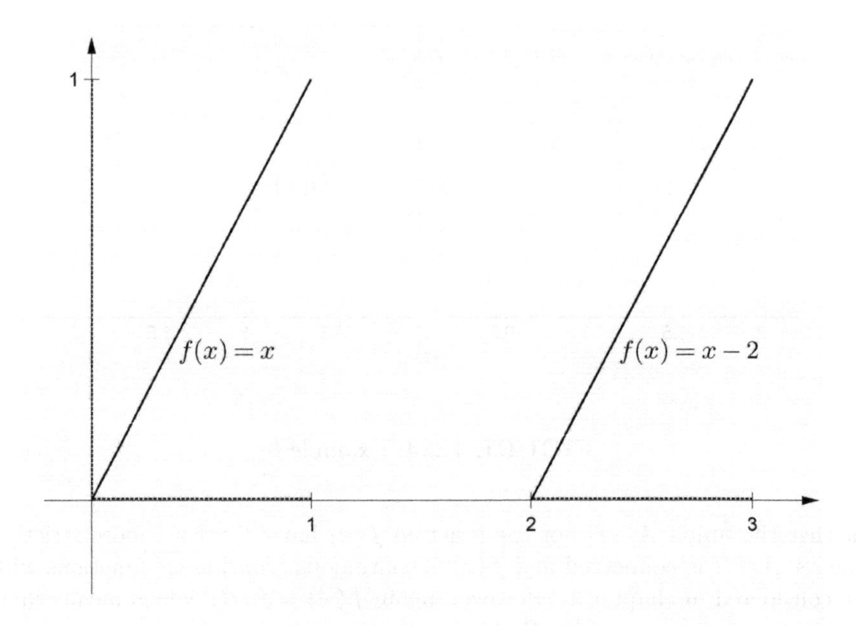

FIGURE 1.2.5: Example 7

Example 9. "If $f(x) : X \to Y$ is surjective and $g(y) : Y \to Z$ is injective, then $g(f(x)) : X \to Z$ is surjective."

Solution.

Let $f(x) = |x|$: $X = [-1, 1] \to Y = [0, 1]$, and $g(y) = y : Y = [0, 1] \to Z = [-1, 1]$. Evidently, the conditions of the statement hold, but the composition $g(f(x)) = |x| : X = [-1, 1] \to Z = [-1, 1]$ is not surjective, because there is no element in $X = [-1, 1]$ that corresponds to any negative number in Z.

Remark 1. The same counterexample can be used for the false statement: "if $f(x) : X \to Y$ is surjective and $g(y) : Y \to Z$ is injective, then $g(f(x)) : X \to Z$ is injective".

Remark 2. The change of the order of the functions does not change the situation, more precisely, the following statement is also false: "if $f(x) : X \to Y$ is injective and $g(y) : Y \to Z$ is surjective, then $g(f(x)) : X \to Z$ is injective or surjective". Consider the injective function $f(x) = x : X = [-1, 1] \to Y = [-2, 2]$, and the surjective $g(y) = y^2 : Y = [-2, 2] \to Z = [0, 4]$. Their composition $g(f(x)) = x^2 : X = [-1, 1] \to Z = [0, 4]$ is not injective neither surjective, because $g(f([-1, 0])) = g(f([0, 1])) = [0, 1]$.

Remark 3. If both $f(x)$ and $g(x)$ are injective then their composition is also injective; and if $f(x)$ and $g(x)$ are surjective then their composition is also surjective.

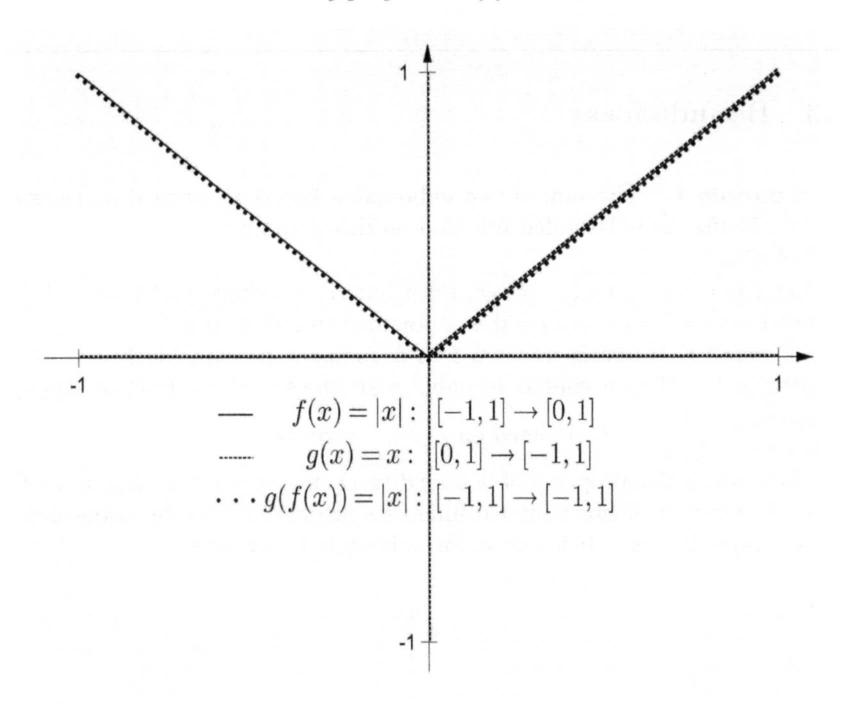

FIGURE 1.2.6: Example 9

Example 10. "If $g(f(x)) : X \to Z$ is bijective composition of $f(x) : X \to Y$ and $g(y) : Y \to Z$, then both $f(x)$ and $g(y)$ are bijective."

Solution.

Let $f(x) = x$: $X = \{0\} \to Y = \mathbb{R}$, and $g(y) = 0 : Y = \mathbb{R} \to Z = \{0\}$. With this choice both functions are not bijective: $f(x)$ is not surjective (although it is injective) and $g(y)$ is not injective (although it is surjective). However, their composition is a bijective function: $g(f(x)) = 0 : X = \{0\} \to Z = \{0\}$.

Another simple counterexample is non-surjective $f(x) = x$: $X = \left[0, \frac{\pi}{2}\right] \to Y = [0, \pi]$, followed by non-injective $g(y) = \sin y : Y = [0, \pi] \to Z = [0, 1]$. However, their composition is a bijective function: $g(f(x)) = \sin x : X = \left[0, \frac{\pi}{2}\right] \to Z = [0, 1]$.

Remark. If the composition $g(f(x))$ is bijective, it implies that $f(x)$ is injective and $g(y)$ is surjective.

1.3 Boundedness

Example 1. "The sum of two unbounded functions defined on the same domain is also an unbounded function on this domain."

Solution.

Let $f(x) = x$ and $g(x) = -x$. Both functions defined and unbounded on $\mathbb{R}$, but $h(x) = f(x) + g(x) = 0$ is a bounded function on $\mathbb{R}$.

Remark 1. If one is interested in a compact domain (such as a closed interval $[a, b]$), then a similar example with the functions $f(x) = -g(x) = \begin{cases} \tan x\,, \ x \in \left(-\frac{\pi}{2}, \frac{\pi}{2}\right) \\ 0\,, \ x = -\frac{\pi}{2}, \frac{\pi}{2} \end{cases}$ defined on $\left[-\frac{\pi}{2}, \frac{\pi}{2}\right]$ works.

Remark 2. The property of boundedness is maintained by the sum, difference and product of two bounded functions (at least for the functions defined on the same domain). It is not so for unbounded functions.

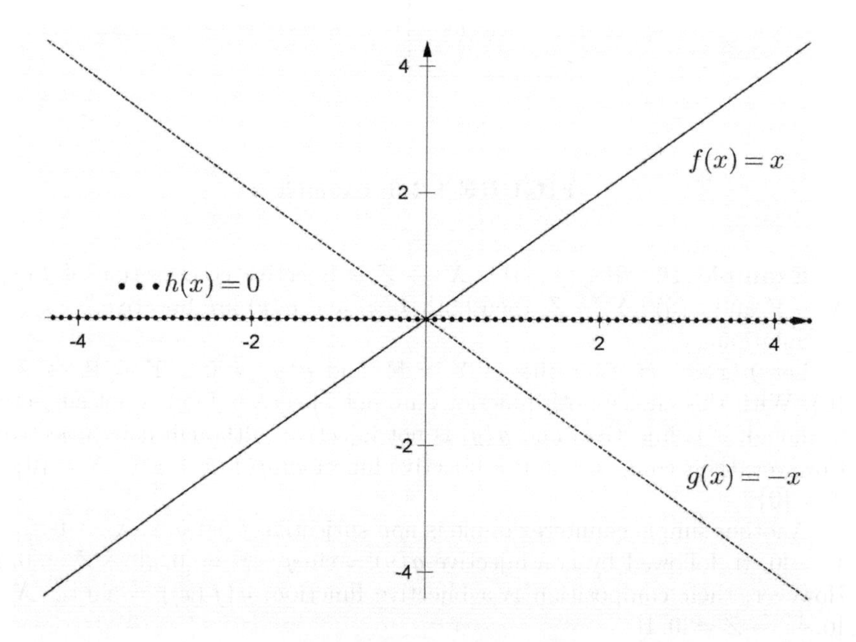

FIGURE 1.3.1: Example 1

Example 2. "A bounded function always attains its global extremum."

Solution.

The function $f(x) = x$ considered on $(0, 1)$ is bounded (for example, with lower bound 0 and upper bound 1), but it does not achieve either its global minimum or global maximum. Indeed, since $f(x)$ assumes positive values, its global maximum, if it existed, would be positive. Since all the values of $f(x)$ are less than 1, any number in $[1, +\infty)$ cannot be its maximum. On the other hand, for any $x_0 \in (0,1)$ the value of the function $f(x_1) = x_1$ at the point $x_1 = \frac{x_0+1}{2} \in (0,1)$ is larger than $f(x_0) = x_0$, implying that any $x_0 \in (0,1)$ cannot be a maximum point. Since we exhausted all options for a global maximum, it remains to conclude that $f(x) = x$ does not attain its maximum on $(0,1)$. A similar reasoning leads to the conclusion about the absence of a global minimum.

Remark 1. Analogous result is true for a compact domain. For instance, in the case of the closed interval $[0,1]$, a slight modification of the above counterexample $f(x) = \begin{cases} x, & x \in (0,1) \\ \frac{1}{2}, & x = 0, x = 1 \end{cases}$ shows a bounded function without global extrema.

Remark 2. In both examples (for open and closed intervals) the chosen bounded functions have no local extremum also.

Example 3. "If a function is bounded in some neighborhood of each point of (a,b), then it is bounded on (a,b)."

Solution.

Consider $f(x) = \tan x$ on $\left(-\frac{\pi}{2}, \frac{\pi}{2}\right)$. For each point $x_0 \in \left(-\frac{\pi}{2}, \frac{\pi}{2}\right)$ there is a neighborhood where $f(x)$ is bounded: for example, in the neighborhood (x_1, x_2), where $x_1 = \frac{x_0 - \pi/2}{2}$, $x_2 = \frac{x_0 + \pi/2}{2}$ all the values of $f(x)$ lie inside the interval $(f(x_1), f(x_2))$, due to strict increasing of $\tan x$ on $\left(-\frac{\pi}{2}, \frac{\pi}{2}\right)$, i.e. $f(x)$ is bounded on (x_1, x_2). However, $\tan x$ is unbounded on $\left(-\frac{\pi}{2}, \frac{\pi}{2}\right)$.

Remark 1. If an open interval (a,b) is substituted by closed $[a,b]$, then the statement will be true.

Remark 2. The boundedness is a "global" property, that is, the property defined on a given set, unlike a "local" property defined in a neighborhood of each point. For "local" properties their extension to the entire set is frequently defined as a fulfillment of "local" property at each point of a set. It is not so for boundedness: local boundedness does not imply global boundedness.

Example 4. "If a function is defined on $\mathbb{R}$, then it is locally bounded at least in one point."

Solution.

Let us suppose below that any rational number is represented in the form $x = \frac{m}{n}$, where m is an integer, n is a natural number, and the fraction $\frac{m}{n}$ is in lowest terms (the last means that m and n have no common factor). In this case the numbers m and n are uniquely determined, and the following function $f(x) = \begin{cases} n, & x \in \mathbb{Q}, x = \frac{m}{n} \\ 0, & x \in \mathbb{I} \text{ and } x = 0 \end{cases}$ is well-defined. It can be shown that this function is unbounded in any neighborhood of an arbitrary real point x_0. In effect, if $f(x)$ was bounded in some neighborhood of x_0, then for all rational points $x = \frac{m}{n}$ in this neighborhood the denominators n would be bounded, and hence the numerators m would be bounded too. But this would permit only finitely many rational numbers in such neighborhood. It leads to a contradiction with the known property that in any neighborhood of an arbitrary real point there exist infinitely many rational numbers.

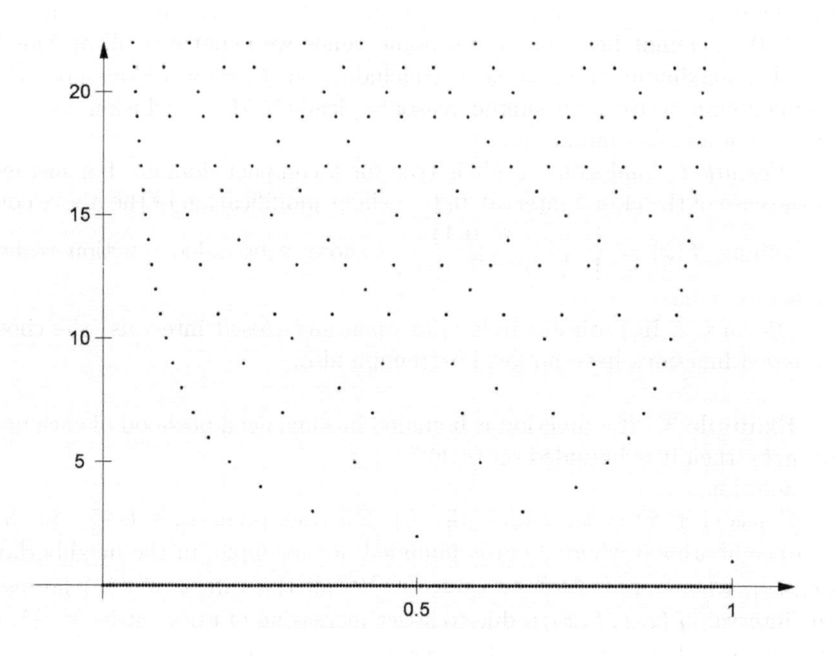

FIGURE 1.3.2: Example 4

1.4 Periodicity

Example 1. "If a non-constant function $f(x)$ is periodic, then it has a fundamental period."

Solution.

Consider Dirichlet's function $D(x) = \begin{cases} 1, & x \in \mathbb{Q} \\ 0, & x \in \mathbb{I} \end{cases}$. Any rational number is a period of this function (due to the properties of the real numbers, the sum of two rational numbers is again rational, and the sum of a rational number with an irrational one is always irrational).

Remark 1. The statement will turn true if additionally $f(x)$ is continuous in at least one point (but proof is not simple and it involves the concept of continuity, which will be considered later).

Remark 2. $D(x)$ is an example of a function for which any rational number is a period and any irrational number is not (the latter is because $D(0) = 1 \neq 0 = D(x)$ for any irrational number x). Contrary to this situation, there is no function for which any irrational number is a period and any rational number is not (supposing, per absurd, that there exists such a function, we pick up two irrational periods $T_1 = \sqrt{2}$ and $T_2 = 1 - \sqrt{2}$ and immediately arrive to contradiction, because their sum $T_1 + T_2 = 1$ should be a period of the same function, but it is a rational number).

Example 2. "If both functions $f(x)$ and $g(x)$ have a fundamental period T, then $f(x) + g(x)$ has the same fundamental period."

Solution.

If $f(x) = \sin x$ and $g(x) = 1 - \sin x$ (both functions have the fundamental period $T = 2\pi$), then $h(x) = f(x) + g(x) = 1$ and it does not have a fundamental period (although $T = 2\pi$ is one of the periods of $h(x)$).

Remark 1. Another simple example when a fundamental period of the sum exists, but is smaller than T: $f(x) = \sin x$ and $g(x) = \sin 2x - \sin x$ have the fundamental period $T = 2\pi$, but $h(x) = f(x) + g(x) = \sin 2x$ has the fundamental period π.

Remark 2. Similar examples can be constructed for other arithmetic operations. For example, the false statement for the product has the form: "if both functions $f(x)$ and $g(x)$ have a fundamental period T , then $f(x) \cdot g(x)$ has the same fundamental period". The corresponding example can be $f(x) = \sin x$, $g(x) = 2\cos x$ (both functions have the fundamental period $T = 2\pi$), and $h(x) = f(x) \cdot g(x) = \sin 2x$ has the fundamental period π.

Remark 3. The correct statement asserts that if $f(x)$ and $g(x)$ have a fundamental period T , then $f(x) + g(x)$ has the same period T. It just may be a non-fundamental period for the sum.

Example 3. "If both functions $f(x)$ and $g(x)$ are periodic, then $f(x) + g(x)$ is also periodic."

Solution.

Consider $f(x) = \sin x$ with the fundamental period $T_f = 2\pi$ and $g(x) = \sin \pi x$ with the fundamental period $T_g = 2$. Let us show that $h(x) = f(x) + g(x) = \sin x + \sin \pi x$ has no period. In fact, if there existed a period T of the function $h(x)$, then the following would hold for an arbitrary x:

$$h(x+T) = \sin(x+T) + \sin \pi (x+T) = \sin x + \sin \pi x = h(x),$$

or

$$\sin(x+T) - \sin x = -\sin \pi (x+T) + \sin \pi x,$$

or

$$\sin \frac{T}{2} \cos\left(x + \frac{T}{2}\right) = -\sin \pi \frac{T}{2} \cos \pi \left(x + \frac{T}{2}\right).$$

Now, we specify the two values of x in the last equation. First, let us set

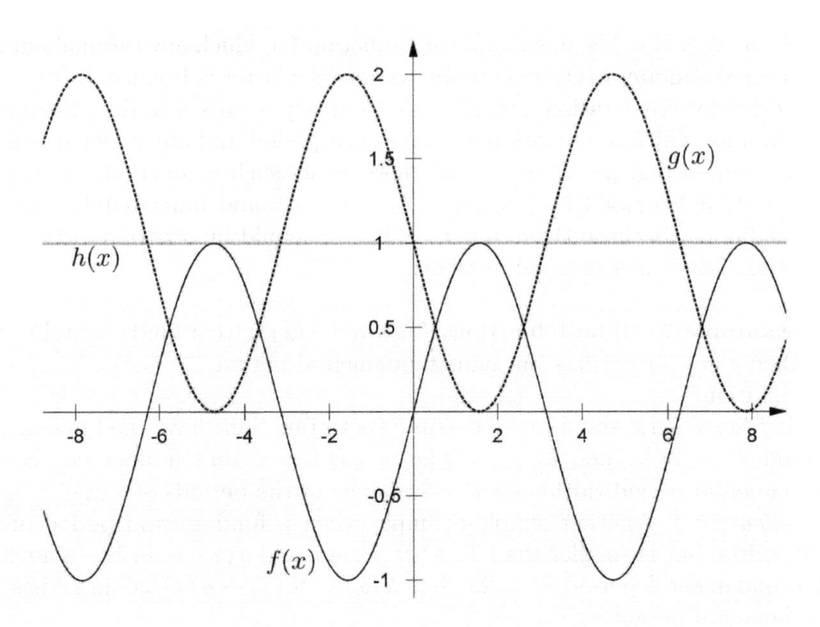

FIGURE 1.4.1: Example 2

$x_1 = \frac{\pi}{2} - \frac{T}{2}$. Then the last equation simplifies to $\sin \pi \frac{T}{2} \cos \pi \frac{\pi}{2} = 0$. Since the second multiplier is not zero ($\cos x = 0$ only at the points $x_k = \frac{\pi}{2} + k\pi$, $k \in \mathbb{Z}$), we have $\sin \pi \frac{T}{2} = 0$, which gives the solutions $T_n = 2n$, $n \in \mathbb{Z} \backslash \{0\}$ ($\sin x = 0$ only at the points $x_n = n\pi$, $n \in \mathbb{Z}$, and zero is eliminated by the definition of a period).

On the other hand, we can set $x_2 = \frac{1}{2} - \frac{T}{2}$. Then the same equation simplifies to $\sin \frac{T}{2} \cos \frac{1}{2} = 0$, imposing the condition $\sin \frac{T}{2} = 0$, that is, $T_m = 2m\pi$, $m \in \mathbb{Z} \backslash \{0\}$. Since all the numbers $T_n = 2n$ are the integers and $T_m = 2m\pi$ are irrational, there is no common number between two sets of solutions, i.e., it does not exist a non-zero number that satisfies the considered equation for a period T both for x_1 and x_2. It means that there is no period for the function $h(x)$.

Remark 1. Another interesting example can be provided with the functions $f(x) = \cos x$ and $g(x) = x - [x]$ (the last has the fundamental period $T_g = 1$). In this case, again supposing, per absurd, that $h(x) = f(x) + g(x) = \cos x + x - [x]$ has a period T, we obtain the relevant equation of periodicity in the form

$$\cos(x + T) + x + T - [x + T] = \cos x + x - [x]$$

or

$$-2 \sin \frac{T}{2} \sin\left(x + \frac{T}{2}\right) = -T + [x + T] - [x].$$

Setting first $x_1 = -\frac{T}{2}$, we obtain $-T + \left[\frac{T}{2}\right] - \left[-\frac{T}{2}\right] = 0$, which implies that T

is an integer, because if it is not so then we readily arrive to a contradiction by rewriting the last equation in the form, $-T+\left[\frac{T}{2}\right]+\left[\frac{T}{2}\right]+1=0$, or $\left[\frac{T}{2}\right]=\frac{T-1}{2}$. On the other hand, by setting $x_2=0$, we obtain $-2\sin^2\frac{T}{2}=-T+[T]$. Since T should be an integer, the right-hand side is zero and, consequently, $\sin\frac{T}{2}=0$, which is not possible for integers. Therefore, the function $h(x)$ does not possess a period.

Remark 2. For the statement to be true, it is necessary to add the condition that the periods of $f(x)$ and $g(x)$ are commensurate (it means that $\frac{T_f}{T_g}\in\mathbb{Q}$).

Remark 3. The sum of two periodic functions with non-commensurate periods still can be a periodic function, but such examples are more complex.

Remark 4. For continuous functions, the condition that periods of two functions are commensurate is necessary and sufficient for the sum to be a periodic function also.

Remark 5. Similar examples can be constructed for other arithmetic operations. For instance, for the false statement: "if both functions $f(x)$ and $g(x)$ are periodic, then $f(x)\cdot g(x)$ is also periodic", the following counterexample works: $f(x)=\sin x,\ g(x)=\sin 2\pi x$.

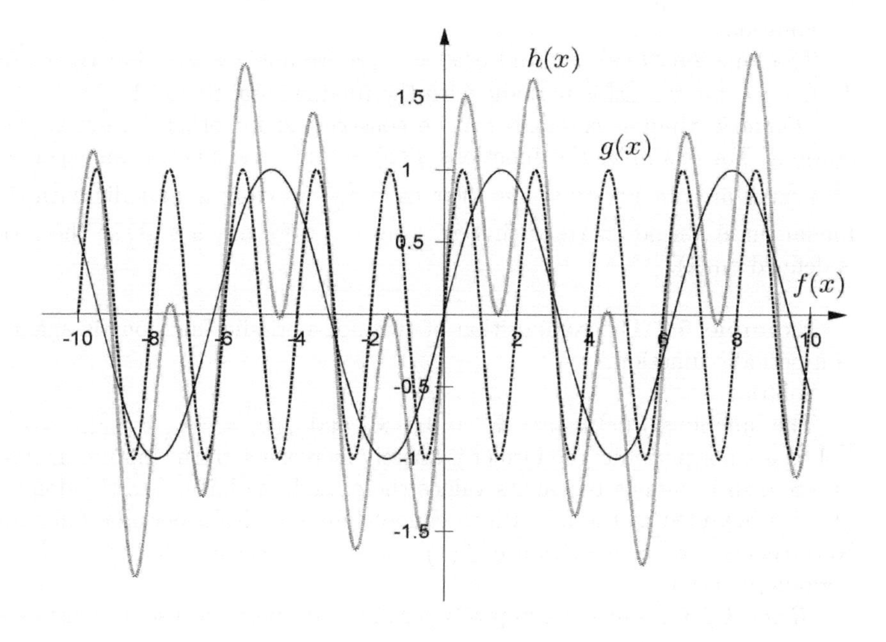

FIGURE 1.4.2: Example 3

Example 4. "A nonconstant function cannot possess both rational and irrational periods."

Solution.

Let us construct a function that has two periods $T_1=1$ and $T_2=\sqrt{2}$. First, we define the following set of real numbers $S = \left\{ x = p + \sqrt{2}q,\ p, q \in \mathbb{Q} \right\}$. Then we define the characteristic function of S, that is,

$$f_S(x) = \begin{cases} 1, & x \in S \\ 0, & x \notin S \end{cases}.$$

It is easy to see that T_1 and T_2 are periods of $f_S(x)$: if $x \in S$ then $(x+1) \in S$ and $\left(x+\sqrt{2}\right) \in S$, and therefore $f_S(x) = f_S(x+1) = f_S(x+\sqrt{2}) = 1$; if $x \notin S$ then $(x+1) \notin S$ and $\left(x+\sqrt{2}\right) \notin S$, and therefore $f_S(x) = f_S(x+1) = f_S(x+\sqrt{2}) = 0$.

Remark 1. This statement becomes true if the condition of the continuity of function is added.

Remark 2. It is worth to notice that in the provided counterexample any rational number and also any number of the form $\sqrt{2}q$, $q \in \mathbb{Q}$ are periods of the presented function.

Example 5. "If the sum of two functions is periodic, then each of the functions is also periodic."

Solution.

The functions $f(x) = x$ and $g(x) = -[x]$ are not periodic, but their sum $f(x) + g(x) = x - [x]$ is periodic with the fundamental period 1.

Remark. Similar examples can be constructed for other arithmetic operations. For example, the functions $f(x) = \left(x^2 + \cos x\right)\cos x$ and $g(x) = x^2 + \cos x$ are not periodic, but their ratio $\frac{f(x)}{g(x)} = \cos x$ is periodic with the fundamental period 2π (note that $x^2 + \cos x \neq 0$ for any $x \in \mathbb{R}$, so the ratio is defined on $\mathbb{R}$).

Example 6. "The composition of two non-periodic functions is again a non-periodic function."

Solution.

The functions $f(x) = x^2 : \mathbb{R} \to [0, +\infty)$ and $g(y) = \cos\sqrt{y} : [0, +\infty) \to [-1, 1]$ are not periodic: the former is strictly increasing on the infinite interval $[0, +\infty)$, so it cannot repeat its values there, and the latter has the domain $[0, +\infty)$ bounded at the left, which cannot happen with a periodic function. Nevertheless, the composition $g(f(x)) = \cos|x| = \cos x : \mathbb{R} \to [-1, 1]$ is a periodic function.

Remark 1. Of course, a composition can be non-periodic also. For instance, just use another order for composition of the above functions to obtain the non-periodic function $f(g(x)) = \left(\cos\sqrt{x}\right)^2 : [0, +\infty) \to [0, 1]$.

Remark 2. Another interesting example is $f(x) = \begin{cases} -1, & x < 0 \\ 1, & x \geq 0 \end{cases}$ and $g(y) = |y|$. In this case, both compositions give the constant function: $g(f(x)) = f(g(x)) = 1$.

Remark 3. For composition of a non-periodic and periodic functions the result depends on the order of composition. For instance, if $f(x)$ is periodic

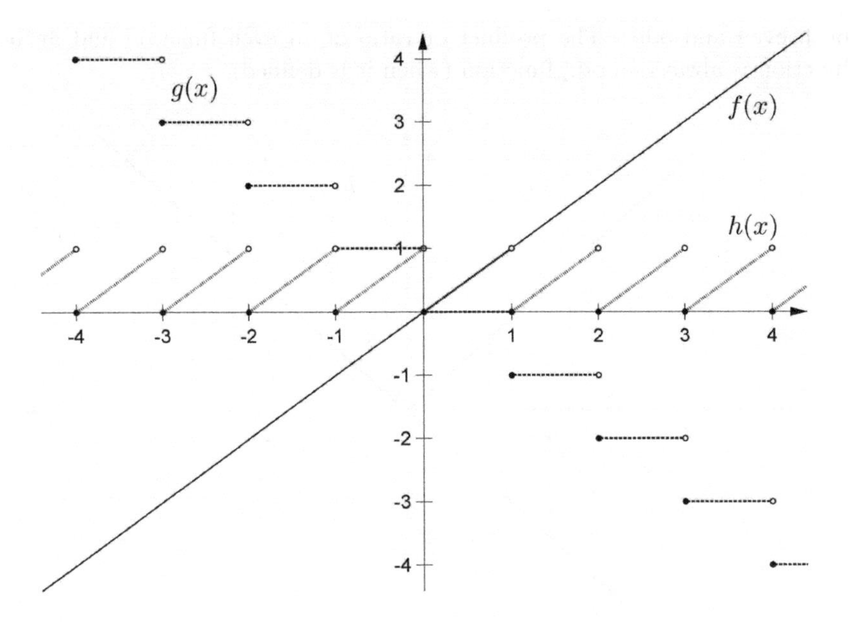

FIGURE 1.4.3: Example 5

and $g(y)$ is not, then the function $g(f(x))$ is always periodic having the same period as that of $f(x)$ (of course, we consider here only the cases when the composition is defined), but the function $f(g(x))$ can be periodic or not, depending on a specific choice of functions. If one chooses $f(x) = \sin x$ and $g(y) = \sqrt{y}$, then $f(g(x)) = \sin\sqrt{x}$ is a non-periodic function, but for $f(x) = \sin x$ and $g(y) = y$ the composition $f(g(x)) = \sin x$ is periodic.

1.5 Even/odd functions

Example 1. "The sum of an even function and an odd function is odd."
Solution.
Consider $f(x) = x$ (odd function) and $g(x) = |x|$ (even function). Then $h(x) = f(x) + g(x) = \begin{cases} 0, \ x < 0 \\ 2x, \ x \geq 0 \end{cases}$ is neither even nor odd.

Remark. A similar simple example can be given for the difference. Actually the sum of an even function and an odd function is always neither even nor odd, unless one of these functions is zero (zero is the only function which is both even and odd). The product or ratio of an even function and an odd function is always an odd function (when it is defined).

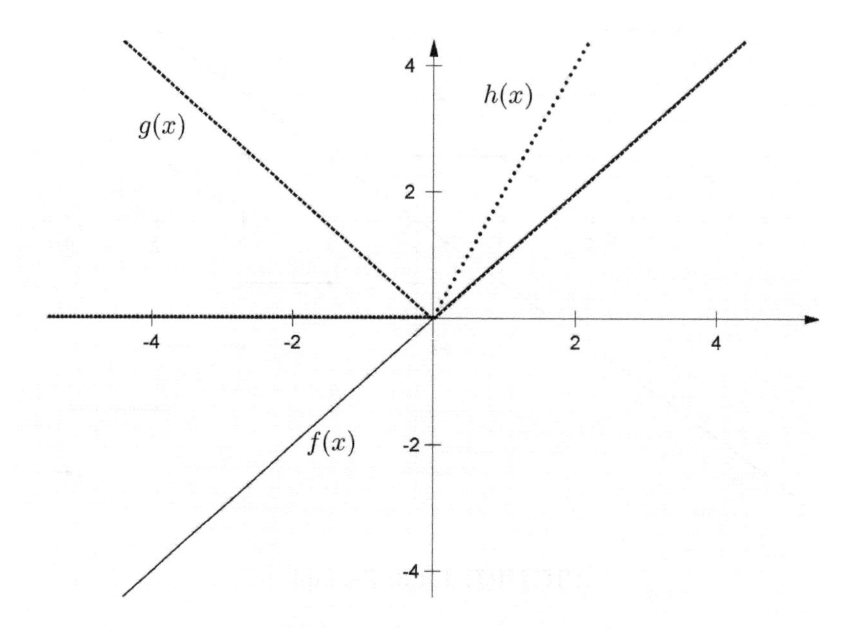

FIGURE 1.5.1: Example 1

Example 2. "If the sum of two functions is even, then each of these two functions is also even."

Solution.

Let $f(x) = x^2 + x$ and $g(x) = x^2 - x$. Both functions are neither even nor odd, but $h(x) = f(x) + g(x) = 2x^2$ is an even function.

Remark 1. Similar simple examples can be given for the difference, product and ratio. For example, the product of the functions $f(x) = x - 1$ and $g(x) = x + 1$, which are neither even nor odd, gives the even function $h(x) = f(x) \cdot g(x) = x^2 - 1$.

Remark 2. The converse is correct: the sum, difference, product and ratio of two even functions is again an even function (when it is defined).

Example 3. "If the sum of two functions is odd, then each of these two functions is also odd."

Solution.

Let $f(x) = \begin{cases} x, \ x < 1 \\ -x, \ x \geq 1 \end{cases}$ and $g(x) = \begin{cases} x, \ x < 1 \\ 3x, \ x \geq 1 \end{cases}$. Both functions are neither even nor odd, but $h(x) = f(x) + g(x) = 2x$ is an odd function.

Remark 1. A similar simple example can be given for the difference.

Remark 2. The converse is correct: the sum and difference of two odd functions is again an odd function (when it is defined).

Remark 3. For the product and ratio, if each of two functions is odd then the result (when defined) is an even function. Therefore, it is hard even to suppose that the product (or the ratio) of two odd functions will give another odd function.

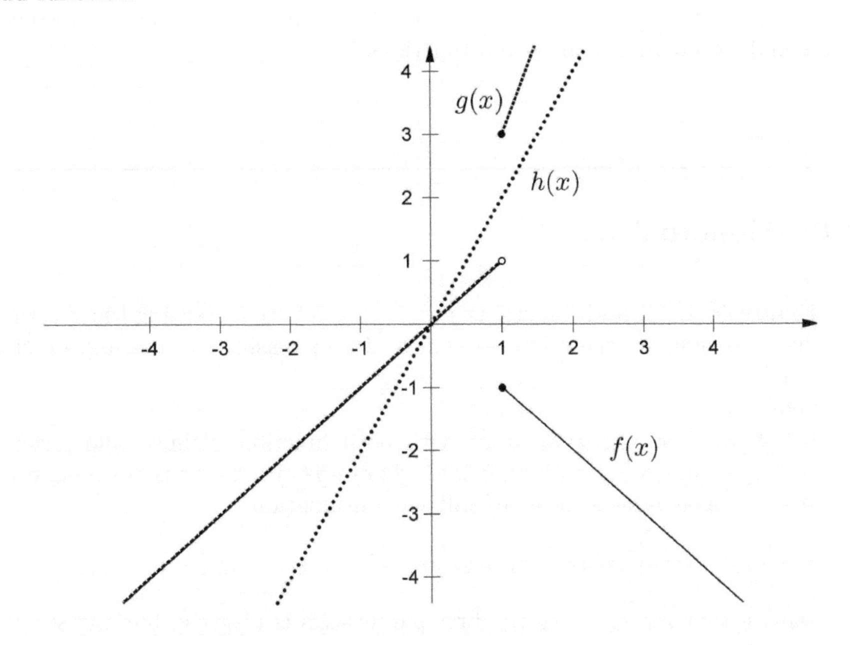

FIGURE 1.5.2: Example 3

Example 4. "If the product of two functions is odd, then one of these functions is even and another is odd."

Solution.

Let $f(x) = \begin{cases} x, \ x < 1 \\ -x, \ x \geq 1 \end{cases}$ and $g(x) = \begin{cases} 1, \ x < 1 \\ -1, \ x \geq 1 \end{cases}$. Both functions are neither even nor odd, but $h(x) = f(x) \cdot g(x) = x$ is an odd function.

Remark 1. A similar simple example can be given for the ratio.

Remark 2. The converse is correct: the product and ratio of an even function and an odd function is an odd function (when it is defined).

Example 5. "If the absolute value of a function is even, then the function itself is also even."

Solution.

If $f(x) = x$ (odd function), then $|f(x)| = |x|$ is even function.

Remark 1. Of course, the original function can be neither even nor odd: $f(x) = \begin{cases} x, & x < 1 \\ -x, & x \geq 1 \end{cases}$ gives the same even function $|f(x)| = |x|$.

Remark 2. The same counterexamples can be used to the following false statement: "if the square of a function is even, then the function itself is also even".

Remark 3. The converse is true: if a function is even, then its absolute value and its square are also even functions.

1.6 Monotonicity

Example 1. "If both functions $f(x)$ and $g(x)$ are increasing (decreasing) on the same domain, then $f(x) - g(x)$ is also increasing (decreasing) on this domain."

Solution.

Let $f(x) = x$ and $g(x) = x^2$ with both functions defined and strictly increasing on $[0, +\infty)$. However, $h(x) = f(x) - g(x) = x - x^2$ is not monotone on $[0, +\infty)$, as it is seen from the following evaluation:

$$h(x_2) - h(x_1) = x_2 - x_1 - (x_2^2 - x_1^2) = (x_2 - x_1)(1 - x_2 - x_1),$$

and notice that for $x_1 < x_2$ the first parenthesis is positive, but the second changes its sign on $[0, +\infty)$ (for $0 \leq x_1 < x_2 \leq \frac{1}{2}$ the second expression is positive, but for $\frac{1}{2} \leq x_1 < x_2$ it is negative).

For the functions defined on $\mathbb{R}$, one can use the counterexample with $f(x) = x$ and $g(x) = x^3$. Evidently, both functions are strictly increasing on $\mathbb{R}$. However, $h(x) = f(x) - g(x)$ is not monotone on $\mathbb{R}$, as it is seen from the following evaluation:

$$h(x_2) - h(x_1) = x_2 - x_1 - (x_2^3 - x_1^3) = (x_2 - x_1)(1 - x_2^2 - x_2 x_1 - x_1^2),$$

and notice that for $x_1 < x_2$ the first parenthesis is positive, but the second changes its sign on $\mathbb{R}$ (for $-\frac{1}{2} < x_1 < x_2 < \frac{1}{2}$ the second expression is positive, but for $1 < x_1 < x_2$ it is negative).

Remark. A similar statement for the sum of two increasing (decreasing) functions is true.

Example 2. "If both functions $f(x)$ and $g(x)$ are increasing (decreasing) on the same domain, then $f(x) \cdot g(x)$ is also increasing (decreasing) on this domain."

Solution.

Let $f(x) = \frac{1}{x}$ and $g(x) = \frac{1}{x}$ be considered on $(-\infty, 0)$. Evidently, both

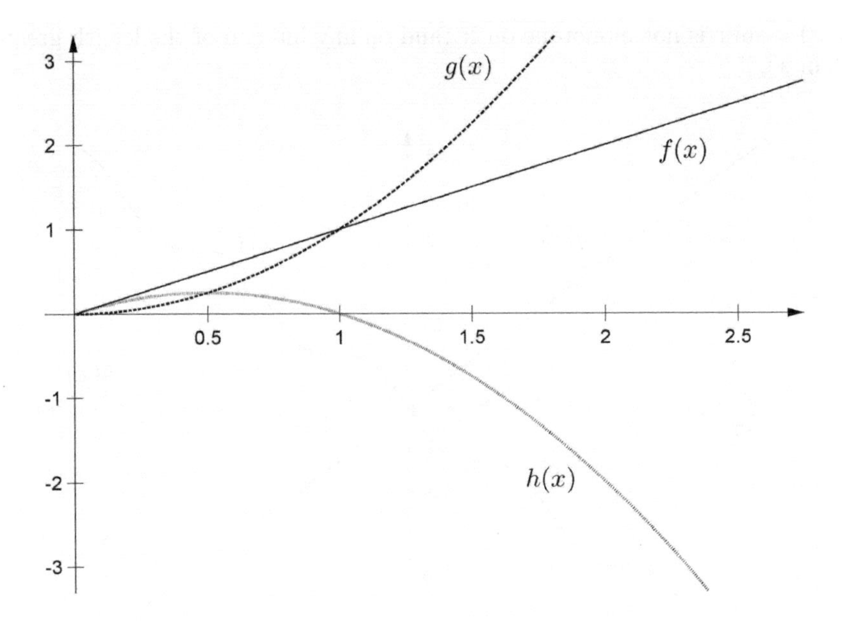

FIGURE 1.6.1: Example 1

function are strictly decreasing on $(-\infty, 0)$. However, $h(x) = f(x) \cdot g(x) = \frac{1}{x^2}$ is strictly increasing on $(-\infty, 0)$.

Another pair of functions can produce non-monotone product. For instance, $f(x) = x$ and $g(x) = x$ are defined and strictly increasing on $\mathbb{R}$. However, $h(x) = f(x) \cdot g(x) = x^2$ is not monotone function on $\mathbb{R}$.

Remark. In the particular case when $f(x) = g(x)$ the above statement is transformed as follows: "if a function $f(x)$ is increasing (decreasing) on its domain, then $f^2(x)$ is also increasing (decreasing) on this domain". The same function $f(x) = x$ disproves the last statement.

Example 3. "If both functions $f(x)$ and $g(x)$ are monotone on the same domain, then $f(x) + g(x)$ is also monotone on this domain."

Solution.

Let $f(x) = x + \sin x$ and $g(x) = -x$ (both defined on $\mathbb{R}$). Evidently, $g(x)$ is strictly decreasing on $\mathbb{R}$, and $f(x)$ is strictly increasing on $\mathbb{R}$, because for an arbitrary pair $x_1, x_2 \in \mathbb{R}$, $x_1 < x_2$, we have

$$f(x_2) - f(x_1) = x_2 - x_1 + \sin x_2 - \sin x_1$$

$$= x_2 - x_1 + 2\sin\frac{x_2 - x_1}{2}\cos\frac{x_2 + x_1}{2} > x_2 - x_1 - 2\frac{x_2 - x_1}{2} = 0$$

(here, we used two elementary properties of trigonometric functions: $\sin\alpha < \alpha$, $\forall\alpha > 0$ and $|\cos\beta| \leq 1$, $\forall\beta$). At the same time, the function $h(x) = f(x) +$

$g(x) = \sin x$ is not monotone on $\mathbb{R}$ (and on any interval of the length greater than π).

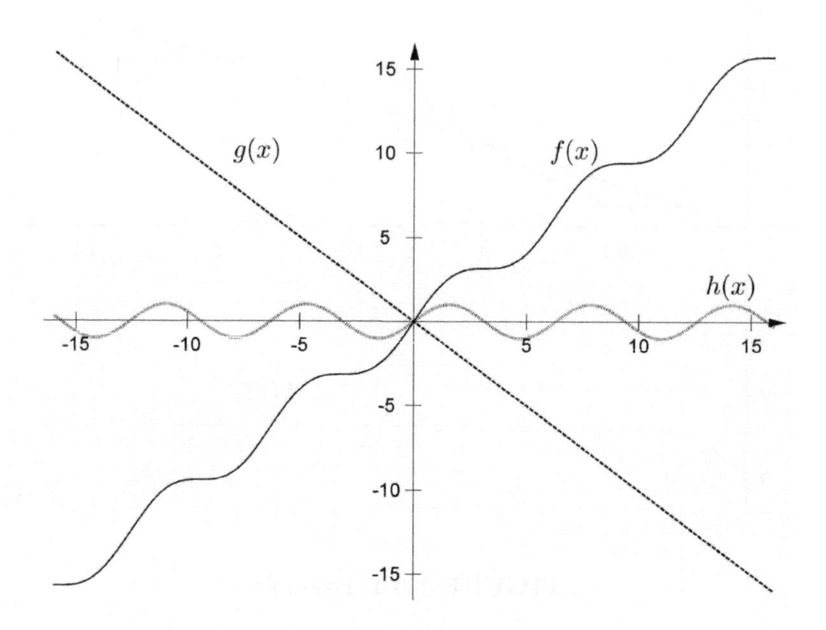

FIGURE 1.6.2: Example 3

Example 4. "If $f(x)$ is not monotone and $g(x)$ is monotone on the same domain, then $f(x) + g(x)$ is monotone on this domain."

Solution.

Let $f(x) = |x|$ and $g(x) = \frac{x}{2}$ be considered on $\mathbb{R}$. Evidently, both functions satisfy the above conditions, but the resulting function $h(x) = f(x) + g(x) = \begin{cases} -\frac{x}{2}, & x < 0 \\ \frac{3x}{2}, & x \geq 0 \end{cases}$ is not monotone on $\mathbb{R}$.

Remark. The statement with the opposite conclusion is also false. The corresponding counterexample can be constructed with the functions $f(x) = |x|$ and $g(x) = 2x$ defined on $\mathbb{R}$. Although $f(x)$ is not monotone and $g(x)$ is monotone on $\mathbb{R}$, the resulting function $h(x) = f(x) + g(x) = \begin{cases} x, & x < 0 \\ 3x, & x \geq 0 \end{cases}$ is monotone on $\mathbb{R}$.

Example 5. "If both functions $f(x)$ and $g(x)$ are non-monotone on the same domain, then $f(x) + g(x)$ is also non-monotone on this domain."

Solution.

Let $f(x) = |2x|$ and $g(x) = x - |2x| = \begin{cases} 3x, & x < 0 \\ -x, & x \geq 0 \end{cases}$ be defined on $\mathbb{R}$.

Evidently, both function are non-monotone on $\mathbb{R}$, but the resulting function $h(x) = f(x) + g(x) = x$ is strictly increasing on $\mathbb{R}$.

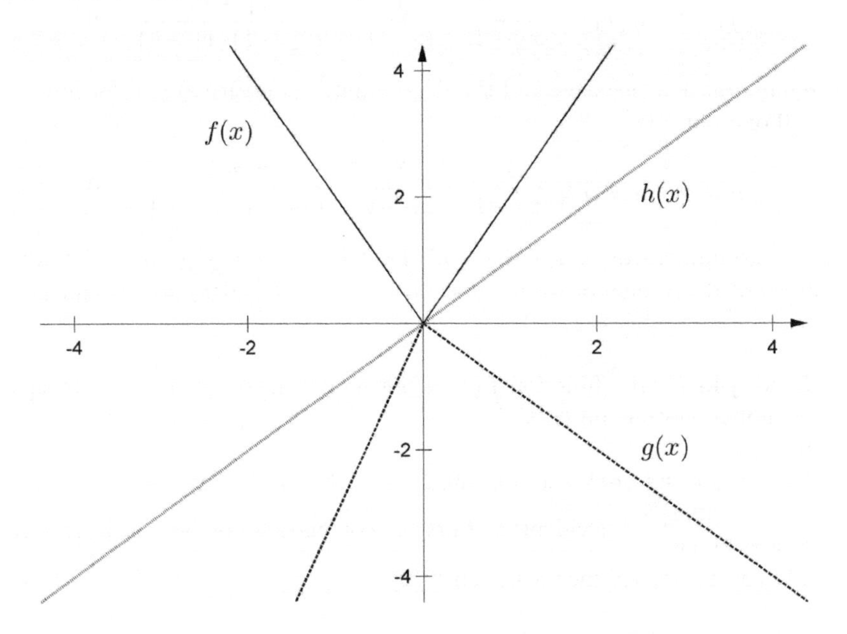

FIGURE 1.6.3: Example 5

Remark to Examples 3-5. All the statements presented above for the sum can also be used for the difference, product and ratio of two functions. For instance, the false statement for the product, similar to Example 5, can be formulated as follows: "if both functions $f(x)$ and $g(x)$ are non-monotone on the same domain, then $f(x) \cdot g(x)$ is also non-monotone on this domain". The counterexample can be constructed using the functions $f(x) = x^2$ and $g(x) = \begin{cases} \frac{1}{x}, \; x \neq 0 \\ 0, \; x = 0 \end{cases}$, both defined and non-monotone on $\mathbb{R}$. Their product $h(x) = f(x) \cdot g(x) = x$ is strictly increasing on $\mathbb{R}$.

Example 6. "If $f(x)$ is monotone on $\mathbb{R}$ and $f(x) \neq 0$, $\forall x \in \mathbb{R}$, then $\frac{1}{f(x)}$ is monotone on $\mathbb{R}$."

Solution.

Let $f(x) = \begin{cases} x-1\,, \; x \leq 0 \\ x+1\,, \; x > 0 \end{cases}$. Evidently, this is a strictly increasing function on $\mathbb{R}$. However, the function $g(x) = \frac{1}{f(x)} = \begin{cases} \frac{1}{x-1}\,, \; x \leq 0 \\ \frac{1}{x+1}\,, \; x > 0 \end{cases}$ is not monotone on $\mathbb{R}$. Indeed, this function is strictly decreasing on separate intervals

$(-\infty, 0]$ and $(0, +\infty)$, since for any $x_1 < x_2 \leq 0$ it follows that

$$g(x_2) - g(x_1) = \frac{1}{x_2 - 1} - \frac{1}{x_1 - 1} = \frac{x_1 - x_2}{(x_2 - 1)(x_1 - 1)} < 0$$

(the numerator is negative and the denominator is positive) and for any $x_2 > x_1 > 0$ one obtains

$$g(x_2) - g(x_1) = \frac{1}{x_2 + 1} - \frac{1}{x_1 + 1} = \frac{x_1 - x_2}{(x_2 + 1)(x_1 + 1)} < 0$$

(again the numerator is negative and the denominator is positive). However, $g(x)$ is not decreasing on $\mathbb{R}$: if one chooses $x_1 = -1$ and $x_2 = 1$, then $g(x_2) - g(x_1) = \frac{1}{1+1} - \frac{1}{-1-1} = 1 > 0$.

Example 7. "If a function $f(x)$ is not monotone on (a, b), then its square also is not monotone on (a, b)."

Solution.

Let $f(x)$ be defined on $[0, 2]$ in the following way: $f(x) = \begin{cases} -x, & x \in [0,1] \\ x, & x \in (1,2] \end{cases}$. Evidently, $f(x)$ is not monotone on $[0, 2]$. However, $f^2(x) = x^2$ is strictly increasing on $[0, 2]$.

Example 8. "The composition of two non-monotone functions is again a non-monotone function."

Solution.

The functions $f(x) = \begin{cases} \frac{1}{x}, & x \neq 0 \\ 0, & x = 0 \end{cases}$, $g(y) \equiv f(y)$ are not monotone on $\mathbb{R}$, but their composition $g(f(x)) = f(g(x)) = x$ is a strictly increasing function on $\mathbb{R}$.

A similar counterexample can be given for a finite interval: the functions $f(x) = \begin{cases} -1 - x, & x \in (-1, 0) \\ 0, & x = 0 \\ 1 - x, & x \in (0, 1) \end{cases}$, $g(y) \equiv f(y)$ are not monotone on $(-1, 1)$, but their composition $g(f(x)) = f(g(x)) = x$ is a strictly increasing function on $(-1, 1)$.

Remark 1. Of course, a composition can give a non-monotone function also. For instance, if $f(x) = x^2$, $g(y) \equiv f(y)$, then $g(f(x)) = f(g(x)) = x^4$. All the functions are non-monotone on $\mathbb{R}$.

Remark 2. The following strengthened version is also false: "The composition of two functions, each of which is non-monotone anywhere on its domain, is again a non-monotone function." In this formulation "non-monotone anywhere" means that a function is not monotone in any neighborhood of an arbitrary point of the domain. The corresponding example is $f(x) = \begin{cases} x, & x \in \mathbb{Q} \\ -x, & x \in \mathbb{I} \end{cases}$, $g(y) \equiv f(y)$ (considered on any infinite or finite interval symmetric with respect to the origin) and $g(f(x)) = f(g(x)) = x$.

Remark 3. A composition of monotone functions is always a monotone (we consider only the cases when the composition is defined). If both functions are strictly monotone, then its composition also is strictly monotone.

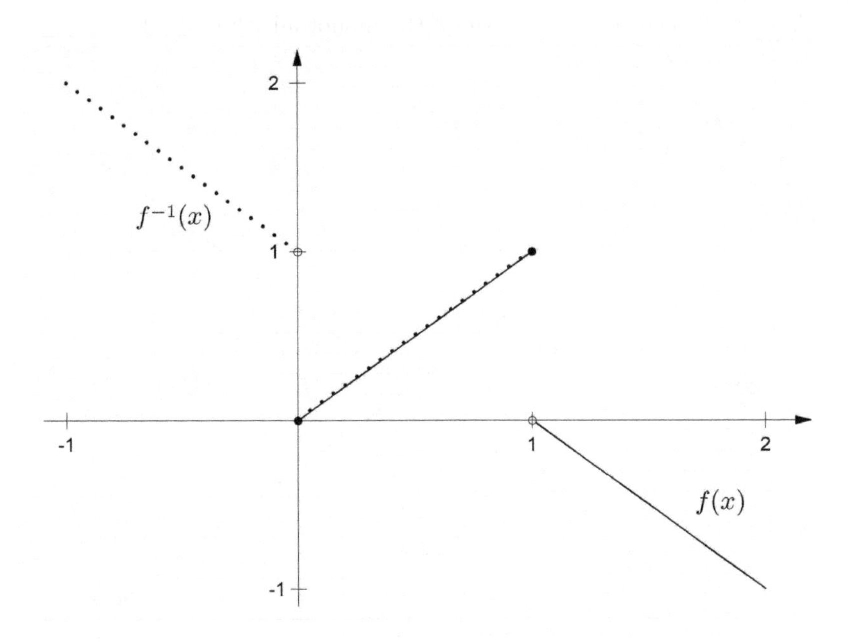

FIGURE 1.6.4: Example 9

Example 9. "If a function $f(x)$ has an inverse on $[a,b]$, then $f(x)$ is monotone on $[a,b]$. "

Solution.

Let $f(x)$ be defined on $[0,2]$ in the following way: $f(x) = \begin{cases} x\,,\ x \in [0,1] \\ 1-x\,,\ x \in (1,2] \end{cases}$. It is easy to see that $f(x)$ is one-to-one correspondence between $[0,2]$ and $[-1,1]$ with the inverse function defined as follows: $x = f^{-1}(y) = \begin{cases} y\,,\ y \in [0,1] \\ 1-y\,,\ y \in [-1,0) \end{cases}$. On the other hand, $f(x)$ is not monotone on $[0,2]$.

Remark 1. The same function can be used if the closed interval is substituted by open. In the last case, if one wishes to obtain an open interval as the image of $f(x)$, then the function should be modified to $f(x) = \begin{cases} x\,,\ x \in (0,1) \\ 1-x\,,\ x \in [1,2) \end{cases}$.

Remark 2. A slightly more complex function $f(x) = \begin{cases} x\,,\ x \in \mathbb{Q} \\ 1-x\,,\ x \in \mathbb{I} \end{cases}$ on

$[0, 1]$ gives the counterexample both for the above statement and the following strengthened statement (also false): "if $f(x)$ has an inverse on $[a, b]$, then there exists a subinterval of $[a, b]$ on which $f(x)$ is monotone" (see explanations in Example 10).

Remark 3. The converse is true if the monotonicity is strict.

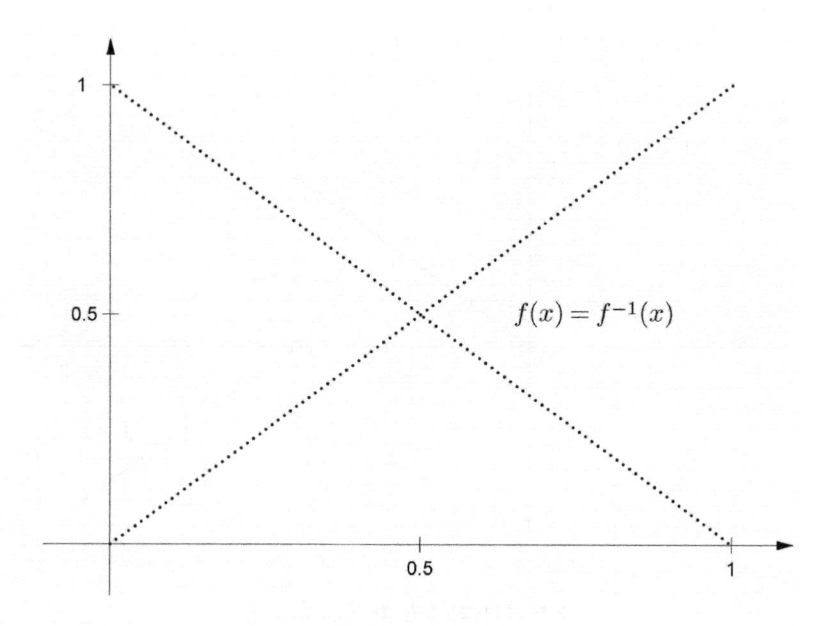

FIGURE 1.6.5: Example 10

Example 10. "If $f(x)$ is a one-to-one correspondence between intervals $[a, b]$ and $[c, d]$, then this function is monotone at least in some subintervals of $[a, b]$."

Solution.

Let $f(x)$ be defined on $[0, 1]$ in the following way: $f(x) = \begin{cases} x, & x \in \mathbb{Q} \\ 1 - x, & x \in \mathbb{I} \end{cases}$. It is easy to see that $f(x)$ is a one-to-one correspondence between $[0, 1]$ and $[0, 1]$ with the inverse function defined as follows: $x = f^{-1}(y) = \begin{cases} y, & y \in \mathbb{Q} \\ 1 - y, & y \in \mathbb{I} \end{cases}$. At the same time, in any interval $(c, d) \subset [0, 1]$ there are both rational and irrational points. If we choose two rational points $x_1, x_2 \in (c, d)$, $x_1 < x_2$ then $f(x_1) < f(x_2)$. On the other hand, for two irrational points $x_3, x_4 \in (c, d)$, $x_3 < x_4$ it follows $f(x_3) > f(x_4)$. Of course, the same considerations are true for closed or half-open intervals. Therefore, $f(x)$ is non-monotone in any subinterval in $[0, 1]$.

Remark 1. This is a finer version of the previous Example 9.

Remark 2. The same function can be used if the closed intervals are substituted by open.

Remark 3. Due to the theorem about an inverse of a strictly monotone function, it may appear that condition of monotonicity is somehow attached to a possibility to find an inverse, but actually there is no requirement on monotonicity for the existence of an inverse function.

1.7 Extrema

Example 1. "A function cannot have infinitely many strict local extrema."

Solution.

The function $f(x) = \cos x$ has infinitely many strict local maxima $x_k = 2k\pi$, $\forall k \in \mathbb{Z}$ and minima $x_n = \pi + 2n\pi$, $\forall n \in \mathbb{Z}$. These strict local extrema are also non-strict global extrema on $\mathbb{R}$.

A more interesting situation occurs when a bounded interval is considered. In this case the counterexample can be provided with the function $f(x) = \cos\frac{1}{x}$ considered on $(0,1)$: its infinite set of strict local maxima consists of the points $x_k = \frac{1}{2k\pi}$, $\forall k \in \mathbb{N}$ and strict local minima are $x_n = \frac{1}{\pi+2n\pi}$, $\forall n \in \mathbb{N} \cup \{0\}$. Again these strict local extrema are also non-strict global extrema on $(0,1)$.

Remark. Although a local extremum is referred to a property of a function on a small interval, one may think that a sufficient finite number of small intervals always covers the given domain (for example, interval) and, consequently, a function can possess only a finite number of local extrema. However, a given interval (domain) can be partitioned in infinitely many small subintervals each of which contains a strict local extremum of a function.

Example 2. "If a function has an infinite set of strict local maxima, then there is the largest function value among these maximum function values."

Solution.

For an infinite interval, we can consider $f(x) = x\cos x$ on $(0,+\infty)$. All the points $x_k = 2k\pi$, $\forall k \in \mathbb{N}$ are strict local maxima and the function values at these points $f(x_k) = x_k = 2k\pi$ increase infinitely as k approaches infinity, so there is no largest number among these values. In the same way, there is no smallest number among the minimum values $f(x_n) = -x_n = -\pi - 2n\pi$ at the minimum points $x_n = \pi + 2n\pi$, $\forall n \in \mathbb{N} \cup \{0\}$.

For a finite interval, we can choose $f(x) = \frac{1}{x}\cos\frac{1}{x}$ defined on $(0,1)$. Its strict local maxima are the points $x_k = \frac{1}{2k\pi}$, $\forall k \in \mathbb{N}$ where the function takes the values $f(x_k) = \frac{1}{x_k} = 2k\pi$, $k \in \mathbb{N}$, and there is no largest value among these values. In the same way, there is no smallest value among the minimum

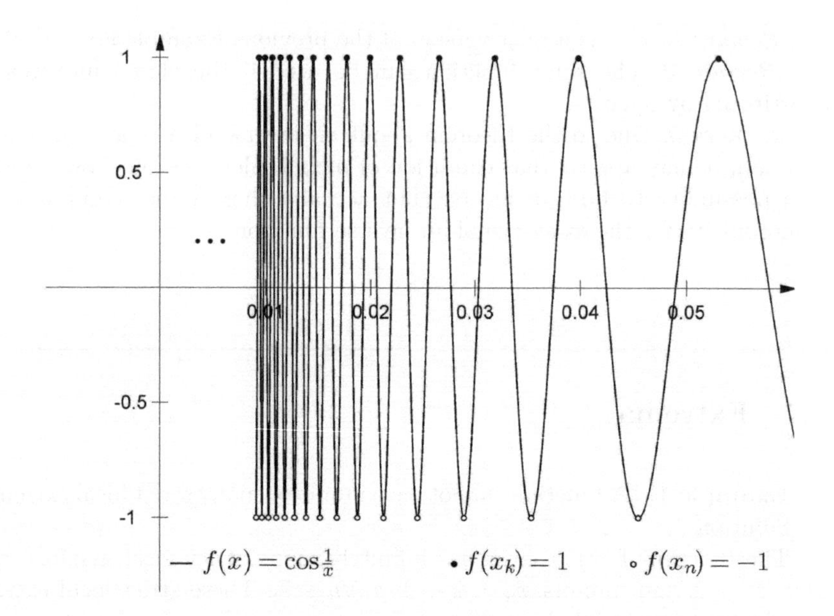

FIGURE 1.7.1: Example 1

values $f(x_n) = -\frac{1}{x_n} = -\pi - 2n\pi$ that the function takes at the minimum points $x_n = \frac{1}{\pi+2n\pi}$, $\forall n \in \mathbb{N} \cup \{0\}$.

Example 3. "If a function $f(x)$ is non-constant on any interval of its domain and has a global maximum at x_0, then there exists a neighborhood of x_0 such that $f(x) < f(x_0)$ for every $x \neq x_0$ in this neighborhood."

Solution.

Consider Dirichlet's function $D(x) = \begin{cases} 1, & x \in \mathbb{Q} \\ 0, & x \in \mathbb{I} \end{cases}$. Evidently, each rational point is a (non-strict) global maximum of $D(x)$ and each irrational point is a (non-strict) global minimum. However, there is no neighborhood of a rational point x_0 such that $D(x) < D(x_0)$ for every $x \neq x_0$ of this neighborhood (in any neighborhood there exist rational points $\tilde{x}$ such that $D(\tilde{x}) = D(x_0) = 1$).

Example 4. "If $f(x)$ has a minimum at x_0, then $\frac{1}{f(x)}$ has a maximum at x_0."

Solution.

First, for local extrema, we can consider the function $f(x) = |x|$, that has a local (and global) minimum at $x_0 = 0$, but the function $\frac{1}{f(x)} = \frac{1}{|x|}$ has no maximum at 0 (it does not defined at 0).

For global extrema, the function $f(x) = x^2 - 1$ has a global minimum at

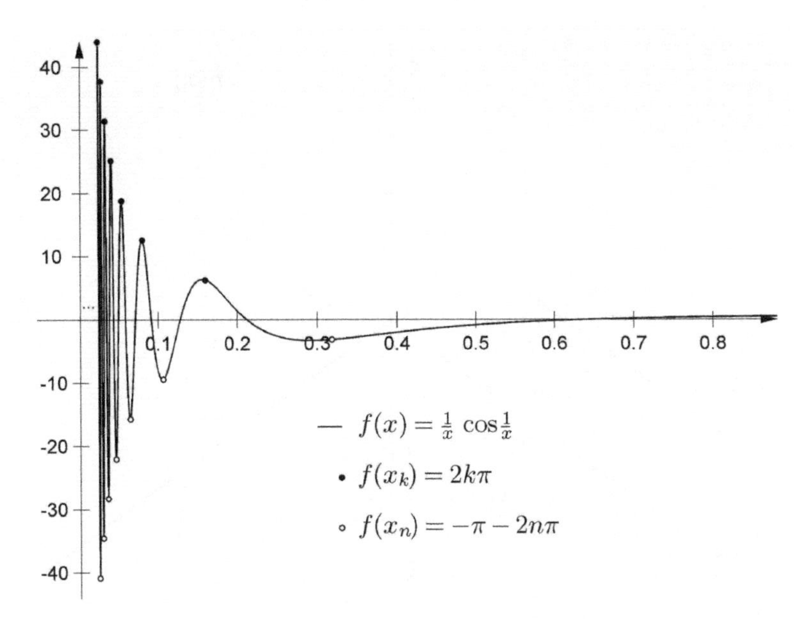

FIGURE 1.7.2: Example 2

$x_0 = 0$, but the function $\frac{1}{f(x)} = \frac{1}{x^2-1}$ has a local, but not global, maximum at $x_0 = 0$.

Example 5. "If $f(x)$ and $g(x)$ have a maximum at x_0, then $f(x) \cdot g(x)$ also has a maximum at x_0."

Solution.

The functions $f(x) = g(x) = -|x|$ have a local (and global) maximum at $x_0 = 0$, but $f(x) \cdot g(x) = x^2$ has a local (and global) minimum at this point.

Another interesting counterexample is $f(x) = -|x|$ and $g(x) = \begin{cases} x+1\,, & x \le 0 \\ -x\,, & x > 0 \end{cases}$. Both functions have a local (and global) maximum at $x_0 = 0$, but their product $h(x) = f(x) \cdot g(x) = \begin{cases} x^2 + x\,, & x \le 0 \\ x^2\,, & x > 0 \end{cases}$ is strictly increasing at $x_0 = 0$, since $x^2 + x$ is strictly increasing on $\left[-\frac{1}{2}, 0\right]$, and x^2 is strictly increasing on $[0, +\infty)$.

Remark. The statement is true for the sum of two functions.

Example 6. "If a function has more than one local minima on an interval, then it is not invertible on this interval."

Solution.

The function $f(x) = x - 2[x]$ has local minimum at each integer point. In fact, on an interval $(n-1, n+1)$, for any fixed $n \in \mathbb{Z}$, the function is defined

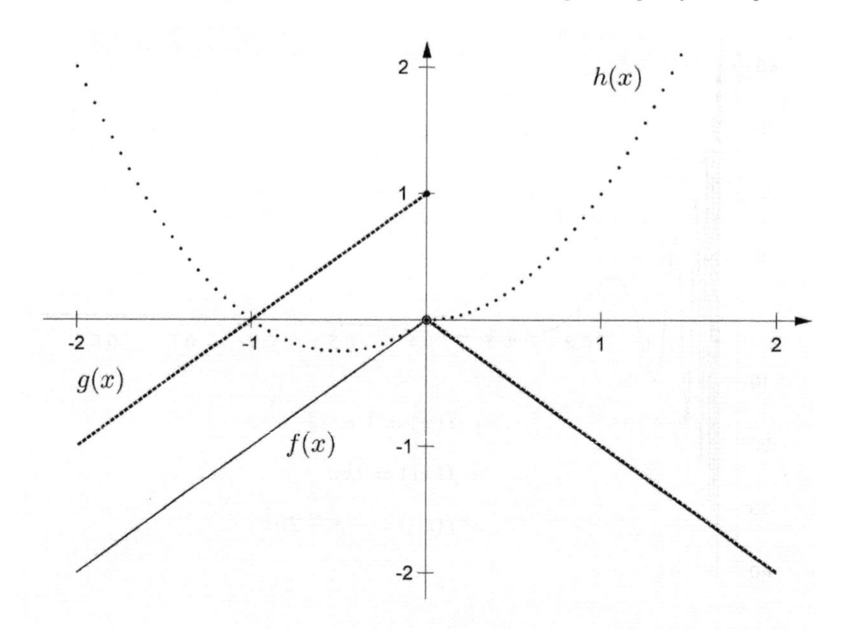

FIGURE 1.7.3: Example 5

as follows: $f(x) = \begin{cases} x - 2(n-1),\ x \in (n-1, n) \\ x - 2n,\ x \in [n, n+1) \end{cases}$, and $f(n) = -n$ is the minimum value on this interval. Therefore, $f(x)$ has infinitely many local minima on $\mathbb{R}$. However, $f(x)$ is invertible with the inverse given by formula $f^{-1}(y) = y + 2[y]$: on each interval $[n, n+1)$ the function is defined by the formula $y = f(x) = x - 2n$, which is simply invertible $x = f^{-1}(y) = y + 2n$, and the image of each interval $[n, n+1)$ is $[-n, -n+1)$, which is different for different values of n; therefore gluing the inverses for each n we obtain the general inverse $f^{-1}(y) = y + 2[y]$.

Remark. Of course, the same is true for local maxima.

Exercises

1. Give one more counterexample to the statement in Example 1, section 1.2.

2. Show that $f(x) = 2x^2 - 1$ is a counterexample to Example 3 in section 1.2 (choose X properly).

3. Give one more counterexample to Example 5, section 1.2 using the function 2^x (choose X properly).

4. Give one more counterexample to the statement in Remark 2 to Example 9, section 1.2.

5. Analyze if the following statement is true: "if $g(f(x)) : X \to Z$ is injective composition of $f(x) : X \to Y$ and $g(y) : Y \to Z$, then both $f(x)$ and $g(y)$ are injective". If not, provide a counterexample. What about surjective functions?

6. Find a counterexample to the following statement: "if $f(x)$ is bounded and non-zero on (a, b), then $\frac{1}{f(x)}$ is bounded on the same interval".

7. It is simple to show that if $f(x)$ is bounded on any open interval contained in $[a, b]$, then it is bounded on $[a, b]$. Show that a similar statement: "if $f(x)$ is bounded on any closed interval contained in (a, b), then it is bounded on (a, b)" is false by providing a counterexample.

8. Show that the following statement "a non-constant function cannot have infinitely many irrational periods" is false by providing a counterexample. Consider a finer version of the above statement: "there is no function which has no rational period, but has infinitely many positive irrational periods smaller then any given positive number" and construct corresponding counterexample. How is this situation related to Remark 2 in Example 1.4.1? (Hint: the first counterexample is simple - just choose a basic trigonometric function, like $\sin x$; but the second is a bit harder - try a version of Dirichlet's function with a scaled variable - $D(x) = \begin{cases} 1, & x = \sqrt{2} \cdot q, \ \forall q \in \mathbb{Q} \\ 0, & otherwise \end{cases}$.)

9. Construct a counterexample to the statement similar to Example 2, section 1.4: "if both functions (defined on the same domain) have the same fundamental period, then their ratio has the same fundamental period".

10. Construct a counterexample to the following statement: "if both functions (defined on the same domain) have the same fundamental period, then their sum has a fundamental period."

11. Prove that the counterexample in Remark 5 to Example 3, section 1.4 works.

12. Give a counterexample to the statement in Example 5, section 1.4, where the sum is substituted by the product.

13. Give a counterexample to statement in Example 2, section 1.5, where the sum is substituted by the ratio.

14. Give a non-trivial example that shows that the situation described in Remark 3 to Example 3 in section 1.5 is feasible, albeit is rare. That is, construct a counterexample to the following false statement: "if the product of two odd functions is odd then both functions are zero" (Hint: Since the product of two odd functions is always an even function and the condition says that the result is an odd function, it follows that the product is zero. Now, to decompose zero into two non-zero odd factors, divide all the domain, say $\mathbb{R}$, into two parts symmetric with respect to zero. Set the first function to be zero and the second to be any non-zero odd function, say x, on the first part, say $(-1,1)$, and vice-versa on the remaining part $\mathbb{R}\backslash(-1,1)$.)

15. Verify if the following statement is true: "if the composite function $g(f(x))$ is odd then at least one of the functions is odd". If not, provide a counterexample. What about even functions?

16. Show that the difference of two functions monotone on the same domain can be both a monotone and non-monotone function. Formulate these results in the form of counterexamples.

17. Provide a counterexample to the statement in Example 4, section 1.6, with the product instead of sum of functions.

18. Provide a counterexample to the statement in Example 5, section 1.6, with the ratio instead of sum of functions.

19. Verify if Dirichlet's function provides a counterexample for Example 1 in section 1.7.

20. Consider a refined version of the statement in Example 4, section 1.7: "if $f(x)$ has a minimum at x_0 and $\frac{1}{f(x)}$ is defined on the same domain as $f(x)$, then $\frac{1}{f(x)}$ has a maximum at x_0." Is this statement still false? If so, provide a counterexample.

21. Consider the following statement: "if $f(x)$ has a maximum at x_0 and $g(x)$ has a minimum at x_0, then $f(x)/g(x)$ has a maximum at x_0." Analyze if the statement is true. If not, provide a counterexample.

22. Consider the following stronger version of the statement in Example 6, section 1.7: "if a function has more than one local minima and maxima on an interval, then it is not invertible on this interval". Verify if the statement is true and construct a counterexample if it is not.

23. Analyze the following statement: "if $f(x)$ and $g(x)$ attain their global maximum values A and B on domain X, then $h(x) = f(x) + g(x)$ has the global maximum value $A+B$ on X". If it is false, provide a counterexample.

Chapter 2

Limits

2.1 Elements of theory

Concepts

Limit (general limit). Let $f(x)$ be defined on X and a be a limit point of X. We say that the *limit* of $f(x)$, as x approaches a, exists and equals A if for every $\varepsilon > 0$ there exists $\delta > 0$ such that for all $x \in X$ such that $0 < |x - a| < \delta$ it follows that $|f(x) - A| < \varepsilon$. The usual notations are $\lim\limits_{x \to a} f(x) = A$ and $f(x) \underset{x \to a}{\to} A$.

Remark. In calculus, a non-essential simplification that $f(x)$ is defined in some deleted neighborhood of a is frequently used.

Partial limit. Let $f(x)$ be defined on X and a be a limit point of X. The *partial limit* of $f(x)$ at a is the limit of $f(x)$, as x approaches a, on a subset of X. Among partial limits, one of the most important for the functions of one variable are the one-sided limits.

One-sided limits. Let $f(x)$ be defined on X and a be a limit point of X. We say that the *right-hand (left-hand) limit* of $f(x)$, as x approaches a, exists and equals A if for every $\varepsilon > 0$ there exists $\delta > 0$ such that for all $x \in X$ such that $0 < x - a < \delta$ $(-\delta < x - a < 0)$ it follows that $|f(x) - A| < \varepsilon$. The standard notations are $\lim\limits_{x \to a_+} f(x) = A$ ($\lim\limits_{x \to a_-} f(x) = A$) and $f(x) \underset{x \to a_+}{\to} A$ ($f(x) \underset{x \to a_-}{\to} A$).

Infinite limit. Let $f(x)$ be defined on X and a be a limit point of X. We say that the limit of $f(x)$, as x approaches a, is $+\infty$ $(-\infty)$, if for every $E > 0$ there exists $\delta > 0$ such that for all $x \in X$ such that $0 < |x - a| < \delta$ it follows that $f(x) > E$ $(f(x) < -E)$. The usual notations are $\lim\limits_{x \to a} f(x) = +\infty$ ($\lim\limits_{x \to a} f(x) = -\infty$) and $f(x) \underset{x \to a}{\to} +\infty$ ($f(x) \underset{x \to a}{\to} -\infty$).

Sometimes it is also considered a "general" infinite limit in the following sense:

The limit of $f(x)$, as x approaches a, is ∞, if for every $E > 0$ there exists $\delta > 0$ such that for all $x \in X$ such that $0 < |x - a| < \delta$ it follows that $|f(x)| > E$. The usual notations are $\lim_{x \to a} f(x) = \infty$ and $f(x) \underset{x \to a}{\to} \infty$.

If any of the above three infinite limits is admitted at the same time the notations used are $\lim_{x \to a} f(x) = (\pm)\infty$.

Limit at infinity. Let $f(x)$ be defined on X, which is unbounded above (below). We say that the limit of $f(x)$, as x approaches $+\infty$ $(-\infty)$, exists and equals A , if for every $\varepsilon > 0$ there exists $D > 0$ such that for all $x \in X$ such that $x > D$ $(x < -D)$ it follows that $|f(x) - A| < \varepsilon$. The usual notations are $\lim_{x \to +\infty} f(x) = A$ ($\lim_{x \to -\infty} f(x) = A$) and $f(x) \underset{x \to +\infty}{\to} A$ ($f(x) \underset{x \to -\infty}{\to} A$).

Elementary properties

Unicity. If the limit exists, then it is unique.

Remark 1. In the properties below it is supposed that $f(x)$ and $g(x)$ are defined on the same domain.

Remark 2. The point a can be finite or infinite.

Comparative properties

1) If $\lim_{x \to a} f(x) = A$, $\lim_{x \to a} g(x) = B$ and $f(x) \le g(x)$ for all $x \in X$ in a deleted neighborhood of a, then $A \le B$.

2) If $\lim_{x \to a} f(x) = A$, $\lim_{x \to a} g(x) = B$ and $A < B$, then $f(x) < g(x)$ for all $x \in X$ in a deleted neighborhood of a.

Arithmetic (algebraic) properties

If $\lim_{x \to a} f(x) = A$, $\lim_{x \to a} g(x) = B$, then

1) $\lim_{x \to a} (f(x) + g(x)) = A + B$

2) $\lim_{x \to a} (f(x) - g(x)) = A - B$

3) $\lim_{x \to a} (f(x) \cdot g(x)) = A \cdot B$

4) $\lim_{x \to a} \frac{f(x)}{g(x)} = \frac{A}{B}$ (under the additional condition $B \ne 0$)

If $\lim_{x \to a} f(x) = A$, then

5) $\lim_{x \to a} |f(x)| = |A|$

6) $\lim_{x \to a} (f(x))^{\alpha} = A^{\alpha}$ (under the condition $\alpha \in N$; or under the condition $\alpha \in Z$ and $A \ne 0$; or under the condition $A > 0$)

7) $\lim_{x \to a} \alpha^{f(x)} = \alpha^{A}$ (under the assumption $\alpha > 0$)

The squeeze theorem. If $\lim_{x \to a} f(x) = \lim_{x \to a} g(x) = A$, and the inequality

$f(x) \le h(x) \le g(x)$ holds for all $x \in X$ in a deleted neighborhood of a, then $\lim\limits_{x\to a} h(x) = A$.

2.2 Concepts

Example 1. "If $f(x)$ is defined on $\mathbb{R}$, then there is at least one point where $\lim\limits_{x\to a} f(x)$ exists."

Solution.

Consider Dirichlet's function $D(x) = \begin{cases} 1, & x \in \mathbb{Q} \\ 0, & x \in \mathbb{I} \end{cases}$. It is easy to show that $D(x)$ has no limit at any real point. In fact, in any deleted neighborhood of an arbitrary point a there exist both rational and irrational points. Therefore, the number 1 cannot be the limit of $D(x)$ at a, because for $\varepsilon_0 = 1$ and any small δ there exist points x (irrational ones) such that $0 < |x - a| < \delta$, but $|f(x) - 1| = |0 - 1| \ge \varepsilon_0$. Similarly, the number 0 cannot be the limit, because in any δ-neighborhood of a there exist points x (rational ones) such that $0 < |x - a| < \delta$, but $|f(x) - 0| = |1 - 0| \ge \varepsilon_0$. Finally, any real number $A \ne 0$ cannot be the limit of $D(x)$, because for $\varepsilon_0 = |A|$ in any δ-neighborhood of a there exist points x (irrational ones) such that $0 < |x - a| < \delta$, but $|f(x) - A| = |0 - A| \ge \varepsilon_0$. Hence, the definition of limit is not satisfied at arbitrary point a, whatever the value of A we try.

Example 2. "If $\nexists \lim\limits_{x\to a} f(x)$ then $\nexists \lim\limits_{x\to a} |f(x)|$."

Solution.

Consider a slight modification of Dirichlet's function $\tilde{D}(x) = \begin{cases} 1, & x \in \mathbb{Q} \\ -1, & x \in \mathbb{I} \end{cases}$.

Evidently, $\tilde{D}(x)$ has no limit at any point just like $D(x)$. However, $\left|\tilde{D}(x)\right| = 1$ and this function has the limit 1 at any point.

Remark 1. This is the false converse to the right statement: if $\exists \lim\limits_{x\to a} f(x)$ then $\exists \lim\limits_{x\to a} |f(x)|$.

Remark 2. The situation with the square is quite similar. The statement: "if $\exists \lim\limits_{x\to a} f^2(x)$, then $\exists \lim\limits_{x\to a} f(x)$" is the false converse to the right statement: if $\exists \lim\limits_{x\to a} f(x)$ then $\exists \lim\limits_{x\to a} f^2(x)$. The corresponding counterexample can be provided by the same function $\tilde{D}(x)$.

Example 3. "If a sequence y_n converges to A and a function $f(x)$ is such that $f(n) = y_n$, $\forall n \in \mathbb{N}$, then $\lim\limits_{x\to+\infty} f(x) = A$."

Solution.

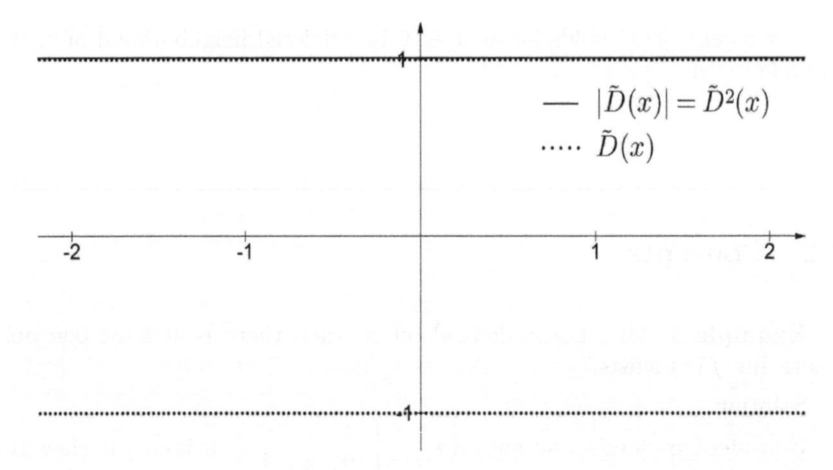

FIGURE 2.2.1: Example 2

Consider the function $f(x) = \sin \pi x$, which in the points $x_n = n$, $\forall n \in \mathbb{N}$ is equal to 0. The corresponding sequence $y_n = f(x_n) = \sin \pi n = 0$, $\forall n \in \mathbb{N}$ converges to 0, but the function has no limit as x approaches infinity ($f(x)$ has partial limits that take any value in $[-1, 1]$, for example, $\lim\limits_{x_n \to +\infty} f(x_n) = \lim\limits_{n \to +\infty} \sin \pi n = 0$ for $x_n = n$, $\forall n \in \mathbb{N}$ and $\lim\limits_{x_k \to +\infty} f(x_k) = \lim\limits_{k \to +\infty} \sin \pi \left(2k + \frac{1}{2}\right) = 1$ for $x_k = 2k + \frac{1}{2}$, $\forall k \in \mathbb{N}$).

Remark. The converse is true: if $\lim\limits_{x \to +\infty} f(x) = A$ and the sequence is defined by the relation $y_n = f(n)$, $\forall n \in \mathbb{N}$, then $\lim\limits_{n \to +\infty} y_n = A$.

Example 4. "If there exists a sequence x_n such that $\lim\limits_{x_n \to a} f(x_n) = A$, then $\lim\limits_{x \to a} f(x) = A$."

Solution.

Let us consider Dirichlet's function $D(x) = \begin{cases} 1, & x \in \mathbb{Q} \\ 0, & x \in \mathbb{I} \end{cases}$ and $a = 0$. Using the rational points in the form $x_n = \frac{1}{n}$, $n \in \mathbb{N}$ we form the sequence such that $x_n \to 0$ when $n \to +\infty$. Calculating the corresponding limit we have $\lim\limits_{x_n \to 0} D(x_n) = \lim\limits_{x_n \to 0} 1 = 1$. However, a general limit does not exist (see Example 1 for details).

Another interesting function is $f(x) = \sin \frac{1}{x}$ defined on $\mathbb{R} \backslash \{0\}$. First, using the points $x_k = \frac{1}{k\pi}$, $k \in \mathbb{N}$ we have the sequence such that $x_k \to 0$ when $k \to +\infty$. For this sequence we get $\lim\limits_{x_k \to 0} f(x_k) = \lim\limits_{k \to +\infty} \sin(k\pi) = \lim\limits_{k \to +\infty} 0 = 0$. Another sequence can be chosen as follows: $x_n = \frac{1}{\pi/2 + 2n\pi}$, $n \in \mathbb{N}$ with the same property $x_n \to 0$ when $n \to +\infty$. For this sequence $\lim\limits_{x_n \to 0} f(x_n) = \lim\limits_{n \to +\infty} \sin\left(\frac{\pi}{2} + 2n\pi\right) = \lim\limits_{n \to +\infty} 1 = 1$. Since two partial limits are different, a general limit does not exist. Actually, it can be easily shown that the set of all the partial limits of this function covers the entire interval $[-1,1]$.

Remark 1. Of course, under the above conditions, if $\lim_{x\to a} f(x)$ exists then it is equal to A.

Remark 2. This is a weakened version of the Cauchy-Heine criterion: $\lim_{x\to a} f(x) = A$ if, and only if, $\lim_{x_n\to a} f(x_n) = A$ for all sequences x_n such that $x_n \underset{n\to+\infty}{\to} a$, $x_n \neq a$.

Remark 3. Evidently, the following more general statement is also false: "if a finite or an infinite number of partial limits exist and coincide, then a general limit also exists". Both counterexamples provided above work in this case too.

Remark 4. The existence and equality of partial limits can guarantee the existence of a general limit only in the case when the subsets of real numbers used in these limits cover all the points of the function domain in a deleted neighborhood of the limit point a. For example, if $f(x)$ is defined in some deleted neighborhood of a, then this is the case of two one-sided limits. Also this is the case of two partial limits based first on all rational numbers and second on all irrational numbers in a deleted neighborhood of a.

Remark 5. In practice, when an evaluation of a general limit is difficult, the comparison of partial limits can be a useful way to get an idea about a general limit and, in the case of inexistence of some partial limits or inequality between two partial limits, it leads to the immediate conclusion about inexistence of a general limit.

Example 5. "If a function has both one-sided limits, then it has a general limit too."

Solution.

Let $f(x) = \operatorname{sgn} x = \begin{cases} -1\,, & x<0 \\ 0\,, & x=0 \\ 1\,, & x>0 \end{cases}$. Then both one-sided limits exist: $\lim_{x\to 0_-} \operatorname{sgn} x = \lim_{x\to 0_-} (-1) = -1$ and $\lim_{x\to 0_+} \operatorname{sgn} x = \lim_{x\to 0_+} 1 = 1$, but since they are different, the general $\lim_{x\to 0} f(x)$ does not exist.

Remark. The well-known in Calculus result states that a general limit exists if, and only if, both one-sided limits exist and coincide (this statement is true both for finite and infinite limits). Notice, however, that this result is true in the context of the Calculus simplified supposition that the function is defined in a deleted neighborhood of the limit point a. In general, this result is not valid as is shown in Example 7 (although it is still true that when both one-sided limits exist and coincide then a general limit exists also and has the same value.)

Example 6. "If the one-sided limits at point a coincide, then there exists a finite limit at a."

Solution.

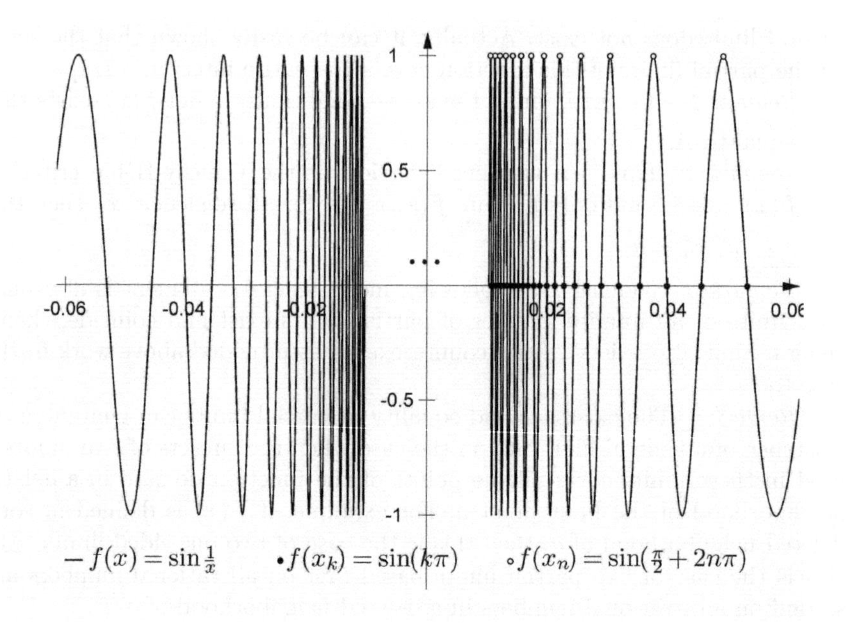

FIGURE 2.2.2: Example 4

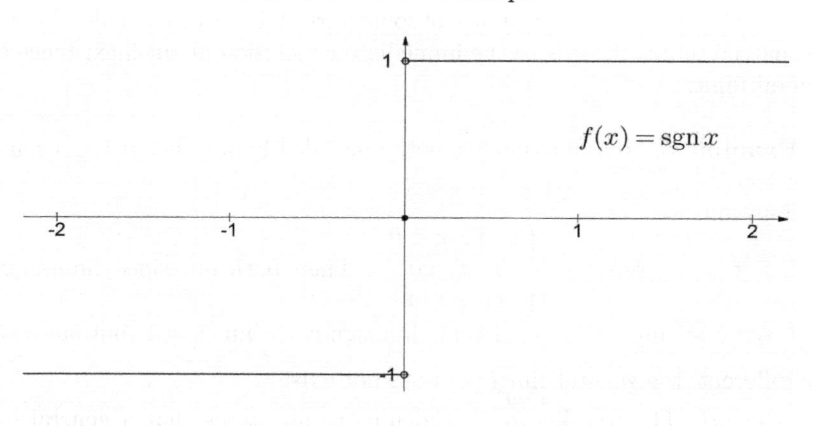

FIGURE 2.2.3: Example 5

The function $f(x) = \frac{1}{x^2}$ has both one-sided limits that are equal: $\lim\limits_{x\to 0_-} \frac{1}{x^2} = \lim\limits_{x\to 0_+} \frac{1}{x^2} = +\infty$. The general limit also exists, but it is infinite: $\lim\limits_{x\to 0} \frac{1}{x^2} = +\infty$.

Example 7. "If $f(x)$ is not defined in the left/right-hand neighborhood of a, then the limit of $f(x)$ at a does not exist."

Solution.

If $f(x)=\sqrt{x}$ and $a=0$ then the function is not defined in any left-hand neighborhood of a. However, applying the general definition of the limit one can easily obtain that $\lim\limits_{x\to 0}\sqrt{x}=\lim\limits_{x\to 0_+}\sqrt{x}=0$. Indeed, for $\forall\varepsilon>0$ choosing $\delta=\varepsilon^2$ we obtain for all x such that $0<x<\delta$ (that is, according to the general definition, for all x of the domain of $f(x)$ that are located in the deleted δ-neighborhood of zero) that $|f(x)-0|=\sqrt{x}<\sqrt{\delta}=\varepsilon$. Therefore, the definition is satisfied.

Remark. In the study of limits in Calculus, the simplified assumption that a function is defined in a deleted neighborhood of the limit point a is frequently used. It leads to the well-known Calculus result about equivalence between existence of a general limit and existence and equality of two one-sided limits. This example shows that it is not so for a general definition of the limit.

Remark to Examples 8-10. To focus on some important elements in the definition of the limit, in the three following Examples we will use the Calculus assumption that a function is defined in a deleted neighborhood of the limit point a.

Example 8. "The definition of the limit can be reformulated as follows: the limit of $f(x)$ as x approaches a is A if for every $\varepsilon>0$ there is a corresponding $\delta>0$ such that if $|f(x)-A|<\varepsilon$ then $|x-a|<\delta$."

Solution.

At first glance, the above "definition" conveys an impression that it has at least some relation to the correct definition, because the inequalities of the correct definition, meaning proximity among function values and among argument values, are involved here. Nevertheless, this statement has nothing to do with the correct definition. In fact, if $f(x)$ is defined on a bounded interval, say (a,b), then the above statement does not imply any restriction on $f(x)$, because by choosing $\delta=b-a$ the required inequality $|x-a|<\delta$ is satisfied for an arbitrary $x\in(a,b)$ and for an arbitrary $\varepsilon>0$. On the other hand, if $f(x)$ is defined on an unbounded domain, say on $\mathbb{R}$, then the above "definition" also fails. In fact, let us consider the function $f(x)=\begin{cases} e^x\,, & x\in\mathbb{Q}\\ -e^x\,, & x\in\mathbb{I}\end{cases}$ on $\mathbb{R}$. Let us choose any $A>0$ and consider ε-neighborhood of A: $|y-A|<\varepsilon$, $\varepsilon\le\frac{A}{2}$. The corresponding δ-neighborhood of the point $a=\ln A$ is defined as $|x-a|<\delta$ with $\delta=\max\{\ln(A+\varepsilon)-a,\,a-\ln(A-\varepsilon)\}=a-\ln(A-\varepsilon)$. It can be checked now that for all the values of $f(x)$ that satisfied the condition $|f(x)-A|<\varepsilon$ it follows that $|x-a|<\delta$ (more precisely, $x\in(\ln(A-\varepsilon),\ln(A+\varepsilon))\subset(a-\delta,a+\delta)$). However, $f(x)$ has no limit at a, because for any $x\in\mathbb{I}$ (in particular for $x\in\mathbb{I}$ in the δ-neighborhood of $a=\ln A$) one obtains $|f(x)-A|=|-e^x-A|>A$.

Remark. For an unbounded domain, say $\mathbb{R}$, the above "definition" can also be too restrictive. For example, if $f(x)=\sin x$, $A=0$ and $\varepsilon=3$, then the inequality $|x-a|<\delta$ holds for an arbitrary x (and a) whatever δ is. It means that the implication of the "definition" is not true even for some continuous functions. (We are acting here in advance, because the concept of continuity will be considered in the next chapter.)

Example 9. "The definition of the limit can be reformulated as follows: the limit of $f(x)$ as x approaches a is A if for every $\delta>0$ there is a corresponding $\varepsilon>0$ such that if $|f(x)-A|<\varepsilon$ then $|x-a|<\delta$."

Solution.

Actually this condition means that there exists a partial limit A of $f(x)$ as x approaches a (that is, there is a sequence of points x_k such that $\lim\limits_{x_k\to a} f(x_k)=A$). However, it is not sufficient to guarantee the existence of the general limit. For example, if $f(x)=\begin{cases} x^2, & x\in\mathbb{Q} \\ 1, & x\in\mathbb{I}\end{cases}$, $a=0$ and $A=0$, then by choosing $\varepsilon=\delta^2$ for any $\delta<1$ one guarantees that for all $f(x)$ such that $|f(x)|=|x^2|<\varepsilon$ we obtain $|x|<\delta$ for the corresponding values of x. However, the general limit does not exist, because there are two different partial limits: $\lim\limits_{x\to 0,\, x\in\mathbb{Q}} f(x)=0$ and $\lim\limits_{x\to 0,\, x\in\mathbb{I}} f(x)=1$.

Remark. If we assume that the inverse function exists, then the above statement is the definition of the limit (equal to a) of the inverse function $x=f^{-1}(y)$ when the argument y approaches A.

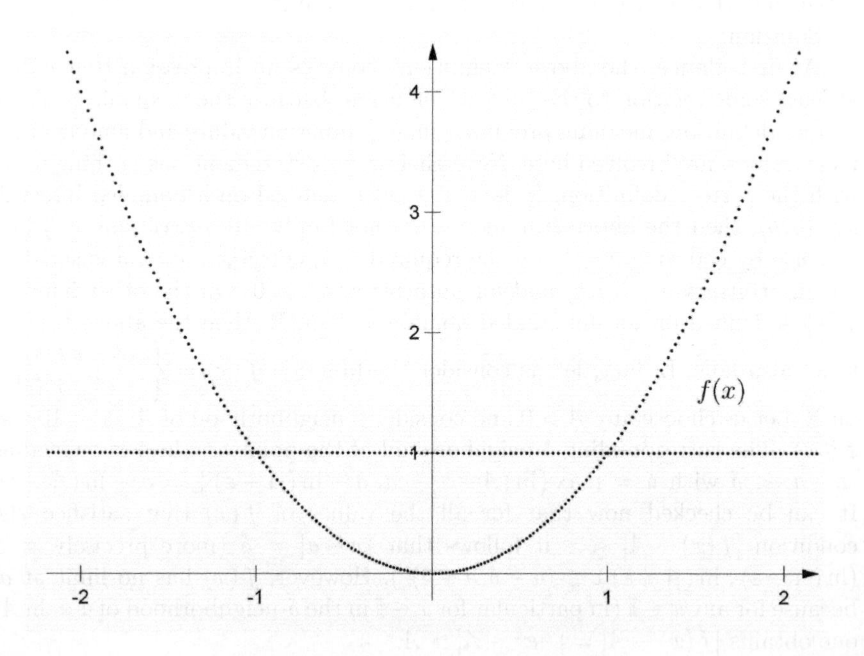

FIGURE 2.2.4: Example 9

Example 10. "The definition of the limit can be reformulated as follows: the limit of $f(x)$ as x approaches a is A if for every $\delta > 0$ there is a corresponding $\varepsilon > 0$ such that if $|x - a| < \delta$ then $|f(x) - A| < \varepsilon$."

Solution.

The above "definition" seems to be rather natural and tempting, because it apparently corresponds to a vague idea about limits: when x approaches a the function values $f(x)$ approach A. However, this similarity is only appearing, because the inversion of the dependence between δ and ε (in comparison with the correct definition) leads to possibility to attribute to ε arbitrary large values, that destroys the requirement that $f(x)$ should approach A. For example, Dirichlet's function $D(x)$, which actually does not have a limit at any point (see Example 1), satisfies the above "definition" with an arbitrary a and $A = 0$ if one chooses $\varepsilon = 2$ for every $\delta > 0$.

Remark. The condition of the statement implies boundedness of $f(x)$ in any deleted neighborhood of the point a.

Example 11. "If for a specific function $f(x)$ and point a the dependence of δ from ε in the limit definition cannot be expressed in the form $\delta = c\varepsilon$, where c is a positive constant, then $f(x)$ does not have a limit at a."

Solution.

For $f(x) = \sqrt[3]{x}$ considered in a neighborhood of the point $a = 0$, if $\delta = \varepsilon^3$ then the limit definition is satisfied: $\lim\limits_{x \to 0} \sqrt[3]{x} = 0$. However, there is no constant $c > 0$ such that $\delta = c\varepsilon$. Indeed, if $0 < |x| < \delta = c\varepsilon$, then we can obtain $|\sqrt[3]{x}| < \sqrt[3]{c\varepsilon}$, but for any fixed c the expression in the right-hand side is greater than ε, if ε is sufficiently small. Therefore, we will not obtain the required evaluation $|\sqrt[3]{x}| < \varepsilon$ if $\delta = c\varepsilon$.

Remark. Of course, similar statements with requirement of any other specific law $\delta(\varepsilon)$ are also false.

Example 12. "If for a specific function $f(x)$ and point a there are two different ways to determine $\delta(\varepsilon)$ in the limit definition, then $f(x)$ does not have a limit at a."

Solution.

Actually, if the limit definition is satisfied, then always there are infinitely many ways to determine $\delta(\varepsilon)$. For example, if $f(x) = x$ and $a = 0$ then an evident choice to satisfy the definition of $\lim\limits_{x \to 0} x = 0$ is $\delta = \varepsilon$. But it means that for any constant $0 < c < 1$ the law $\delta = c\varepsilon$ is also suitable, as well as $\delta = \min\left\{\varepsilon, \varepsilon^2\right\}$, or $\delta = \min\left\{\varepsilon, \varepsilon^3\right\}$, or more exotic $\delta = \ln(1 + \varepsilon)$ and $\delta = 1 - e^{-\varepsilon}$, or many others.

2.3 Elementary properties (arithmetic and comparative)

Example 1. "If $\nexists \lim\limits_{x\to a} f(x)$ and $\nexists \lim\limits_{x\to a} g(x)$ then $\nexists \lim\limits_{x\to a} (f(x)+g(x))$."

Solution.

Let $f(x) = -g(x) = D(x)$. Then for an arbitrary a the limit of $f(x)$ and $g(x)$ does not exist (see Example 1, section 2.2 for details). However, $h(x) = f(x) + g(x) = 0$ has zero limit at any point a.

Remark 1. Similar simple examples can be given for other arithmetic operations. For example, for the product we can use the following modification of Dirichlet's function: $f(x) = -g(x) = \tilde{D}(x) = \begin{cases} 1, & x \in \mathbb{Q} \\ -1, & x \in \mathbb{I} \end{cases}$, with $h(x) = f(x) \cdot g(x) = -1$.

Remark 2. This is a false "negative" version of the sum rule for limits.

Remark 3. This "negative" formulation is equivalent to the direct converse to the sum rule, that is the following statement is equally false: "if $\exists \lim\limits_{x\to a} (f(x)+g(x))$ then $\exists \lim\limits_{x\to a} f(x)$ and $\exists \lim\limits_{x\to a} g(x)$".

Example 2. "If both limits $\lim\limits_{x\to a} f(x)$ and $\lim\limits_{x\to a} g(x)$ exist and the former equals zero, then $\lim\limits_{x\to a} \frac{f(x)}{g(x)} = 0$."

Solution.

Let $f(x) = g(x) = x$ and $a = 0$. Then $\lim\limits_{x\to 0} f(x) = \lim\limits_{x\to 0} g(x) = 0$, but $\lim\limits_{x\to 0} \frac{f(x)}{g(x)} = \lim\limits_{x\to 0} \frac{x}{x} = \lim\limits_{x\to 0} 1 = 1$. (The function $\frac{x}{x}$ is defined in a deleted neighborhood of zero, which is sufficient to consider the concept of the limit, and in any such neighborhood $\frac{x}{x}$=1, which justify the equality $\lim\limits_{x\to 0} \frac{x}{x} = \lim\limits_{x\to 0} 1$).

Remark 1. This is a false extension of the ratio rule for limits.

Remark 2. It is important (and interesting) to note that if $\lim\limits_{x\to a} f(x) = 0$ and $\lim\limits_{x\to a} g(x) = 0$ then the result for $\lim\limits_{x\to a} \frac{f(x)}{g(x)}$ is unpredictable in a general case, and it can lead to a bunch of different conclusions depending on a specific choice of functions $f(x)$ and $g(x)$. This is why this general situation is called an indeterminate form and is denoted by symbol $\frac{0}{0}$. To specify different situations that can occur under the general assumption that $\lim\limits_{x\to a} f(x) = 0$ and $\lim\limits_{x\to a} g(x) = 0$, let us briefly consider the following examples with the limit point $a = 0$:

1) if $f(x) = x^2$ and $g(x) = x$ then $\lim\limits_{x\to 0} f(x) = \lim\limits_{x\to 0} g(x) = 0$ and $\lim\limits_{x\to 0} \frac{f(x)}{g(x)} = \lim\limits_{x\to 0} x = 0$.

2) if $f(x) = cx$ $(c = const)$ and $g(x) = x$ then $\lim\limits_{x\to 0} f(x) = \lim\limits_{x\to 0} g(x) = 0$ and $\lim\limits_{x\to 0} \frac{f(x)}{g(x)} = \lim\limits_{x\to 0} c = c$.

3) if $f(x) = x$ and $g(x) = x^3$ then $\lim_{x\to 0} f(x) = \lim_{x\to 0} g(x) = 0$ and $\lim_{x\to 0} \frac{f(x)}{g(x)} = \lim_{x\to 0} \frac{1}{x^2} = +\infty$.

4) if $f(x) = -x$ and $g(x) = x^3$ then $\lim_{x\to 0} f(x) = \lim_{x\to 0} g(x) = 0$ and $\lim_{x\to 0} \frac{f(x)}{g(x)} = \lim_{x\to 0} \frac{-1}{x^2} = -\infty$.

5) if $f(x) = \begin{cases} x, \ x \in \mathbb{Q} \\ 0, \ x \in \mathbb{I} \end{cases}$ and $g(x) = x$ then $\lim_{x\to 0} f(x) = \lim_{x\to 0} g(x) = 0$ and $\frac{f(x)}{g(x)} = D(x)$ in $\mathbb{R}\setminus\{0\}$; hence $\lim_{x\to 0} \frac{f(x)}{g(x)}$ does not exist.

Therefore, depending on a specific choice of functions $f(x)$ and $g(x)$, an indeterminate form $\frac{0}{0}$ can result in an arbitrary constant (including zero), infinity or even nonexistence of the limit. This makes false any statement about a specific property of $\lim_{x\to a} \frac{f(x)}{g(x)}$ - existence/non-existence, assuming a specific value or be infinite - and the above specifications of $f(x)$ and $g(x)$ are counterexamples to such statements.

Remark 3. The indeterminate form $\frac{0}{0}$ is one of seven indeterminate forms, others being denoted by $0\cdot\infty$, $\frac{\infty}{\infty}$, $\infty-\infty$, 1^∞, ∞^0, 0^0. It is easy to specify what each symbol means just comparing with the description for $\frac{0}{0}$. For example, an indeterminate form $0\cdot\infty$ means that $\lim_{x\to a} f(x) = 0$ and $\lim_{x\to a} g(x) = (\pm)\infty$ and we are considering the limit $\lim_{x\to a} f(x)\cdot g(x)$. Of course, similar statements and corresponding examples can be constructed for each of indeterminate forms.

Remark 4. All the above considerations are true for both finite and infinite limit points a.

Remark 5. The condition of non-zero limit of denominator is the important condition for the validity of the ratio rule. At the same time, a mere fact that the denominator has zero limit does not mean that the limit of the ratio does not exist. This last observation is of extreme importance for one of the central concepts in Calculus and applications - derivative, which is dealing with the indeterminate form $\frac{0}{0}$.

Example 3. "If $\lim_{x\to a} f(x) = A$, $\lim_{x\to a} g(x) = B$ and $A \le B$, then there is a deleted neighborhood of a where $f(x) \le g(x)$."

Solution.

Let $f(x) = x$, $g(x) = 2x$ and $a = 0$. Then $\lim_{x\to 0} f(x) = 0 \le \lim_{x\to 0} g(x) = 0$, but in any left-hand neighborhood of $a = 0$ it holds $f(x) > g(x)$.

Remark 1. This statement is false for both finite and infinite limit points a.

Remark 2. This is a wrong version of the correct property: if $\lim_{x\to a} f(x) = A$, $\lim_{x\to a} g(x) = B$ and $A < B$, then there is a deleted neighborhood of a where $f(x) < g(x)$.

Example 4. "If $\lim_{x\to a} f(x) = A$, $\lim_{x\to a} g(x) = B$ and there is a deleted neighborhood of a where $f(x) < g(x)$ then $A < B$."

Solution.

Let $f(x) = -x^2$, $g(x) = x^2$ and $a = 0$. Then in any deleted neighborhood of zero $f(x) < g(x)$ and both limits exist, but $\lim_{x\to 0} f(x) = 0 = \lim_{x\to 0} g(x)$.

Remark 1. This statement is false for both finite and infinite limit points a. For example, for $a = +\infty$ this statement assumes the form: "if $\lim_{x\to+\infty} f(x) = A$, $\lim_{x\to+\infty} g(x) = B$ and $f(x) < g(x)$ for all $x > 0$ then $A < B$". The corresponding counterexample is: $f(x) = \frac{1}{x}$, $g(x) = \frac{2}{x}$ with $f(x) < g(x)$ for all $x > 0$ and $\lim_{x\to+\infty} f(x) = \lim_{x\to+\infty} g(x) = 0$.

Remark 2. This is a wrong version of the correct property: if $\lim_{x\to a} f(x) = A$, $\lim_{x\to a} g(x) = B$ and there is a deleted neighborhood of a where $f(x) \le g(x)$ then $A \le B$.

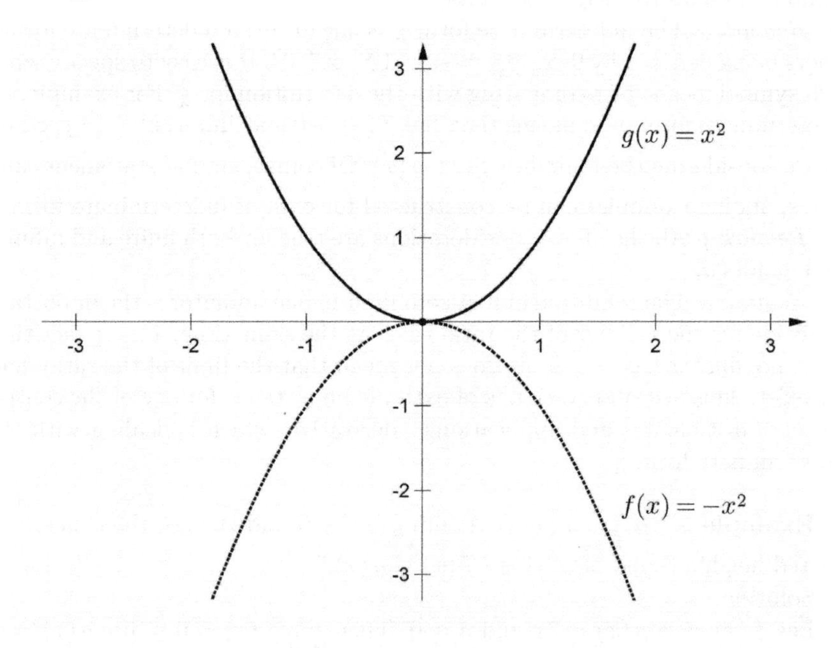

FIGURE 2.3.1: Example 4

Remark to Examples 3 and 4. The two last Examples show that slight modifications of the signs in formulations of the comparative properties (changing strong inequality to non-strong and vice-versa) lead to wrong results.

Example 5. "If $\lim_{x\to a} f(x) = A$ and $\lim_{x\to a} g(x) = (\pm)\infty$, then $\lim_{x\to a} f(x) \cdot g(x) = (\pm)\infty$."

Solution.

Let $f(x) = x^2$, $g(x) = \frac{1}{x^2}$ and $a = 0$. Then $\lim_{x\to 0} f(x) = 0$, $\lim_{x\to 0} g(x) = +\infty$, but $\lim_{x\to 0} f(x) \cdot g(x) = 1$.

Remark 1. This statement is false for both finite and infinite limit points a.

Remark 2. Of course the simplest example is with $f(x) \equiv 0$, but it can give a wrong impression that $f(x) \equiv 0$ is the only function when the limit of the product exists.

Remark 3. The statement will be true if to add that $A \neq 0$.

Remark 4. For the sum the corresponding statement is true: if $\lim_{x\to a} f(x) = A$ and $\lim_{x\to a} g(x) = (\pm)\infty$, then $\lim_{x\to a} (f(x) + g(x)) = (\pm)\infty$. The same is for the difference.

Example 6. "If $\lim_{x\to a} f(x) = A$ and $\lim_{x\to a} g(x)$ does not exist, then $\lim_{x\to a} f(x) \cdot g(x)$ does not exist."

Solution.

Let $f(x) = x$, $g(x) = D(x)$ and $a = 0$. Then $\lim_{x\to 0} f(x) = 0$ and $\lim_{x\to 0} g(x)$ does not exist (see Example 1, section 2.2), but $h(x) = f(x) \cdot g(x) = \begin{cases} x, & x \in \mathbb{Q} \\ 0, & x \in \mathbb{I} \end{cases}$ and $\lim_{x\to 0} h(x) = 0$ (since $\lim_{x\to 0, x\in\mathbb{Q}} h(x) = \lim_{x\to 0, x\in\mathbb{Q}} x = 0$ and $\lim_{x\to 0, x\in\mathbb{I}} h(x) = \lim_{x\to 0, x\in\mathbb{I}} 0 = 0$ and $\mathbb{Q} \cup \mathbb{I} = \mathbb{R}$).

Remark 1. This statement is false for both finite and infinite limit points a.

Remark 2. Of course the simplest example is with $f(x) \equiv 0$, but it can give a wrong impression that $f(x) \equiv 0$ is the only case when the limit of the product exists.

Remark 3. The statement will be true if to add that $A \neq 0$.

Remark 4. The existence of finite limits of both functions $f(x)$ and $g(x)$ guarantees the existence of the limit of their product (and also sum and difference). On the other hand, the existence of one of the limits and non-existence of another one assures the non-existence of the limit of the sum and difference, but not of the product.

Remark 5. The corresponding statement for the ratio is also false.

Example 7. "If $\lim_{x\to a} f(x) = A$ and $\lim_{x\to A} g(x) = B$, then $\lim_{x\to a} g(f(x)) = B$."

Solution.

If $f(x) = 0$, $\forall x \in \mathbb{R}$ and $a = 0$ then $\lim_{x\to 0} f(x) = 0 = A$. Further, if $g(x) = \begin{cases} 0, & x \neq 0 \\ 1, & x = 0 \end{cases}$ then $\lim_{x\to 0} g(x) = 0 = B$. However, $g(f(x)) = 1$, $\forall x \in \mathbb{R}$ and hence $\lim_{x\to 0} g(f(x)) = 1 \neq B$.

Remark 1. The statement will be true if the following condition is added: $f(x) \neq A$ when $x \neq a$.

Remark 2. In chapter 3 we will consider continuous functions and their properties, but acting in advance we can recall the following Composite function theorem: if $\lim\limits_{x\to a} f(x) = b$, $g(x)$ is continuous at b and the composite function $g(f(x))$ is defined in a deleted neighborhood of a, then $\lim\limits_{x\to a} g(f(x)) = g(b)$. It is quite tempting to relax a bit the condition of continuity in this theorem by using instead the condition of the existence of a finite limit. But this Example shows that the result of this simple substitution is disastrous.

Example 8. "If a function $f(x)$ is non-negative and unbounded in a neighborhood of a, then $f(x)$ does not vanish in a deleted neighborhood of a."

Solution.

Consider $f(x) = \begin{cases} \left|\frac{1}{x}\sin\frac{1}{x}\right|, & x \neq 0 \\ 0, & x = 0 \end{cases}$. This non-negative function is unbounded in a neighborhood of $a = 0$, since for $x_k = \frac{1}{\pi/2+2k\pi}$, $k \in \mathbb{N}$ we have $\lim\limits_{x_k\to 0} f(x_k) = \lim\limits_{k\to+\infty} \left(\frac{\pi}{2}+2k\pi\right)\sin\left(\frac{\pi}{2}+2k\pi\right) = \lim\limits_{k\to+\infty} \left(\frac{\pi}{2}+2k\pi\right) = +\infty$. On the other hand, for $x_n = \frac{1}{n\pi}$, $n \in \mathbb{N}$ we have $x_n \underset{n\to\infty}{\to} 0$ and $f(x_n) = n\pi\sin n\pi = 0$ for any $n \in \mathbb{N}$.

Remark 1. Of course the point zero has nothing special for this example. A similar counterexample $f(x) = \begin{cases} \left|\frac{1}{x-a}\sin\frac{1}{x-a}\right|, & x \neq a \\ 0, & x = a \end{cases}$ can be given for any a. Moreover, a can be infinite. In the last case, $f(x) = x^2(1-\cos x)$ is a suitable function for the counterexample.

Remark 2. If $f(x)$ is non-constant on any interval, non-negative and bounded on $\mathbb{R}$, the statement still remains false. The counterexample with the function $f(x) = \begin{cases} 1+\sin\frac{1}{x}, & x \neq 0 \\ 0, & x = 0 \end{cases}$ and $a = 0$ works in this case.

Example 9. "If $f(x)$ is not bounded in any neighborhood of a point a, then at least one of the limits $\lim\limits_{x\to a_-} |f(x)|$ or $\lim\limits_{x\to a_+} |f(x)|$ is infinite."

Solution.

Let us consider the function $f(x) = \frac{1}{x}\cos\frac{1}{x}$ defined in $\mathbb{R}\backslash\{0\}$. Since for $x_k = \frac{1}{2k\pi}$, $k \in \mathbb{Z}\backslash\{0\}$ we get $\lim\limits_{x_k\to 0} f(x_k) = \lim\limits_{k\to\pm\infty} 2k\pi\cos(2k\pi) = \lim\limits_{k\to\pm\infty} 2k\pi = \pm\infty$, that is, the function is unbounded in any neighborhood of zero. At the same time, the general limit does not exist, because for $x_n = \frac{1}{\pi/2+n\pi}$, $n \in \mathbb{Z}$ we obtain $\lim\limits_{x_n\to 0} f(x_n) = \lim\limits_{n\to\pm\infty} \left(\frac{\pi}{2}+n\pi\right)\cos\left(\frac{\pi}{2}+n\pi\right) = \lim\limits_{n\to\pm\infty} 0 = 0$.

Remark. Under the conditions of the statement one can only guarantee that there exists a partial infinite limit at a.

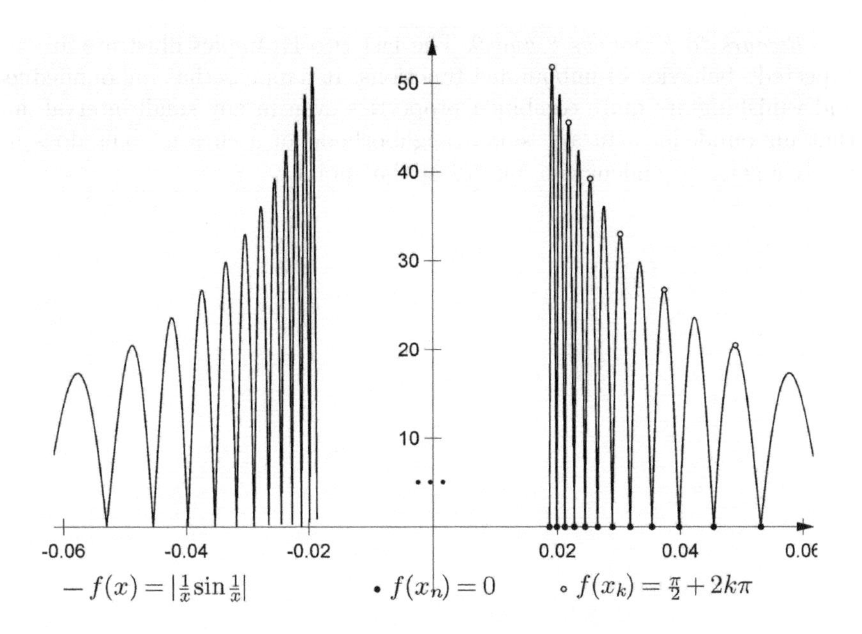

FIGURE 2.3.2: Example 8

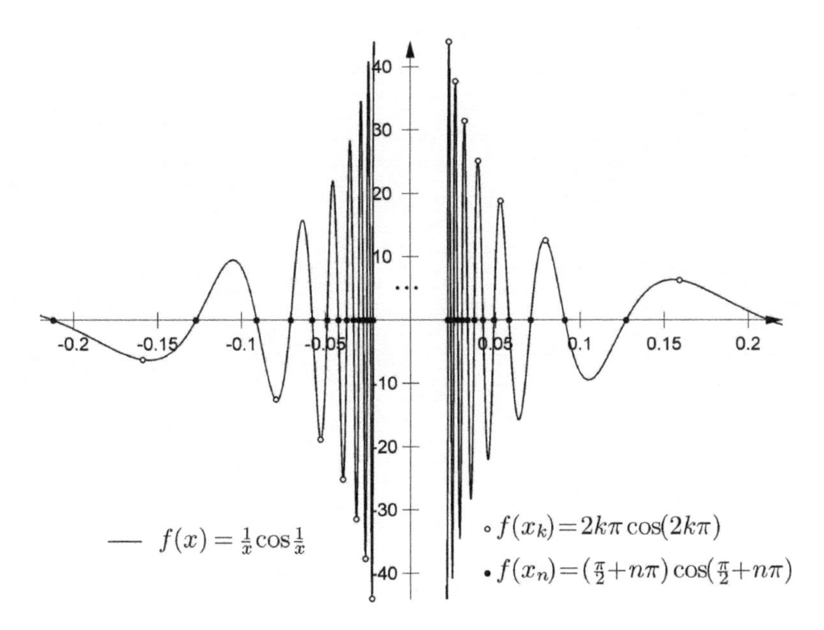

FIGURE 2.3.3: Example 9

Remark to Examples 8 and 9. The last two Examples illustrate an "unexpected" behavior of unbounded functions. It happens that unboundedness and vanishing are quite combined properties even in any small interval, and that unboundedness in any small neighborhood of a chosen point does not imply a general tendency to infinity at that point.

Exercises

1. Show that $\cos\frac{1}{x}$ considered at the limit point $a = 0$ is another counterexample to the statement in Example 4 in section 2.2. What about the $e^{\frac{1}{x}}$ at $a = 0$?

2. Provide a counterexample to the statement: "if $\lim\limits_{x\to a} f(x) = \infty$ and $\lim\limits_{x\to a} g(x)$ exists then $\lim\limits_{x\to a} (f(x) + g(x)) = \infty$."

3. Provide a counterexample to the statement: "if $f(x)$ is bounded in a neighborhood of a and $\lim\limits_{x\to a_+} f(x) = A$, then left-hand limit also exists".

4. Show that the following "definition" of $\lim\limits_{x\to+\infty} f(x) = A$ is not correct: "for every $D > 0$ there exists a corresponding $\varepsilon > 0$ such that $x > D$ implies that $|f(x) - A| < \varepsilon$ ".

5. Show that the following "definition" of $\lim\limits_{x\to a} f(x) = +\infty$ is not correct: "for every $\delta > 0$ there exists a corresponding $E > 0$ such that $0 < |x - a| < \delta$ implies that $f(x) > E$ ".

6. Verify if the following "definition" of the limit is correct: "for every $\varepsilon > 0$ there exists δ such that for all $x \in X$ such that $0 < |x - a| < \delta$ it follows that $|f(x) - A| < \varepsilon$". If not, provide a counterexample.

7. Suppose $f(x)$ is defined on X, a is a limit point of X, and $k \in \mathbb{N}$ is a fixed number. Show that the following definition of the limit is false: "if $\forall\varepsilon \geq 10^{-k}$ there exists $\delta(\varepsilon) > 0$ such that $\forall x \in X,\ 0 < |x - a| < \delta$ it follows that $|f(x) - A| < \varepsilon$, then $\lim\limits_{x\to a} f(x) = A$". (Hint: use the function $f(x) = 10^{-(k+1)}[x], X = \mathbb{R}, a = 1, A = 10^{-(k+1)}$.)

8. Give a counterexample for the statement in Example 1, section 2.3 with the difference and ratio of functions instead of the sum.

9. Show that $0 \cdot \infty$ is the indeterminate form, that is, under the condition $\lim\limits_{x\to a} f(x) = 0$ and $\lim\limits_{x\to a} g(x) = \infty$ any statement about existence/non-existence or (in the case of the existence) about a specific value (0, non-zero constant, infinity) of the $\lim\limits_{x\to a} f(x) \cdot g(x)$ is not valid. (Hint: follow the construction of counterexamples in Remark 2 to Example 2, section 2.3)

Do the same for the indeterminate form 0^0.

10. Give a counterexample to Example 3, section 2.3 such that $f(x) > g(x), \forall x \in \mathbb{R}\backslash\{a\}$.

11. Construct a counterexample to the following statement "if $\sin f(x)$ has a limit at a, then $f(x)$ also has a limit at this point". (Notice that the converse is true.)

12. Provide the counterexample to the statement: "if $\lim\limits_{x\to a} f(x) = A$, $\lim\limits_{x\to a} g(x) = B$ and $A^2 < B^2$, then there is a deleted neighborhood of a where $f(x) < g(x)$".

13. Suppose that $f(x)$ and $g(x)$ are defined on X, a is a limit point of X, and $\lim_{x\to a} f(x)$ does not exist (either finite or infinite) while $\lim_{x\to a} g(x) = \infty$. Show that under the above conditions the following statements are false:
1) "$f(x) + g(x)$ has no limit at a"
2) "$f(x) + g(x)$ has the infinite limit at a"
3) "$f(x) \cdot g(x)$ has no limit at a"
4) "$f(x) \cdot g(x)$ has the infinite limit at a"

14. Show that the functions $f(x) = x^2 \cos \frac{\pi}{x}$ and $g(y) = \operatorname{sgn}^2 y$ with the limit point $a = 0$ provide another counterexample to the statement in Example 7, section 2.3.

15. Use the function in Remark 2 to Example 8 in section 2.3 to disprove the statement in that Remark.

16. Give a counterexample to the following statement: "if $f(x)$ is defined on $(a, +\infty)$, bounded on any finite interval (a, b), and $\lim_{x\to+\infty} \frac{f(x)}{x} = A$, then $\lim_{x\to+\infty} (f(x+1) - f(x)) = A$". (The converse is true: if $f(x)$ is defined on $(a, +\infty)$, bounded on any finite interval (a, b), and $\lim_{x\to+\infty} (f(x+1) - f(x)) = A$, then $\lim_{x\to+\infty} \frac{f(x)}{x} = A$.)

17. Provide a counterexample to the following false modification of the squeeze theorem: "if $\lim_{x\to a} f(x) = A \le B = \lim_{x\to a} g(x)$, and the inequality $f(x) \le h(x) \le g(x)$ holds for all $x \in X$ in a deleted neighborhood of a, then $A \le \lim_{x\to a} h(x) \le B$".

Chapter 3

Continuity

General Remark. Due to the close connection between the concepts of limit and continuity, the majority of the examples of chapter 2 can be reformulated for continuous functions. A few of them are described in this chapter to exemplify the modifications required. Notice that all these properties are of a local character, and they appear in section 3.2 together with some additional examples of local properties. The greatest attention in this chapter is paid to global properties of functions related to continuity, because these issues were not discussed so far and they are important both in mathematics and applications.

3.1 Elements of theory

Concepts

Continuity at a point. A function $f(x)$ defined on X is *continuous* at a point $a \in X$ if for every $\varepsilon > 0$ there exists $\delta > 0$ such that whenever $x \in X$ and $|x-a| < \delta$ it follows that $|f(x) - f(a)| < \varepsilon$.

Remark. In calculus, a non-essential simplification that $f(x)$ is defined in some neighborhood of a is frequently used. In this case the continuity can be rewritten in the form $\lim\limits_{x \to a} f(x) = f(a)$ or $f(x) \underset{x \to a}{\to} f(a)$.

Actually the original definition of continuity and the definition by limit differ only in one "pathological" situation of an isolated point a of X, because the limit definition requires a to be a limit point of X, while the general definition of continuity does not contain such a requirement. It means that at any isolated point a of X, a function $f(x)$ is continuous according to the general definition, but not continuous according to the definition by limit. Since the behavior of a function at an isolated point is hardly of any interest, we will usually consider the definition of continuity by limit as a complete definition.

One-sided continuity. A function $f(x)$ defined on X is *right-hand (left-hand) continuous* at a point $a \in X$ if for every $\varepsilon > 0$ there exists $\delta > 0$ such that whenever $x \in X$ and $0 \le x - a < \delta$ $(-\delta < x - a \le 0)$ it follows that $|f(x) - f(a)| < \varepsilon$.

Continuity on a set. A function $f(x)$ defined on X is *continuous on a set* $S \subset X$, if $f(x)$ is continuous at every point of S.

Remark. It is worth to notice that in the case when endpoints belong to the interval, for example a closed interval $[a, b]$, continuity on $[a, b]$ means continuity at every point $c \in (a, b)$ plus right-hand continuity at a and left-hand continuity at b.

Discontinuity point. Let a be a limit point of the domain of $f(x)$. If $f(x)$ is not continuous at a, then a is a point of *discontinuity* of $f(x)$ (or equivalently, $f(x)$ has a discontinuity at a).

Remark. Sometimes it is required that a should be a point of the domain of $f(x)$. We will not impose this restriction.

Classification of discontinuities.

1) Removable discontinuity. $f(x)$ has a removable discontinuity at a, if the finite limit of $f(x)$, as x approaches a, exists, but the function is not defined at a or has the value different from the value of the limit. It is called removable, because the proper redefinition of the original function just at the point a removes this discontinuity.

2) Discontinuity of the first kind (or jump discontinuity). $f(x)$ has a jump discontinuity at a, if both one-sided limits of $f(x)$, as x approaches a, exist and finite, but they have different values (that is, a general limit does not exist).

3) Discontinuity of the second kind (or essential discontinuity). $f(x)$ has an essential discontinuity at a, if at least one of the one-sided limits does not exist or is infinite.

Remark. There are other variants (more and less detailed) of the classification of discontinuities.

Local properties

Remark. The properties of functions that characterize the function behavior in an arbitrary small neighborhood of a point are called the *local properties* as opposed to the *global properties* related to the function behavior on a chosen fixed set (in particular, on the entire domain).

Comparative properties

1) If $f(x)$ and $g(x)$ are continuous at a, and $f(x) \le g(x)$ for all $x \in X$ in a deleted neighborhood of a, then $f(a) \le g(a)$.
2) If $f(x)$ and $g(x)$ are continuous at a, and $f(a) < g(a)$, then $f(x) < g(x)$ for all $x \in X$ in a neighborhood of a.

Arithmetic (algebraic) properties

If $f(x)$ and $g(x)$ are continuous at a, then
1) $f(x) + g(x)$ is continuous at a
2) $f(x) - g(x)$ is continuous at a
3) $f(x) \cdot g(x)$ is continuous at a
4) $\frac{f(x)}{g(x)}$ is continuous at a (under the additional condition $g(a) \ne 0$)
If $f(x)$ is continuous at a, and $\alpha \in \mathbb{R}$, then
5) $|f(x)|$ is continuous at a
6) $(f(x))^\alpha$ is continuous at a (under the condition $\alpha \in \mathbb{N}$; or under the condition $f(a) > 0$)
7) $\alpha^{f(x)}$ is continuous at a (under the assumption $\alpha > 0$)

Composite function theorem. If $\lim_{x \to a} f(x) = b$, $g(x)$ is continuous at b and the composite function $g(f(x))$ is defined in a deleted neighborhood of a, then $\lim_{x \to a} g(f(x)) = g(b)$.

Corollary. If $f(x)$ is continuous at a, $g(x)$ is continuous at $f(a)$, and the composite function $g(f(x))$ is defined in a neighborhood of a, then $g(f(x))$ is continuous at a.

Global properties

Remark. The properties of functions that characterize the function behavior on a chosen set (in particular, on the entire domain) are called the global properties as opposed to the local properties that describe the behavior in a small neighborhood of a point.

Open Set Characterization. $f(x)$ is continuous on a domain X if, and only if, the inverse image $f^{-1}(S)$ of any open set $S \subset f(X)$ is open.

Closed Set Characterization. $f(x)$ is continuous on a domain X if, and only if, the inverse image $f^{-1}(S)$ of any closed set $S \subset f(X)$ is closed.

The Compact Set Theorem. If $f(x)$ is continuous on a compact set S, then its image $f(S)$ is also a compact set. In particular, if $f(x)$ is continuous on $[a,b]$, then its image is $[f_{\min}, f_{\max}]$, where $f_{\min} = \min_{x \in S} f(x)$, $f_{\max} = \max_{x \in S} f(x)$.

The First Weierstrass Theorem. If $f(x)$ is continuous on a compact set, then $f(x)$ is bounded on this set. In particular, if $f(x)$ is continuous on $[a, b]$, then it is bounded on this interval.

The Second Weierstrass Theorem. If $f(x)$ is continuous on a compact set, then $f(x)$ attains its global maximum and minimum values on this set. In particular, if $f(x)$ is continuous on $[a, b]$, then it attains its global maximum and minimum values on this interval.

Remark. Evidently, the Second Weierstrass Theorem includes the result of the First Theorem. However, by historical reasons, these two theorems are frequently considered in the given above sequence.

Intermediate Value property. A function $f(x)$ satisfies the *intermediate value property* on $[a, b]$, if for any pair $x_1, x_2 \in [a, b]$, $x_1 < x_2$, the image of $[x_1, x_2]$ under $f(x)$ contains $[f(x_1), f(x_2)]$ if $f(x_1) \leq f(x_2)$ or $[f(x_2), f(x_1)]$ if $f(x_1) \geq f(x_2)$ (that is, if $f(x)$ takes on two values somewhere on $[a, b]$, it also takes on every value in between).

The Intermediate Value Theorem. If $f(x)$ is continuous on a connected set S, then $f(x)$ satisfies the intermediate value property on S. In particular, if $f(x)$ is continuous on a connected set S, then its image $f(S)$ is also a connected set.

Uniform continuity

Uniform Continuity. $f(x)$ is *uniformly continuous* on a set S if for every $\varepsilon > 0$ there exists $\delta > 0$ such that whenever $x_1, x_2 \in S$ and $|x_1 - x_2| < \delta$ it follows that $|f(x_1) - f(x_2)| < \varepsilon$.

Nonuniform Continuity. $f(x)$ fails to be uniformly continuous on a set S if there exists $\varepsilon_0 > 0$ such that for any $\delta > 0$ can be found two points x_δ and $\tilde{x}_\delta$ in S such that $|x_\delta - \tilde{x}_\delta| < \delta$, but $|f(x_\delta) - f(\tilde{x}_\delta)| \geq \varepsilon_0$.

Criterion for Nonuniform Continuity. $f(x)$ fails to be uniformly continuous on a set S if there exists $\varepsilon_0 > 0$ and two sequences x_n and $\tilde{x}_n$ in S such that $|x_n - \tilde{x}_n| \underset{n\to\infty}{\to} 0$, but $|f(x_n) - f(\tilde{x}_n)| \geq \varepsilon_0$.

The Cantor Theorem. If $f(x)$ is continuous on a compact set S, then $f(x)$ is uniformly continuous on S.

3.2 Local properties

Example 1. "If a function has a limit at a, then it is continuous at a."

Solution.

Let $f(x)=\frac{\sin x}{x}$, $a=0$. In this case, $\lim\limits_{x\to 0} f(x)=1$, but $f(x)$ is not defined at $a=0$, so $f(x)$ is not continuous at $a=0$. This kind of discontinuity is called removable, because the redefinition of the original function just at a single point can remove discontinuity: $\tilde{f}(x)=\begin{cases}\frac{\sin x}{x}, & x\neq 0\\ 1, & x=0\end{cases}$ is a continuous function at $a=0$ (and at any other real point).

Remark 1. Another example of the same removable discontinuity is when $f(x)$ is defined at a, but $\lim\limits_{x\to a} f(x)\neq f(a)$, for instance (in line with the above example), $\hat{f}(x)=\begin{cases}\frac{\sin x}{x}, & x\neq 0\\ 0, & x=0\end{cases}$, $a=0$ has a removable discontinuity at $a=0$.

Remark 2. Besides a removable discontinuity presented in the above counterexamples, there are two more kinds of discontinuities (actually there are variations of the classification of discontinuities). Let us briefly list these situations below:

1)$f(x)=\operatorname{sgn} x$, $a=0$. In this case, $\lim\limits_{x\to 0} f(x)$ does not exist, and consequently $f(x)$ is not continuous at $a=0$, but both one-sided limits exist: $\lim\limits_{x\to 0_-} f(x)=-1$ and $\lim\limits_{x\to 0+} f(x)=1$. This is an example of a discontinuity of the first kind (or jump discontinuity).

2) $f(x)=\frac{1}{x^2}$, $a=0$. In this case, $\lim\limits_{x\to 0} f(x)=+\infty$ and this is an example of a discontinuity of the second kind (or essential discontinuity). The same kind of discontinuity occurs when the limit does not exist (either finite or infinite), as for the function $f(x)=\begin{cases}\sin\frac{1}{x}, & x\neq 0\\ 0, & x=0\end{cases}$ at the point $a=0$.

Example 2. "If a function is continuous at a, then it has a finite limit at a."

Solution.

For $f(x)=\begin{cases}\sqrt{x}, & x\geq 0\\ 0, & x=-1\end{cases}$ the point $a=-1$ is isolated (that is, there is a deleted neighborhood of $a=-1$ where there is no point of the domain of $f(x)$). For this reason, $a=-1$ is not a limit point for the domain of $f(x)$, and hence the limit of $f(x)$ does not exist at $a=-1$. On the other hand, $f(x)$ is continuous at $a=-1$ by definition.

Remark to Examples 1 and 2. While the fact that the existence of a finite limit does not imply a continuity at the same point is pointed out in all

Calculus books (see Example 1), the converse statement presented in Example 2 is generally accepted as correct, although there are some nuances regarding this.

Example 3. "Whatever function $f(x)$ is considered, if only an information on its domain is available, it is not possible to make any conclusion about continuity of this function."

Solution.

By definition, any function with the domain $\mathbb{Z}$ is continuous at every point of its domain (every point of $\mathbb{Z}$ is isolated and the definition of the continuity is satisfied).

Remark. Of course, any function with the domain $\mathbb{Z}$ has no limit due to the absence of limit points in the domain.

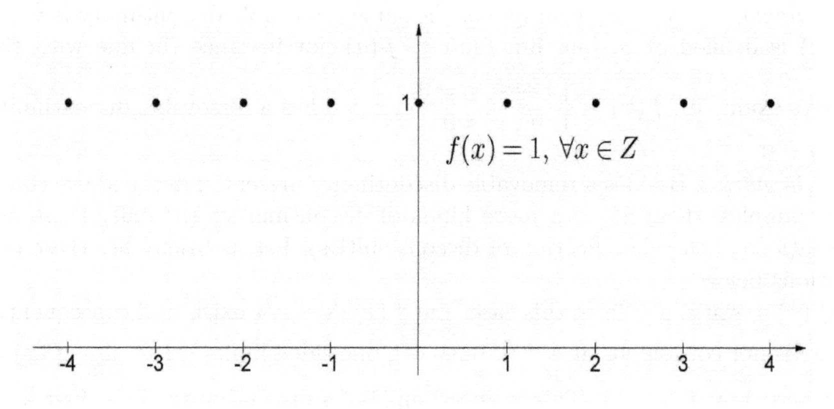

FIGURE 3.2.1: Example 3

Remark to Examples 2 and 3. The situation with an isolated point of the domain of $f(x)$ is a kind of pathological situation with no importance for understanding of the continuity, and apparently with no interest in applications. For this reason, in calculus this situation is not even considered and it usually assumed that a function $f(x)$ is continuous at a if , and only if, $\lim\limits_{x \to a} f(x) = f(a)$. So usually we will avoid this hardly useful specific situation of an isolated point and will focus on the cases when a is a limit point of the domain of a function $f(x)$.

Example 4. "If $f^2(x)$ is continuous at a point a, then $f(x)$ is also continuous at this point."

Solution.

We can use the same example as for the limits (Example 2, section 2.2):

$f(x) = \tilde{D}(x) = \begin{cases} 1, & x \in \mathbb{Q} \\ -1, & x \in \mathbb{I} \end{cases}$ is discontinuous (essential discontinuity) at every real point, while $f^2(x) = 1$ is continuous on $\mathbb{R}$.

Remark 1. Other simple examples with different kinds of discontinuity can be provided:

1) $f(x) = \begin{cases} 1, & x \neq 0 \\ -1, & x = 0 \end{cases}$ has a removable discontinuity at $a = 0$, but $f^2(x) = 1$ is continuous on $\mathbb{R}$.

2) $f(x) = \begin{cases} 1, & x < 0 \\ -1, & x \geq 0 \end{cases}$ has a jump discontinuity at $a = 0$, but $f^2(x) = 1$ is continuous on $\mathbb{R}$.

Remark 2. This is the wrong converse to the correct statement: if $f(x)$ is continuous at a point a, then $f^2(x)$ is also continuous at this point.

Remark 3. The three last counterexamples can also be used to disprove a similar false statement about the absolute value: "if $|f(x)|$ is continuous at a point a, then $f(x)$ is also continuous at this point". The converse is correct: if $f(x)$ is continuous at a point a, then $|f(x)|$ is also continuous at this point.

Example 5. "If $f(x)$ is continuous, positive and unbounded in any right-hand side neighborhood of a, then $\frac{1}{f(x)}$ is bounded in some right-hand side neighborhood of a."

Solution.

The function $f(x) = \frac{1}{x}\left|\sin\frac{1}{x}\right| + 2x$ is positive, continuous on $(0, 1)$ (check it by applying the arithmetic and composition rules) and unbounded in any right-hand side neighborhood of $a = 0$, because for $x_n = \frac{2}{(1+2n)\pi}$, $n \in \mathbb{N}$ we have

$$\lim_{x_n \to 0_+} f(x_n) = \lim_{n \to +\infty} \left(\frac{(1+2n)\pi}{2} + \frac{4}{(1+2n)\pi} \right) = +\infty.$$

At the same time, $g(x) = \frac{1}{f(x)} = \frac{x}{|\sin 1/x| + 2x^2}$ is also unbounded, because for $x_k = \frac{1}{k\pi}$, $k \in \mathbb{N}$ we obtain

$$\lim_{x_k \to 0_+} g(x_k) = \lim_{k \to +\infty} \frac{x_k}{2x_k^2} = \lim_{k \to +\infty} \frac{k\pi}{2} = +\infty.$$

Remark. A similar example can be given for an infinite point a. The false statement: "if $f(x)$ is continuous, positive and unbounded on $(0, +\infty)$, then $\frac{1}{f(x)}$ is bounded on $(0, +\infty)$", can be disproved by using counterexample with function $f(x) = x \sin x + x + \frac{1}{x}$.

Example 6. "If $f(x)$ is continuous at 0 and $f(0) = 0$, then the inequality $|f(x)| \leq |x|$ holds at least in some neighborhood of 0."

Solution.

If $f(x) = \sqrt[3]{x}$ then $f(x)$ is continuous at 0 (since $\lim\limits_{x \to 0} \sqrt[3]{x} = 0 = \sqrt[3]{0}$), however $\sqrt[3]{x} > x$ for any $x \in (0, 1)$.

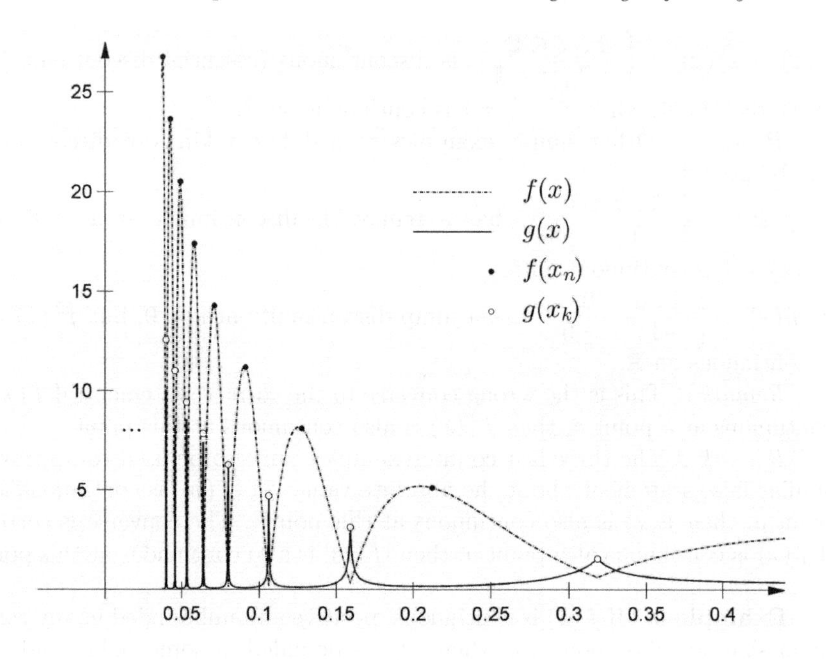

FIGURE 3.2.2: Example 5

Remark 1. This Example shows that the continuity does not prescribe any rate of approximation of a function to its value at the continuity point. The only thing that is guaranteed is the existence of approximation (but not its form).

Remark 2. The converse statement is true.

Example 7. "If $f(x)$ satisfies $\lim_{h\to 0}[f(a+h)-f(a-h)]=0$, then $f(x)$ is continuous at a."

Solution.

If $f(x)=\frac{1}{x^2}$ and $a=0$, then the condition of the statement is satisfied, because $\frac{1}{x^2}$ is an even function. However, $f(x)$ is not continuous at zero (it is not even defined there and besides $\lim_{x\to 0}\frac{1}{x^2}=+\infty$).

Remark 1. The continuity condition is satisfied if a function approaches its value $f(a)$ whatever the form of tendency of x to a is chosen. Therefore, any specification of the way that x approaches a can destroy the concept of continuity. The above Example shows how it happens when the pairs of points approaching a are symmetric with respect to a.

Remark 2. The converse is true.

Example 8. "If $f(x)$ and $g(x)$ are discontinuous at a point a, then $f(x)+g(x)$ is also discontinuous at this point."

Solution.

We can use the same example as for the limits (Example 1, section 2.3): $f(x) = -g(x) = D(x)$. Since Dirichlet's function does not have a limit at any point, it is discontinuous at any point. However, $h(x) = f(x) + g(x) = 0$ has zero limit at any point a.

Remark 1. This is a false "negative" version of the sum rule for continuous functions.

Remark 2. Since there are different kinds of discontinuity, it is worth noting that a counterexample can be constructed with discontinuity of any kind:

1) if $f(x) = \begin{cases} \frac{\sin x}{x}, & x \neq 0 \\ 0, & x = 0 \end{cases}$ and $g(x) = \begin{cases} 0, & x \neq 0 \\ 1, & x = 0 \end{cases}$, both with a removable discontinuity at $a = 0$, then the function $h(x) = f(x) + g(x) = \begin{cases} \frac{\sin x}{x}, & x \neq 0 \\ 1, & x = 0 \end{cases}$ is continuous at the origin.

2) if $f(x) = \begin{cases} x^2 + 1, & x < 0 \\ x, & x \geq 0 \end{cases}$ and $g(x) = \begin{cases} x, & x < 0 \\ x^2 + 1, & x \geq 0 \end{cases}$, both with a jump discontinuity at $a = 0$, then the function $h(x) = f(x) + g(x) = x^2 + x + 1$ is continuous at zero.

3) if $f(x) = \begin{cases} \frac{1}{x^2}, & x \neq 0 \\ 0, & x = 0 \end{cases}$ and $g(x) = \begin{cases} x - \frac{1}{x^2}, & x \neq 0 \\ 0, & x = 0 \end{cases}$, both having the infinite limits at $a = 0$, that is, an essential discontinuity at $a = 0$, then the function $h(x) = f(x) + g(x) = x$ is continuous at zero.

Remark 3. If $f(x)$ and $g(x)$ have different kinds of discontinuities at a point a then the statement of this Example is true.

Remark 4. Similar simple examples can be given for other arithmetic operations. For instance, for the product we can use the same modified Dirichlet function as in Example 2, section 2.2 of the limits:
$f(x) = -g(x) = \tilde{D}(x) = \begin{cases} 1, & x \in \mathbb{Q} \\ -1, & x \in \mathbb{I} \end{cases}$, with $h(x) = f(x) \cdot g(x) = -1$.

Example 9. "If $f(x)$ is continuous at a point a and $g(x)$ is discontinuous at a, then $f(x) \cdot g(x)$ is discontinuous at this point."

Solution.

We can use an example similar to that applied in limits (Example 6, section 2.3): $f(x) = x$ is continuous and $g(x) = 1 - D(x) = \begin{cases} 0, & x \in \mathbb{Q} \\ 1, & x \in \mathbb{I} \end{cases}$ is discontinuous at $a = 0$ (an essential discontinuity). However, $h(x) = f(x) \cdot g(x) = \begin{cases} 0, & x \in \mathbb{Q} \\ x, & x \in \mathbb{I} \end{cases}$ is continuous at $a = 0$, because $\lim\limits_{x \to 0} h(x) = 0 = h(0)$.

Remark 1. Other simple examples can be provided with other types of discontinuity:

1) $f(x) = x$ is continuous and $g(x) = \begin{cases} \sin x, & x \neq 0 \\ 1, & x = 0 \end{cases}$ has a removable discontinuity at $a = 0$, but $f(x) \cdot g(x) = x \sin x$ is continuous at $a = 0$.

2) $f(x) = x$ is continuous and $g(x) = \operatorname{sgn} x$ has a jump discontinuity at $a = 0$, but $f(x) \cdot g(x) = |x|$ is continuous at $a = 0$.

3) $f(x) = x^2$ is continuous and $g(x) = \begin{cases} 1/x, & x \neq 0 \\ 1, & x = 0 \end{cases}$ has an essential discontinuity at $a = 0$, but $f(x) \cdot g(x) = x$ is continuous at $a = 0$.

Remark 2. The statement will be true if to add that $f(a) \neq 0$.

Remark 3. The corresponding statement for the ratio is also false, but the statement for the sum and difference is true.

Example 10. "If $f(x)$ is defined in a neighborhood of a point a, is increasing on the left and decreasing on the right of a, then a is a local maximum point of $f(x)$."

Solution.

The function $f(x) = \begin{cases} \frac{1}{x^2}, & x \neq 0 \\ 0, & x = 0 \end{cases}$ satisfies the conditions of the statement in a neighborhood of $a = 0$, but $f(0) = 0 < \frac{1}{x^2} = f(x)$ for $\forall x \neq 0$.

Remark. This statement will be true if the condition of continuity at the point a would be added.

3.3 Global properties: general results

Example 1. "If $f^2(x)$ is continuous on $[a, b]$, then $f(x)$ is also continuous on this interval."

Solution.

The modified Dirichlet function $\tilde{D}(x) = \begin{cases} 1, & x \in \mathbb{Q} \\ -1, & x \in \mathbb{I} \end{cases}$ is not continuous at any point (it does not even have a limit - see Example 2, section 2.2), but $\tilde{D}^2(x) = 1$ is continuous on an arbitrary interval $[a, b]$.

Remark 1. The same counterexample works also for (a, b).

Remark 2. The same counterexample works also for following false statement about the absolute value of a function: "if $|f(x)|$ is continuous on $[a, b]$ (or (a, b)) then $f(x)$ is also continuous on this interval".

Example 2. "If $f(x)$ is continuous on $[a, b]$ and is not anywhere constant, then it can not take its global maximum and minimum values infinitely many times."

Solution.

Let us consider $f(x) = \begin{cases} -x + x\cos\frac{1}{x}, & x \neq 0 \\ 0, & x = 0 \end{cases}$ on $[0, 1]$. Due to arithmetic and composition rules this function is continuous on $(0, 1]$. It is also continuous at 0, because $\lim_{x \to 0} f(x) = 0 = f(0)$ (the result for the limit follows from the following evaluation $|f(x)| = |-x| \cdot \left|1 - \cos\frac{1}{x}\right| \leq 2|x| \underset{x \to 0}{\to} 0$). Therefore, $f(x)$ is continuous on $[0, 1]$. Also there is no interval inside $[0, 1]$ where

$f(x)$ is constant. However, this function takes its global maximum infinitely many times. Indeed, $f(x) = x\left(-1+\cos\frac{1}{x}\right) \le 0$ for $\forall x \in [0,1]$ and $f(0)=0$, so 0 is the global maximum value of $f(x)$. It is easy to see that the function takes this value in all points $x_n = \frac{1}{2n\pi}$, $n \in \mathbb{N}$. Since all these points lie in $[0,1]$, $f(x)$ takes its global maximum infinitely many times.

Remark 1. A similar example can be constructed for a global minimum or for both minimum and maximum. In the former case one can just use the function symmetric to the shown above with respect to the x-axis: $g(x) = -f(x) = \begin{cases} x - x\cos\frac{1}{x}, & x \ne 0 \\ 0, & x = 0 \end{cases}$ on $[0,1]$. In the latter case, the function $h(x) = \begin{cases} -x + x\cos\frac{\pi}{x}, & x \in (0,1] \\ -x + (2-x)\cos\frac{\pi}{2-x}, & x \in (1,2) \\ 0, \ x=0\ ; \ -2, & x = 2 \end{cases}$ considered on $[0,2]$ shows that continuous anywhere non-constant function can take both the global maximum and minimum values infinitely many times.

Remark 2. Since the behavior of the continuous on an interval function is supposed to be sufficiently "smooth", it may appear that it cannot have too many oscillations within a finite interval in order to attain infinitely many extreme values. However, it is possible.

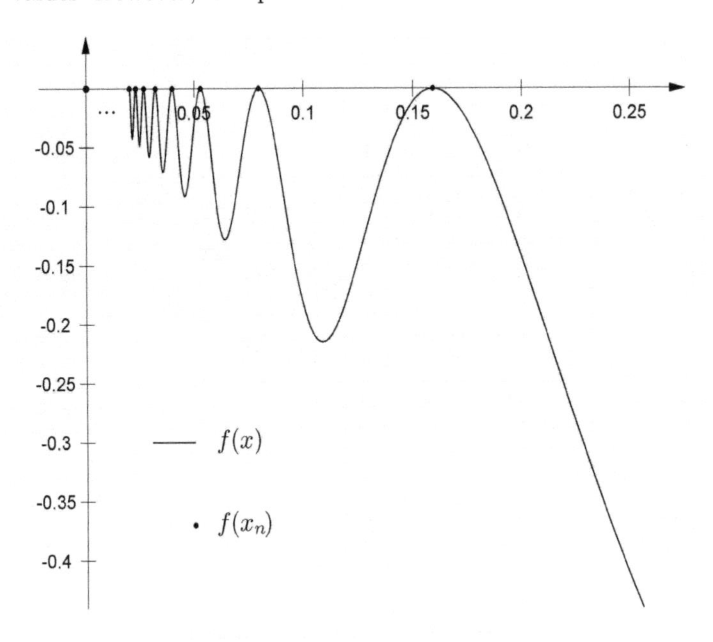

FIGURE 3.3.1: Example 2

Example 3. "If a function is defined on $\mathbb{R}$, it cannot be continuous at only one point."

Solution:

The function $f(x) = \begin{cases} x, & x \in \mathbb{Q} \\ -x, & x \in \mathbb{I} \end{cases}$ is discontinuous at any point, except for 0. In fact, in any neighborhood of every point x_0 there are rational and irrational points. Hence, we can consider two partial limits at x_0:

$$\lim_{x \to x_0,\, x \in \mathbb{Q}} f(x) = \lim_{x \to x_0,\, x \in \mathbb{Q}} x = x_0$$

and

$$\lim_{x \to x_0,\, x \in I} f(x) = \lim_{x \to x_0,\, x \in \mathbb{I}} (-x) = -x_0.$$

Since these partial limits are different for any $x_0 \neq 0$, it follows that the limit of $f(x)$ does not exist at any $x_0 \neq 0$, and therefore, $f(x)$ is not continuous at any $x_0 \neq 0$. At the point $x_0 = 0$ we get $\lim_{x \to 0} f(x) = 0 = f(0)$, so $f(x)$ is continuous at $x_0 = 0$.

Remark. Of course, the statement is also false for an arbitrary domain. For example, if the domain is the interval $(-a, a)$, then one can choose the same function in the counterexample.

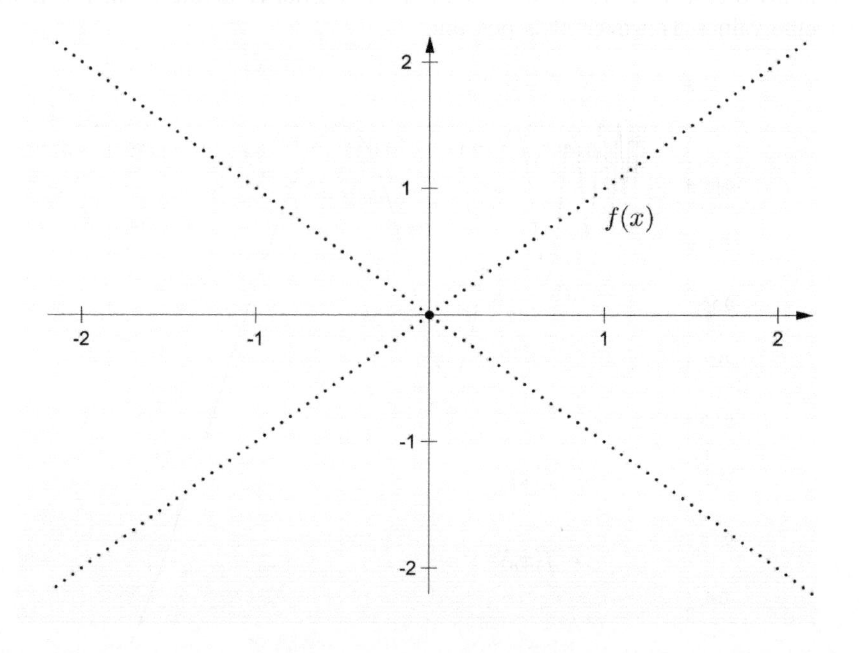

FIGURE 3.3.2: Example 3

Example 4. "A function cannot have infinitely many continuity points and infinitely many discontinuity points at the same time."

Solution.

A trivial counterexample can be constructed by just gluing a continuous and discontinuous functions defined on different sets. For instance, the function $f(x) = \begin{cases} 1, x \in [-1, 0] \\ D(x), x \in [0, 1] \end{cases}$ defined on $[-1, 1]$ is continuous at every point in $[-1, 0)$ and discontinuous at every point in $[0, 1]$.

Much more interesting counterexample arises if we interpret the inexact term "at the same time" in a more strict sense: on any interval in the function domain. In this case, one can use the *Riemann function* defined as follows. Suppose that any rational number is represented in the form $x = \frac{m}{n}$, where m is integer, n is natural, and the fraction $\frac{m}{n}$ is in lowest terms (the last means that m and n have no common factor). In this case the numbers m and n are uniquely determined, and the Riemann function $f(x) = R(x) = \begin{cases} \frac{1}{n}, x \in \mathbb{Q}, x = \frac{m}{n} \\ 0, x \in \mathbb{I} \\ 1, x = 0 \end{cases}$ is well-defined on $\mathbb{R}$. It can be shown that this function is continuous at every irrational point and discontinuous at every rational point. Indeed, in any neighborhood of a rational point $x_1 \in \mathbb{Q}$ there exist irrational ones. Therefore, the partial limit $\lim\limits_{x \to x_1, x \in \mathbb{I}} f(x) = 0$ is different from $f(x_1)$ which is equal $\frac{1}{n}$ if $x_1 = \frac{m}{n}$ or 1 if $x_1 = 0$. It means that $f(x)$ is discontinuous at every rational point in $\mathbb{R}$. Now, let us consider an arbitrary irrational point $x_2 \in \mathbb{I}$. First, approaching x_2 by irrational points we readily obtain $\lim\limits_{x \to x_2, x \in \mathbb{I}} f(x) = 0 = f(x_2)$. If we choose any set of rational points $x = \frac{m}{n}$ approaching x_2, then we can note that the denominator n should approach infinity, because if it would not be so, that is, if the denominator n would be bounded, say $n \leq n_0$, then it will be only finitely many numbers of the form $x = \frac{m}{n}$, $n \leq n_0$ in a neighborhood of the point x_2. However, it will contradict the well-known fact that in any neighborhood of an arbitrary real point there are infinitely many rational points. But if the condition that the rational points $x = \frac{m}{n}$ approach x_2 implies that $n \to +\infty$, then it follows that $\lim\limits_{x \to x_2, x \in \mathbb{Q}} f(x) = \lim\limits_{n \to +\infty} \frac{1}{n} = 0 = f(x_2)$. Since the two considered partial limits involve all the real points ($\mathbb{Q} \cup \mathbb{I} = \mathbb{R}$), we can conclude that $\lim\limits_{x \to x_2} f(x) = 0 = f(x_2)$, i.e., $f(x)$ is continuous at every irrational point in $\mathbb{R}$.

Remark. It can be shown that there does not exist a function continuous at every rational point and discontinuous at every irrational point.

Remark to Examples 3 and 4. The number of points where a defined on $\mathbb{R}$ function can be continuous and discontinuous is largely variable. Examples 3 and 4 illustrate two possible distributions of continuity and discontinuity points, the first being simple while the second being quite non-trivial.

Example 5. "If $f(x)$ is continuous on $\mathbb{R}$ and $\lim\limits_{n \to +\infty} f(n) = A$, then $\lim\limits_{x \to +\infty} f(x) = A$."

Solution.

The function $f(x) = \sin \pi x$ is continuous on $\mathbb{R}$ and $\lim_{n \to +\infty} f(n) = \lim_{n \to +\infty} \sin \pi n = 0$, but this is only one of the partial limits when x approaches $+\infty$. Another partial limit $\lim_{n \to +\infty} f\left(2n + \frac{1}{2}\right) = \lim_{n \to +\infty} \sin\left(2\pi n + \frac{\pi}{2}\right) = 1$ gives a different result, so $\lim_{x \to +\infty} f(x)$ does not exist.

Remark 1. The converse is true: if $\lim_{x \to +\infty} f(x) = A$ and $f(x)$ is defined on $\mathbb{N}$, then $\lim_{n \to +\infty} f(n) = A$.

Remark 2. This Example is a strengthened form of Example 3 in section 2.2: it shows that even continuity on $\mathbb{R}$ is not sufficient to make a conclusion about the behavior at infinity by knowing only a specific partial limit at infinity.

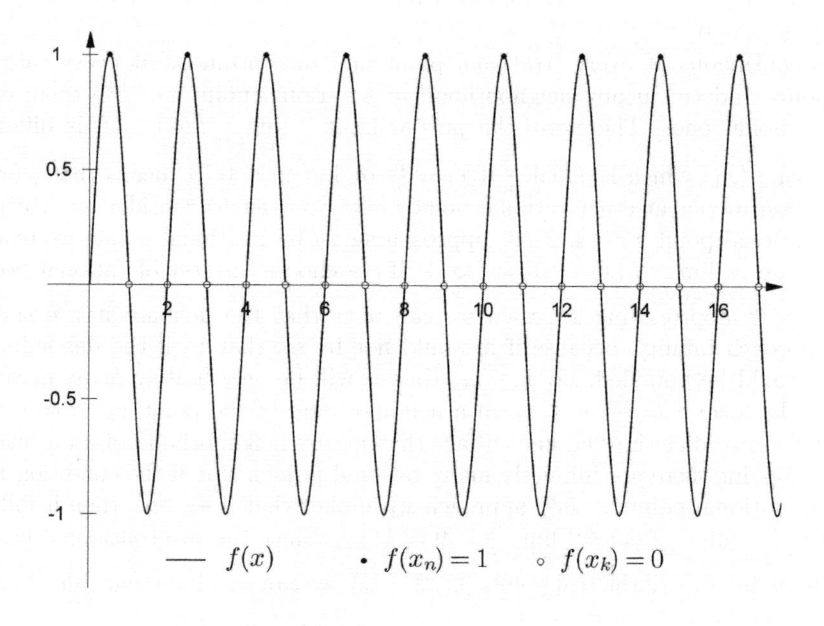

FIGURE 3.3.3: Example 5

Example 6. "If $f(x)$ is continuous on an interval A and on an interval B, then it is also continuous on $A \cup B$."

Solution.

The function $f(x) = \begin{cases} x\,, & x \in [0,1] \\ x+1\,, & x \in (1,2) \end{cases}$ is continuous on $A = [0,1]$ and on $B = (1,2)$, but it is not continuous on $A \cup B = [0,1] \cup (1,2) = [0,2)$, because $f(x)$ has a jump discontinuity at the point 1.

Remark. The statement will be true if $A \cap B \neq \varnothing$.

Example 7. "If $f(x)$ satisfies the condition $\lim_{h\to 0}[f(x+h)-f(x-h)] = 0$, $\forall x \in \mathbb{R}$, then $f(x)$ is continuous on $\mathbb{R}$."

Solution.

If $f(x) = \begin{cases} 1, & x \in \mathbb{Z} \\ 0, & x \notin \mathbb{Z} \end{cases}$ then the statement property is satisfied: for $\forall x \notin \mathbb{Z}$ there is a neighborhood where all the points are not integers, so $\lim_{h\to 0}[f(x+h)-f(x-h)] = 0$; for $\forall x \in \mathbb{Z}$ in a deleted neighborhood of radius 1 all the points are not integers, so again $\lim_{h\to 0}[f(x+h)-f(x-h)] = 0$. However, for $\forall x \in \mathbb{Z}$ it follows that $\lim_{h\to 0} f(x+h) = 0 \neq 1 = f(x)$, that is, $f(x)$ is not continuous at any integer.

Remark 1. The converse is true.

Remark 2. This Example is a replica of Example 7 in the previous section reformulated for an entire domain of a function. A simple even function does not provide a counterexample in this case, but the symmetry in the choice of points in the suggested "definition" still lead to violation of the continuity.

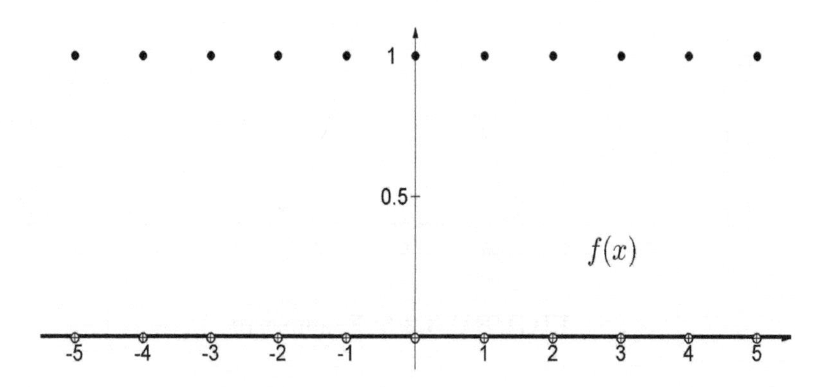

FIGURE 3.3.4: Example 7

Example 8. "If function $f(x)$ is continuous on an interval, then $\frac{1}{f(x)}$ is also continuous on this interval."

Solution.

The function $f(x) = \sin x$ is continuous on $\mathbb{R}$, but $\frac{1}{f(x)} = \frac{1}{\sin x}$ is discontinuous at all points $x_k = k\pi$, $k \in \mathbb{Z}$.

Remark. The statement will be true with the additional condition $f(x) \neq 0$ for all points of the interval.

Example 9. "If $f(x)$ is continuous and bounded on an interval and $f(x) \neq 0$ for every x in this interval, then $\frac{1}{f(x)}$ is also continuous and bounded on this interval."

Solution.

Consider the continuous function $f(x) = x$ which is bounded and different from zero on $(0,1)$. Evidently, $\frac{1}{f(x)} = \frac{1}{x}$ is unbounded on $(0,1)$.

Remark. Under the statement conditions, the function $\frac{1}{f(x)}$ maintains continuity on the chosen interval.

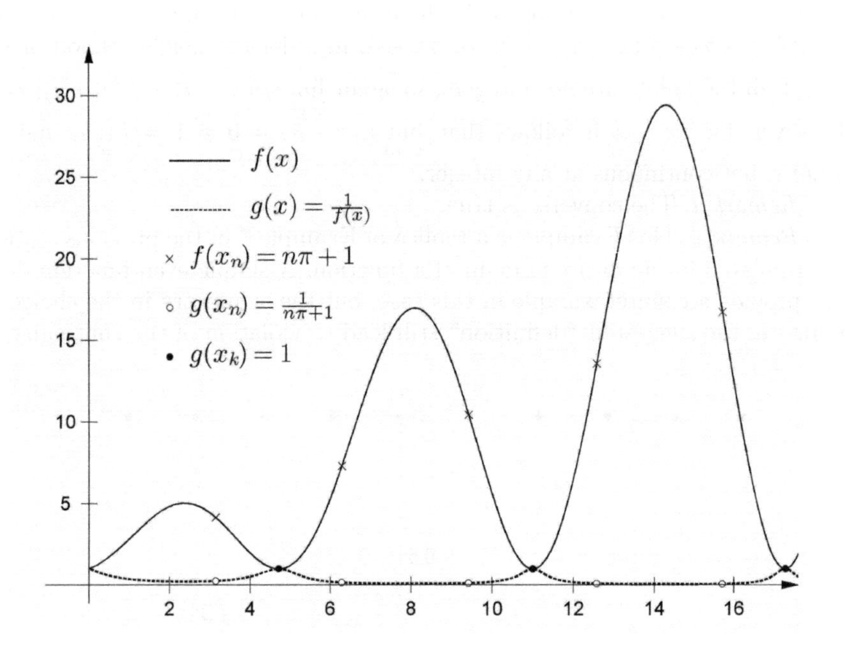

FIGURE 3.3.5: Example 10

Example 10. "If $f(x)$ is continuous and unbounded on $(0,+\infty)$, and $f(x) > 0$, $\forall x \in (0,+\infty)$, then $\lim\limits_{x\to+\infty} \frac{1}{f(x)} = 0$."

Solution.

Consider $f(x) = x\sin x + x + 1$ on $(0,+\infty)$. Evidently, $f(x)$ is continuous on $(0,+\infty)$ due to the arithmetic properties of continuous functions and $f(x) = x(\sin x + 1) + 1 > 0$ for $\forall x \in (0,+\infty)$. Also, choosing $x_n = n\pi$, $n \in \mathbb{N}$ one obtains the following partial limit

$$\lim_{x_n\to+\infty} f(x_n) = \lim_{n\to+\infty} (n\pi(\sin n\pi + 1) + 1) = +\infty,$$

implying that $f(x)$ is unbounded on $(0,+\infty)$. On the other hand, for the same sequence $x_n = n\pi$ we obtain

$$\lim_{x_n\to+\infty} \frac{1}{f(x_n)} = \lim_{n\to+\infty} \frac{1}{n\pi(\sin n\pi + 1) + 1} = 0,$$

while for $x_k = -\frac{\pi}{2} + 2k\pi$, $k \in \mathbb{N}$ the partial limit is different:

$$\lim_{x_k \to +\infty} \frac{1}{f(x_k)} = \lim_{k \to +\infty} \frac{1}{(2k\pi - \pi/2)(\sin(2k\pi - \pi/2) + 1) + 1}$$

$$= \lim_{k \to +\infty} \frac{1}{0+1} = 1.$$

It means that the limit $\lim\limits_{x \to +\infty} \frac{1}{f(x)}$ does not exist.

Remark 1. We can also consider a modified statement: "if $f(x)$ is continuous and unbounded on $(0, +\infty)$, and $f(x) > 0$, $\forall x \in (0, +\infty)$, then $\frac{1}{f(x)}$ is bounded on $(0, +\infty)$". A similar counterexample is provided by the function $f(x) = x \sin x + x + \frac{1}{x}$ on $(0, +\infty)$.

Remark 2. Similar statements for bounded intervals are also false.

3.4 Global properties: the famous theorems

3.4.1 Mapping sets

General Remarks. In this subsection the question of what properties of the sets are preserved under a continuous mapping is considered. One knows that a continuous function preserves openness and closeness of any set in its image. Naturally the question arises if analogous property is true for sets contained in its domain. The answer to this question is given in the first two Examples. On the other hand, it is well-known that a continuous function preserves compactness of any set from its domain, but it does not do so for sets from its image, as shown in Example 3. Further, boundedness is one of the components of compactness, but this property alone is not preserved by a continuous function as shown in Example 4.

Example 1. "If $f(x)$ is continuous on (a, b), then its image is also an open interval."

Solution.

Evidently, the function $f(x) = x^2$ is continuous on $(-1, 1)$, but its image $f((-1, 1))$ is the interval $[0, 1)$, which is not open.

Remark 1. This false statement can be reformulated in a more general form (which is also false): "if $f(x)$ is continuous on an open set S then its image is also an open set".

Remark 2. This is a wrong reformulation of the following result (frequently called the Open Set Characterization of continuous functions): $f(x)$ is continuous on a domain X if, and only if, the inverse image $f^{-1}(S)$ of any open set S contained in the image $f(X)$ is open. (Recall that for a function $f(x)$ with domain X and image $Y = f(X)$ the inverse image $f^{-1}(S)$ of a set $S \subset Y$ is the set of all points $x \in X$ such that $f(x) \in S$).

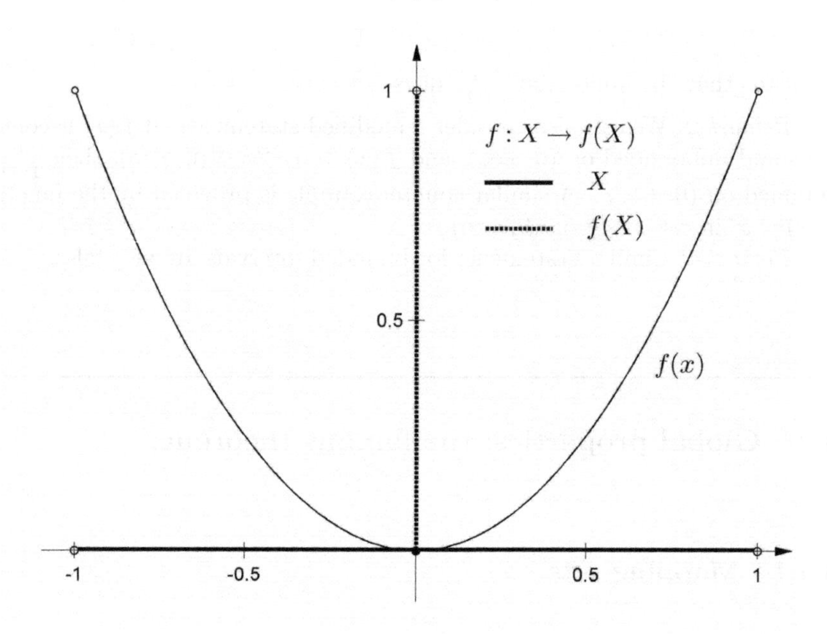

FIGURE 3.4.1: Example 1

Example 2. "If $f(x)$ is continuous on a closed set $[a, +\infty)$, then its image is also a closed set."

Solution.

The function $f(x) = \frac{1}{1+x^2}$ is continuous on $[0, +\infty)$ (just apply the arithmetic properties), but its image $f([0, +\infty))$ is the interval $(0, 1]$, which is not closed (to determine the image notice that $f(0) = 1$, $f(x)$ is strictly decreasing and positive on $[0, +\infty)$ and $\lim\limits_{x \to +\infty} f(x) = 0$).

Remark 1. This false statement can be reformulated in a more general form (which is also false): "if $f(x)$ is continuous on a closed set S then its image is also a closed set".

Remark 2. This is a wrong reformulation of another characterization of continuous functions: $f(x)$ is continuous on a domain X if, and only if, the inverse image $f^{-1}(S)$ of any closed set S contained in the image $f(X)$ is closed.

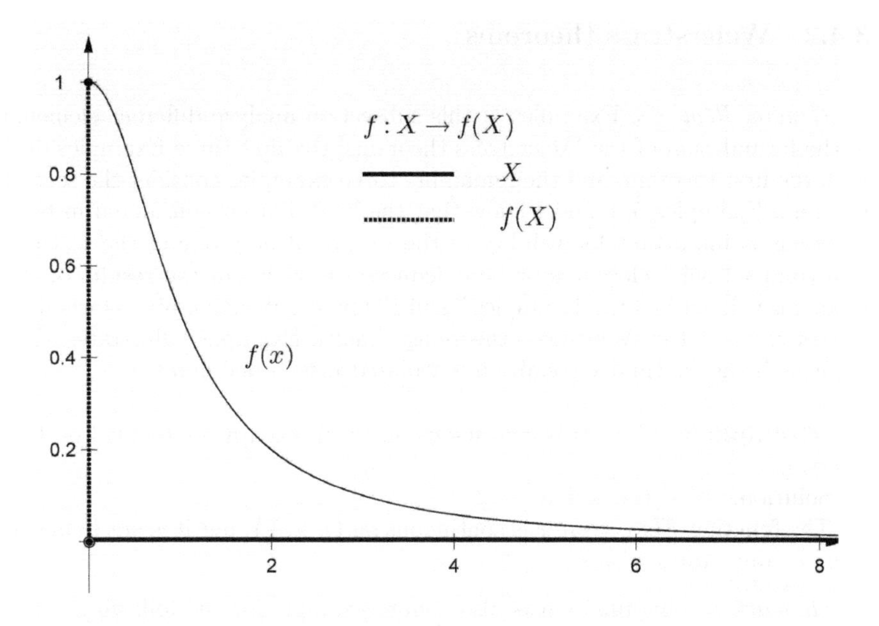

FIGURE 3.4.2: Example 2

Example 3. "If $f(x)$ is continuous on S and its image $f(S)$ is a compact set, then S is also a compact set."

Solution.

The function $f(x) = \sin x$ is continuous on $S = (0, 2\pi)$ and the image $f(S) = [-1, 1]$ is the compact, but the original set S is not a compact set.

Example 4. "If $f(x)$ is continuous on a bounded set, then its image is also a bounded set."

Solution.

The function $f(x) = \tan x$ is continuous on $\left(0, \frac{\pi}{2}\right)$, but its image $f\left(\left(0, \frac{\pi}{2}\right)\right)$ is the interval $(0, +\infty)$, which is not bounded.

Remark to Examples 1-4. These statements can be viewed as wrong modifications of the correct statement on the preservation of compactness under a continuous mapping: if $f(x)$ is continuous on a compact set, then its image is also a compact set. (Recall that in $\mathbb{R}$ a compact set is a bounded and closed set, for example $[a, b]$, where a and b are finite points.) This important global property sometimes is called the Compact Set theorem.

3.4.2 Weierstrass theorems

General Remarks. Examples of this subsection analyze different elements in the formulation of the Weierstrass theorems: the first three Examples deal with the first theorem, and the remaining three examples consider the second theorem. Examples 5, 6 and 8 show that the kind of a set considered in both theorems is important for validity of the results: if only one of the features of a compact set - closeness or boundedness - is violated, the results of the theorems will not be true. Examples 7 and 10 investigate attempts to formulate the converses to the Weierstrass theorems. Finally, Example 9 illustrates what happens if the continuity condition is violated just in one point.

Example 5. "If $f(x)$ is continuous on (a,b), then it is bounded on this interval."

Solution.

The function $f(x) = \tan x$ is continuous on $\left(-\frac{\pi}{2}, \frac{\pi}{2}\right)$, but it is not bounded there ($\lim\limits_{x\to\pi/2_-} \tan x = +\infty$).

Remark 1. This function is also counterexample for the following statement: "if $f(x)$ is continuous on (a,b) then it attains its global minimum and maximum on this interval".

Remark 2. Of course, the interval (a,b) can be substituted by $[a,b)$ or $(a,b]$ with the same false statement. For instance, for $[a,b)$ the same function considered on $\left[0, \frac{\pi}{2}\right)$ works.

Example 6. "If $f(x)$ is continuous on a closed set, then it is bounded on this set."

Solution.

The function $f(x) = x$ is continuous on the closed set $[0,+\infty)$, but it is not bounded there.

Remark to Examples 5 and 6. These statements are wrong modifications of the correct result representing an important global property of continuous functions: if $f(x)$ is continuous on $[a,b]$, then it is bounded on this interval. The same result holds for any compact set in $\mathbb{R}$. This result is frequently called the first Weierstrass theorem.

Example 7. "If $f(x)$ is bounded on $[a,b]$, then it is continuous on this interval."

Solution.

Dirichlet's function is bounded on any interval $[a,b]$, but it does not even have a limit at any point (see Example 1, section 2.2 for details).

Remark 1. The same example works also for (a,b).

Remark 2. Another simple example with a less "pathological" function is $f(x) = \operatorname{sgn} x$ considered on such $[a, b]$, that $0 \in (a, b)$ (it has a jump discontinuity at the point zero).

Remark 3. The converse is true and represents the first Weierstrass theorem.

Example 8. "If $f(x)$ is continuous and bounded on (a, b), then it attains its global minimum and maximum on this interval."

Solution.

The function $f(x) = x$ is continuous and bounded on $(-1, 1)$, but it does not attain either minimum or maximum values.

Remark. For an infinite interval, for example $(-\infty, +\infty)$, the function $f(x) = \arctan x$ provides a counterexample.

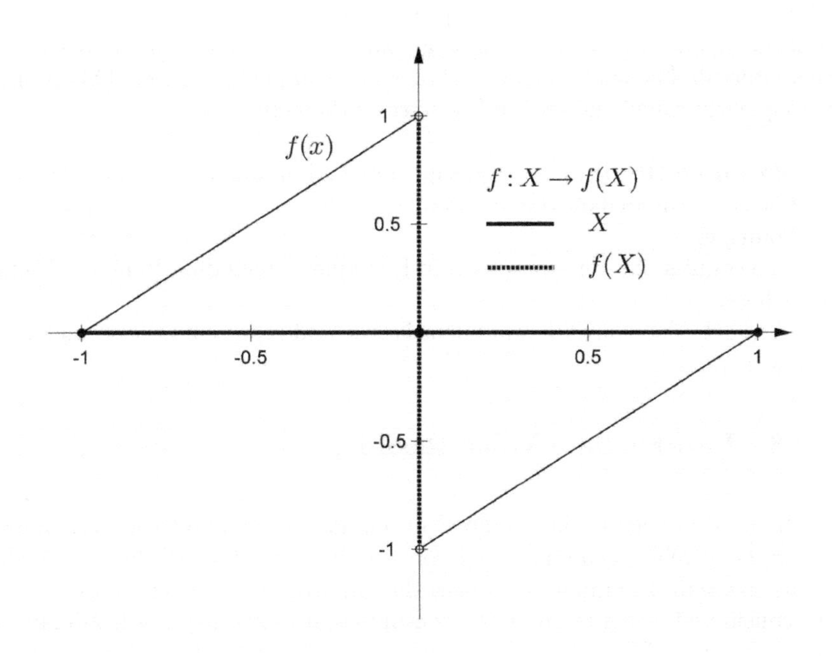

FIGURE 3.4.3: Example 9, Remark 1

Example 9. "If $f(x)$ is defined on $[a, b]$ and continuous on $[a, b]$ except for one point $c \in [a, b]$, then it attains its global extrema on $[a, b]$."

Solution.

The function $f(x) = \begin{cases} |x|, & x \in [-1, 0) \cup (0, 1] \\ 1, & x = 0 \end{cases}$ is defined on $[-1, 1]$ and continuous on $[-1, 0)$ and $(0, 1]$, but not at 0 (it has a removable discontinuity at 0: $\lim\limits_{x \to 0} f(x) = 0$, but $f(0) = 1$). At the same time this function does not attain its minimum, because $\lim_{x \to 0} f(x) = 0$, but all the function values are positive.

Remark 1. Another counterexample when function, which satisfies the above conditions, does not attain either global minimum or maximum is provided by the function $f(x) = \begin{cases} x+1, \, x \in [-1,0) \\ x-1, \, x \in (0,1] \\ 0, \, x = 0 \end{cases}$ on $[-1,1]$.

Remark 2. Of course, the discontinuity point c can be one of the endpoints. For instance, the function of the above counterexample considered on $[0,1]$, has a discontinuity at the left-hand endpoint.

Remark to Examples 8 and 9. These statements are wrong modifications of the following important global property of continuous functions: if $f(x)$ is continuous on $[a,b]$ then it attains global maximum and minimum values on this interval. The same result holds for any compact set in $\mathbb{R}$. This property is frequently called the second Weierstrass theorem.

Example 10. "If $f(x)$ attains its global minimum and maximum on $[a,b]$, then it is continuous on this interval."

Solution.

The counterexamples proposed in Example 7 (including Remarks 1 and 2) work here.

Remark. The converse statement is true and represents the second Weierstrass theorem.

3.4.3 Intermediate Value theorem

General Remarks. The examples in this subsection aim to analyze different attempts to relax, generalize and convert the conditions of the Intermediate Value theorem. Example 11 stresses the importance of the continuity over all the considered set. A failure to be continuous at only one point invalidates the theorem result. Example 12 shows the importance of connectedness of the set: the theorem is not true even on the union of two closed bounded intervals. The next Example examines if the known corollary to the Intermediate Value theorem remains true when the condition of continuity is omitted. Finally, Examples 14-16 analyze attempts to construct the converse statement to the theorem. It happens, that simple converse is very far to be true (Example 14), and even together with the additional conditions of extreme properties the converse fails to assure a continuity (Remark 2 to Example 14). Other versions of supposed converses fail to guarantee continuity even in a single point (Examples 15 and 16).

Example 11. "If $f(x)$ is defined on $[a,b]$ and continuous on (a,b), then for any constant C between $f(a)$ and $f(b)$ there is $c \in (a,b)$ such that $f(c) = C$."

Solution.

The function $f(x) = \operatorname{sgn} x$ considered on $[0,1]$ is continuous on $(0,1)$, and consequently satisfies the conditions. However, for $C = 1/2$, $0 = f(0) < C < f(1) = 1$, there is no point in $(0,1)$ (neither in $[0,1]$) such that $f(c) = C$.

Remark. Of course, continuity on (a,b) can be substituted by continuity on $[a,b)$ or $(a,b]$ with the same effect. For instance, for $(a,b] = (0,1]$ the above function provides a counterexample.

Example 12. "If $f(x)$ is continuous on $[a,b] \cup [\alpha,\beta]$, where $b < \alpha$, then for any constant C between $f(a)$ and $f(\beta)$ there is a point $c \in [a,b] \cup [\alpha,\beta]$ such that $f(c) = C$."

Solution.

Let us consider $f(x) = \operatorname{sgn} x$ on the set $S \equiv [-2,-1] \cup [1,2]$. Evidently, this function is continuous on S, but for $C = 0$, $-1 = f(-2) < C < f(2) = 1$, there is no point in S such that $f(c) = C$.

Remark 1. The values $f(a)$ and $f(\beta)$ can be substituted for any other pair with the same effect.

Remark 2. The closed intervals can be substituted for any other kind of intervals (and more generally by non-connected set). For instance, for the set $(-2,1) \setminus \{0\} = (-2,0) \cup (0,1)$ the same functions and $C = 0$ provide a counterexample.

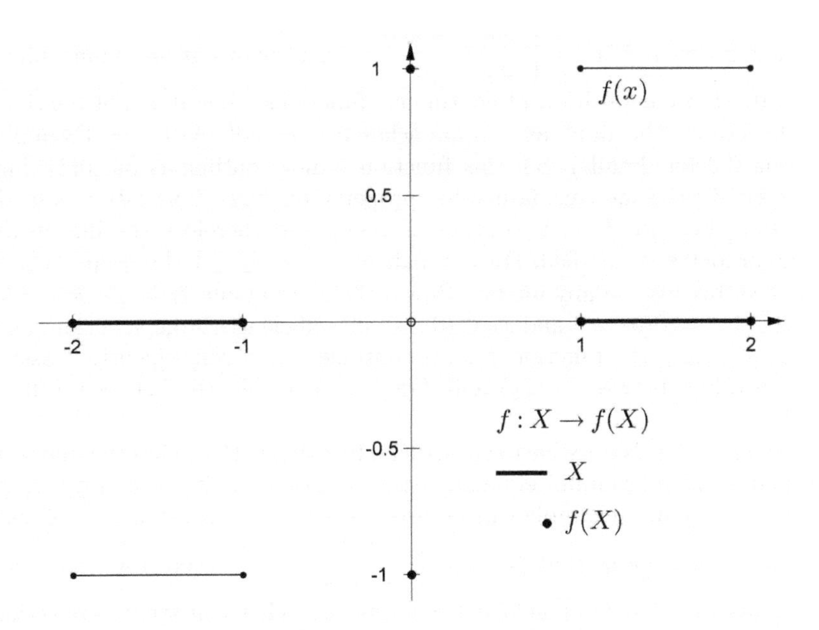

FIGURE 3.4.4: Example 12

Remark to Examples 11 and 12. These statements are wrong modifications of the following important global property called the Intermediate Value theorem: if $f(x)$ is continuous on $[a,b]$, then $f(x)$ satisfies the intermediate value property on $[a,b]$ (that is, if $f(x)$ takes on two values somewhere on $[a,b]$, it also takes on every value in between). In more general form this theorem states: if $f(x)$ is continuous on a connected set, then its image is also a connected set (recall that the connected sets on $\mathbb{R}$ are intervals, that is the result holds for $[a,b]$, $[a,b)$, $(a,b]$ and (a,b), where a and b can be finite or infinite).

Example 13. "Between two local minima there is a local maximum."

Solution.

The function $f(x) = \sec^2 x$ has infinitely many local minima on $\mathbb{R}$ and no local maximum.

Remark 1. Another simple counterexample with a bounded function is $f(x) = x - [x]$ on $\mathbb{R}$.

Remark 2. The statement is true for a continuous function.

Example 14. "If $f(x)$ satisfies the intermediate value property on $[a,b]$ (that is, if $f(x)$ takes on two values somewhere on $[a,b]$, it also takes on every value in between), then $f(x)$ is continuous on $[a,b]$."

Solution.

The function $f(x) = \begin{cases} \sin\frac{1}{x}, & x \neq 0 \\ 0, & x = 0 \end{cases}$ is continuous at any point different from 0 (as a composition of continuous functions), but it is not continuous at 0, because the limit as x approaches 0 does not exist (see Example 4, section 2.2 for details). So, this function is not continuous on $[0,1]$, but it does satisfy the intermediate value property on $[0,1]$. The latter is evident for $(0,1]$, because $f(x)$ is continuous there, and therefore the intermediate value property is satisfied. For any pair $0 = x_1 < x_2 \leq 1$ the property is also satisfied, because on any interval $(0,x_2)$ there are points $\bar{x}_k = \frac{1}{k\pi}$, $k \in \mathbb{N}$ such that $f(\bar{x}_k) = f(0) = 0$ and $\bar{x}_k \in (0,x_2)$ for sufficiently large k. Again, on the interval $[\bar{x}_k, x_2]$ the function $f(x)$ is continuous and consequently it assumes all the values between $f(\bar{x}_k)$ and $f(x_2)$, or equivalently, between $f(0)$ and $f(x_2)$.

Remark 1. It is possible to construct a function with the intermediate value property, which has infinitely many points of discontinuity on a finite interval. The above counterexample can be used as a base for construction of such a function. First notice that $f_a(x) = \begin{cases} \sin\frac{\pi}{x-a}, & x \neq a \\ 0, & x = a \end{cases}$ keeps the properties of the above function $f(x)$ on $[a, a+\delta]$ for $\forall \delta > 0$. Then construct a collection of such functions $f_n(x) = \begin{cases} \sin\frac{\pi}{x-1/(n+1)}, & x \neq \frac{1}{n+1} \\ 0, & x = \frac{1}{n+1} \end{cases}$ each of which is defined on its own domain $X_n = \left[\frac{1}{n+1}, \frac{1}{n}\right]$, $\forall n \in \mathbb{N}$. Finally notice that $\bigcup_{n\in\mathbb{N}} X_n = (0,1]$

and define the resulting function on $(0,1]$ as follows: $\tilde{f}(x) = f_n(x)$, $x \in X_n$, $\forall n \in \mathbb{N}$. Using the reasoning for the function $f(x)$ it is straightforward to show that $\tilde{f}(x)$ satisfies the intermediate value property on $(0,1]$. Indeed, for an arbitrary pair of points $0 < \tilde{x}_1 < \tilde{x}_2 \leq 1$ there are two possibilities: one of the points $x_n = \frac{1}{n+1}$, $n \in \mathbb{N}$ lies in $[\tilde{x}_1, \tilde{x}_2)$ or none of these points lies in this interval. In the former case, the image of $\tilde{f}(x)$ is $[-1,1]$ already on the interval $(x_n, \tilde{x}_2)$; and in the latter case $\tilde{f}(x)$ is continuous on $[\tilde{x}_1, \tilde{x}_2]$. So in both cases $\tilde{f}(x)$ satisfies the intermediate value property. At the same time, $\tilde{f}(x)$ is not continuous at the points $x_n = \frac{1}{n+1}$, $\forall n \in \mathbb{N}$.

Remark 2. The same counterexample works even for the strengthened false statement: "if $f(x)$ satisfies the intermediate value property on $[a,b]$ and attains its global minimum and maximum on $[a,b]$, then $f(x)$ is continuous on this interval". This is actually a strengthened version for this Example and for Example 10 as well, with the conditions including the conclusions of the Weierstrass and Intermediate Value theorems. It shows that these theorems are not convertible: even when a function satisfies the resulting properties of both theorems, it is not sufficient to guarantee the function continuity. It is necessary to add some additional condition to arrive to correct "converse" result. One of the versions of such "converse" theorems is: if $f(x)$ satisfies the intermediate value property on $[a,b]$ and if additionally $f(x)$ takes on each value only finitely many times, then $f(x)$ is continuous on $[a,b]$. Another version is as follows: if $f(x)$ is monotone on an interval I and the image of I is also an interval, then $f(x)$ is continuous on I.

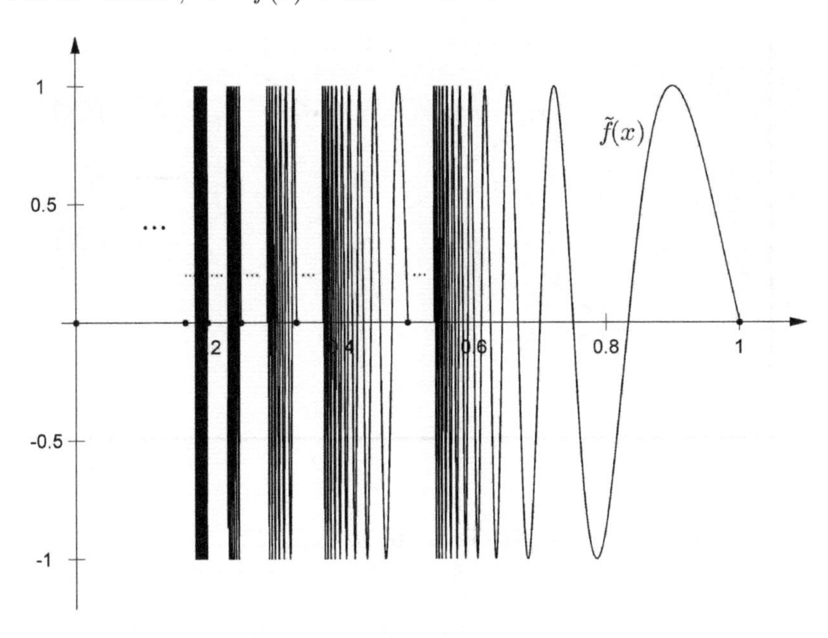

FIGURE 3.4.5: Example 14, Remark 1

Example 15. "If $f(x)$ attains its global minimum and maximum on $[a,b]$ and the image of this interval is $[f(a), f(b)]$, then $f(x)$ is continuous on $[a,b]$."

Solution

The function $f(x) = \begin{cases} x, \ x \in [0,1] \cup [2,3] \\ 3-x \ , \ x \in (1,2) \end{cases}$ attains the minimum ($f(0) = 0$) and maximum ($f(3) = 3$) at the endpoints of the domain $[0,3]$. It also assumes any value between 0 and 3, because the image of $[0,1]$, defined by the function $f(x) = x$, is $[0,1]$, the image of $(1,2)$, defined by the function $f(x) = 3-x$, is $(1,2)$, and the image of $[2,3]$, defined by the function $f(x) = x$, is $[2,3]$. So the total image is the interval $[0,3] = [f(0), f(3)]$. However, $f(x)$ has a jump discontinuity at the points 1 and 2: $\lim\limits_{x\to 1_-} f(x) = 1$, $\lim\limits_{x\to 1_+} f(x) = 2$, $\lim\limits_{x\to 2_-} f(x) = 1$, $\lim\limits_{x\to 2_+} f(x) = 2$.

Remark. Another example of a function, which satisfies the above conditions and has an essential discontinuity, is $f(x) = \begin{cases} \sin\frac{\pi}{x}, \ x \neq 0 \\ 0 \ , \ x = 0 \end{cases}$ on the interval $[-2,2]$.

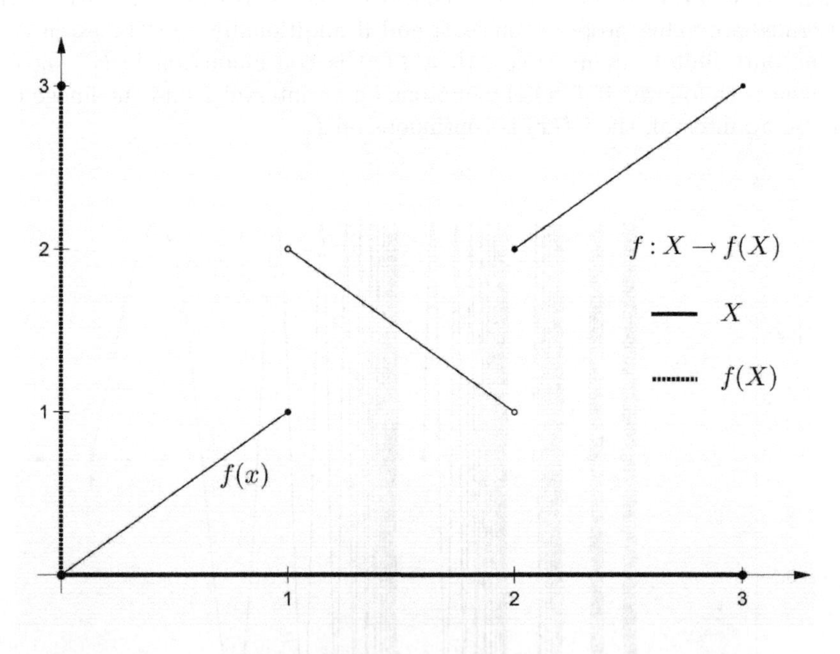

FIGURE 3.4.6: Example 15

Example 16. "If $f(x)$ attains its global minimum A and maximum B on $[a,b]$ and its image is $[A,B]$, then $f(x)$ is continuous at least at one point on $[a,b]$."

Solution

Let us consider $f(x) = \begin{cases} x, \, x \in \mathbb{Q}, \, x \neq 0, \, x \neq 1 \\ -x, \, x \in \mathbb{I} \\ 1\,, \, x = 0\,; \; 0\,, \; x = 1 \end{cases}$ on $[-1, 1]$. Evidently, all function values are located between -1 and 1, this function achieves its global minimum $(f(-1) = -1)$ and maximum $(f(0) = 1)$ and it assumes every value in $[-1, 1]$, that is its image is the interval $[-1, 1]$. Therefore, the conditions of the statement are satisfied. However, there is no point in $[-1, 1]$ where function is continuous. In fact, in any neighborhood of a point x_0 there are rational points different from 0 and 1 (denote them $Q_1 \equiv \mathbb{Q} \setminus \{0, 1\}$) and irrational points. Let us consider the two partial limits at x_0: $\lim\limits_{x \to x_0,\, x \in Q_1} f(x) = \lim\limits_{x \to x_0,\, x \in Q_1} x = x_0$ and $\lim\limits_{x \to x_0,\, x \in \mathbb{I}} f(x) = \lim\limits_{x \to x_0,\, x \in \mathbb{I}} (-x) = -x_0$. Since these partial limits are different for any $x_0 \neq 0$, it follows that the limit of $f(x)$ does not exist at any $x_0 \neq 0$, and therefore, $f(x)$ is not continuous at any $x_0 \neq 0$. Finally, at the point $x_0 = 0$ the limit exists $\lim\limits_{x \to 0} f(x) = 0$, but it is different from the function value $f(0) = 1$. Therefore,$f(x)$ is not continuous at $x_0 = 0$ also. Hence, we have exhausted all possibilities and there is no point in $[-1, 1]$ where $f(x)$ is continuous.

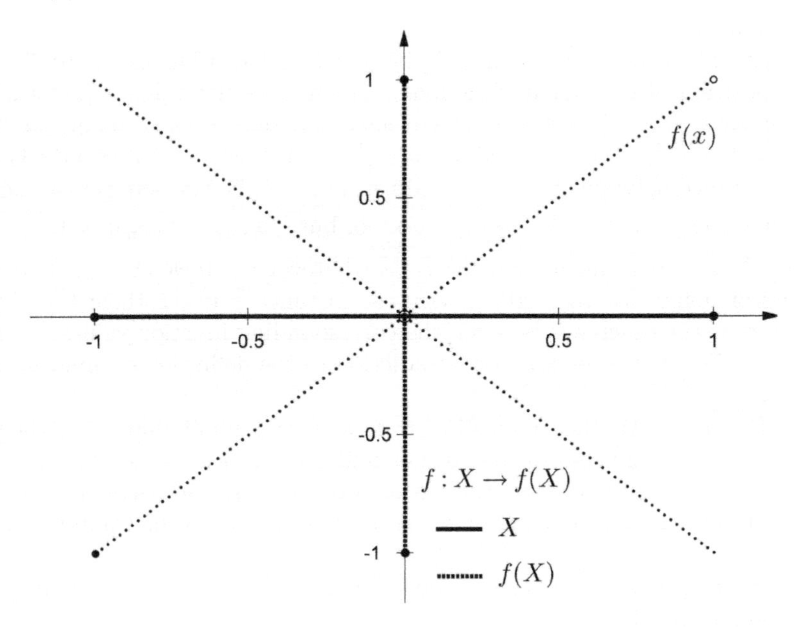

FIGURE 3.4.7: Example 16

Remark 1 to Examples 15 and 16. The last two Examples are strengthened versions of Example 10, with the conditions including the conclusions of the

Weierstrass and Compact Set theorems. It shows that these theorems are not convertible: even if a function satisfies the resulting properties of both theorems, this is not sufficient to guarantee the function continuity, even at a single point.

Remark 2 to Examples 15 and 16. Note that the conditions on images in these Examples do not mean that a function satisfies the intermediate value property, because these conditions require only that the function should assume all the values between the minimum and maximum on the entire interval, but there is no requirement for its subintervals, which is important in the intermediate value property.

3.5 Uniform continuity

Example 1. "If $f(x)$ is continuous and bounded on (a,b), then $f(x)$ is uniformly continuous on (a,b)."

Solution.

The function $f(x) = \sin\frac{\pi}{x}$ is bounded and continuous on $(0,1)$ (as a composition of continuous functions), but it does not possess uniform continuity on $(0,1)$. Indeed, let us consider two sequences of points in $(0,1)$: $x_n = \frac{1}{n}$, $\forall n \in \mathbb{N}$, $n > 1$ and $\tilde{x}_n = \frac{2}{1+4n}$, $\forall n \in \mathbb{N}$. For sufficiently large n the distance between the corresponding points of the two sets gets as close as we wish: $|x_n - \tilde{x}_n| = \left|\frac{1}{n} - \frac{2}{1+4n}\right| \underset{n\to+\infty}{\to} 0$, but $|f(x_n) - f(\tilde{x}_n)| = |0-1| = 1$, $\forall n \in \mathbb{N}$, $n > 1$. It means that for $\varepsilon < 1$ whatever $\delta > 0$ we choose, there exist pairs of points x_n, $\tilde{x}_n \in (0,1)$ with the distance between them less than δ, such that the difference between the corresponding function values is greater than ε. Therefore, we have a contradiction to the definition of uniform continuity.

Remark 1. Another interesting example for an infinite interval is the function $f(x) = \sin x^2$ continuous but not uniformly continuous on $\mathbb{R}$.

Remark 2. If $f(x)$ is continuous on a compact set, in particular, on $[a,b]$, then $f(x)$ is uniformly continuous on $[a,b]$ according to the Cantor theorem.

Example 2. "If $f(x)$ is unbounded on $\mathbb{R}$, then $f(x)$ is not uniformly continuous on $\mathbb{R}$."

Solution.

Consider the function $f(x) = x$, which is unbounded on $\mathbb{R}$. According to the definition of uniform continuity, for $\forall\varepsilon > 0$ we can choose $\delta = \varepsilon$ such that for any pair of points x_1, x_2 such that $|x_1 - x_2| < \delta$ it follows that $|f(x_1) - f(x_2)| = |x_1 - x_2| < \varepsilon$. Hence, the definition is satisfied.

Remark 1. The conditions of the Cantor theorem guarantee, in particular,

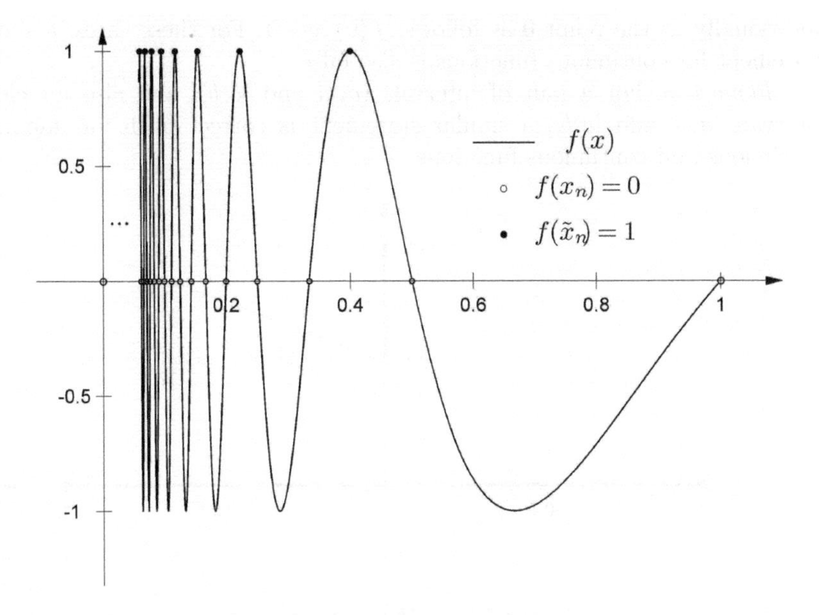

FIGURE 3.5.1: Example 1

that the function is bounded on the considered compact set. However, it is shown above that the boundedness of a function is not a necessary condition for its uniform continuity.

Remark 2. For a finite interval (a, b) this statement is true.

Example 3. "If $f(x)$ is uniformly continuous on (a, c) and (c, b), then $f(x)$ is uniformly continuous on $(a, c) \cup (c, b)$."

Solution.

Let $f(x) = \begin{cases} x-1, \, x \in (-1,0) \\ x+1, \, x \in (0,1) \end{cases}$. This function is uniformly continuous on $(-1, 0)$ and $(0, 1)$, but it does not keep uniform continuity on $(-1, 0) \cup (0, 1)$. In fact, for $(0, 1)$ we get: for an arbitrary pair x_1, x_2 such that $|x_1 - x_2| < \delta$ it follows that $|f(x_1) - f(x_2)| = |x_1 - x_2| < \delta = \varepsilon$, that is the definition holds. The same is true for $(-1, 0)$. However, for $(-1, 0) \cup (0, 1)$, if $x_1 \in (0, 1)$ and $x_2 \in (-1, 0)$, then we obtain $|f(x_1) - f(x_2)| = |2 + x_1 - x_2| > 2$ whatever close points x_1, x_2 are chosen.

Remark 1. A similar statement for continuous functions is also false (see Example 6 in section 3.3).

Remark 2. The statement is also false for a pair of intervals $(a, c]$ and (c, b), and for a pair of intervals (a, c) and $[c, b)$. For instance, to construct a counterexample for the first pair it is sufficient to define the above function additionally at the point 0 as follows: $f(0) = -1$. For these cases, a similar statement for continuous functions is also false.

Remark 3. For a pair of intervals $(a, c]$ and $[c, b)$, and also for closed intervals $[a, c]$ and $[c, b]$, a similar statement is correct both for uniformly continuous and continuous functions.

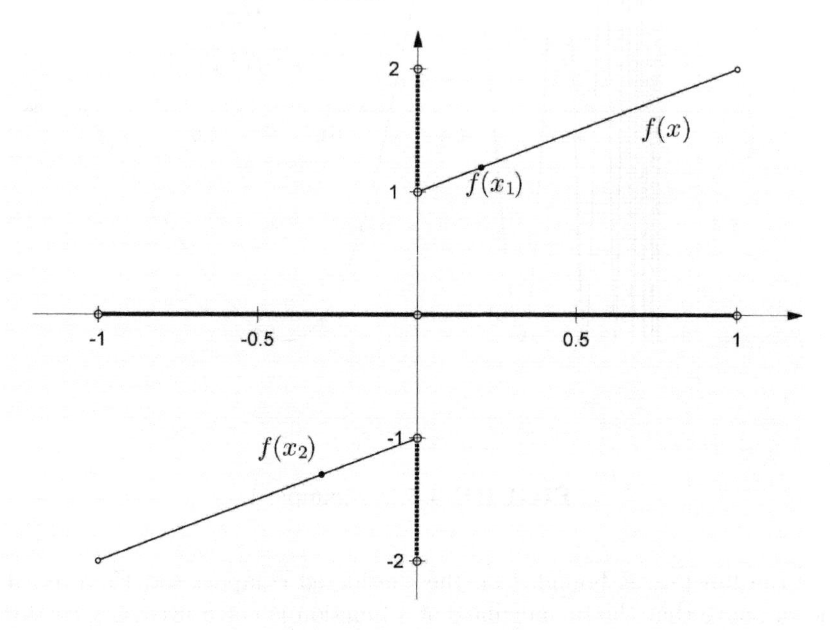

FIGURE 3.5.2: Example 3

Example 4. "If $f(x)$ is uniformly continuous on $\mathbb{R}$, then $f^2(x)$ is also uniformly continuous on $\mathbb{R}$."

Solution:

Evidently, the function $f(x) = x$ is uniformly continuous on $\mathbb{R}$ (see Example 2 for details), but $f^2(x) = x^2$ is not, since for the two sequences $x_n = n + \frac{1}{n}$, $\forall n \in \mathbb{N}$ and $\tilde{x}_n = n$, $\forall n \in \mathbb{N}$ the distance between the corresponding points of the two sets gets as close as we wish for sufficiently large n: $|x_n - \tilde{x}_n| = \frac{1}{n} \underset{n\to+\infty}{\to} 0$, but $\left|f^2(x_n) - f^2(\tilde{x}_n)\right| = 2 + \frac{1}{n^2} > 2$, $\forall n \in \mathbb{N}$.

Example 5. "If $f(x)$ and $g(x)$ are uniformly continuous on a set S, then $f(x) \cdot g(x)$ is also uniformly continuous on S."

Solution.

Evidently, the function $f(x) = x$ is uniformly continuous on $\mathbb{R}$ (see Example 2 for details). The uniform continuity of $g(x) = \sin x$ on $\mathbb{R}$ is also of an elementary proof:

$$|\sin x_1 - \sin x_2| = \left|2\sin\frac{x_1 - x_2}{2}\cos\frac{x_1 + x_2}{2}\right| \le |x_1 - x_2| < \delta = \varepsilon,$$

for $\forall x_1, x_2 \in \mathbb{R}$ such that $|x_1 - x_2| < \delta$ (here, we used two elementary properties of the trigonometric functions: $|\sin\alpha| \le |\alpha|$, $\forall\alpha$ and $|\cos\beta| \le 1$, $\forall\beta$). However, for the function $h(x) = f(x) \cdot g(x) = x\sin x$ we can choose $x_n = 2n\pi + \frac{1}{n}$, $\forall n \in \mathbb{N}$ and $\tilde{x}_n = 2n\pi$, $\forall n \in \mathbb{N}$ such that $|x_n - \tilde{x}_n| = \frac{1}{n} \underset{n\to+\infty}{\to} 0$, but

$$|h(x_n) - h(\tilde{x}_n)| = \left|\left(2n\pi + \frac{1}{n}\right)\sin\frac{1}{n}\right| > \left(2n\pi + \frac{1}{n}\right)\frac{2}{\pi}\frac{1}{n} = 4 + \frac{2}{\pi n^2} > 4,\ \forall n \in \mathbb{N}$$

(here the trigonometric inequality $\sin x > \frac{2}{\pi}x$ for $x \in \left(0, \frac{\pi}{2}\right)$ was applied). It means that $h(x)$ is not uniformly continuous on $\mathbb{R}$.

Remark 1. If S is a bounded domain, then the statement is true.

Remark 2. The sum of uniformly continuous functions on a set S is uniformly continuous function on S.

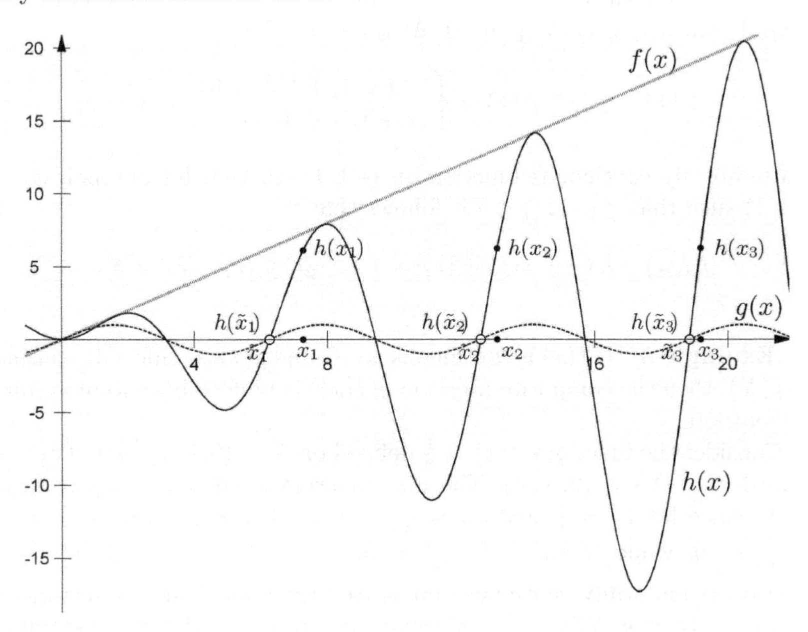

FIGURE 3.5.3: Example 5

Example 6. "If $f(x)$ is uniformly continuous and $f(x) \neq 0$ on a set S, then $\frac{1}{f(x)}$ is also uniformly continuous on S."

Solution.

The function $f(x) = x$ satisfies the conditions of the statement on $(0,1)$: $f(x) \neq 0$ for $x \in (0,1)$ and it is uniformly continuous on $\mathbb{R}$ (see the proof in Example 2). However, the function $g(x) = \frac{1}{f(x)} = \frac{1}{x}$ is not uniformly continuous on $(0,1)$. In fact, choosing the points in $(0,1)$ in the form $x_n = \frac{1}{n}$, $\tilde{x}_n = \frac{1}{n+1}$, $\forall n \in \mathbb{N}\backslash\{1\}$, we obtain: $|x_n - \tilde{x}_n| = \left|\frac{1}{n} - \frac{1}{n+1}\right| \underset{n\to\infty}{\to} 0$, but $|g(x_n) - g(\tilde{x}_n)| = |n - (n+1)| = 1$.

Remark to Examples 4-6. Notice that analogous properties for continuous functions are true.

Example 7. "If $f(x)$ and $g(x)$ are not uniformly continuous on a set S, then $f(x) \cdot g(x)$ also is not uniformly continuous on S."

Solution.

Let us consider a slight modification to the function of Example 3: $f(x) = \begin{cases} x-1, \ x \in (-1,0) \\ x+1, \ x \in [0,1) \end{cases}$, and also the function $g(x) = \begin{cases} -1, \ x \in (-1,0) \\ 1, \ x \in [0,1) \end{cases}$. These functions are uniformly continuous on $(-1,0)$ and $[0,1)$, but they are not uniformly continuous on $(-1,0) \cup [0,1) = (-1,1)$, because both have a jump discontinuity at the point 0. At the same time,

$$h(x) = f(x) \cdot g(x) = \begin{cases} -x+1, \ x \in (-1,0) \\ x+1, \ x \in [0,1) \end{cases} = 1 + |x|$$

is a uniformly continuous function on $(-1,1)$. In fact, for any pair $x_1, x_2 \in (-1,1)$ such that $|x_1 - x_2| < \delta$ it follows that

$$|h(x_1) - h(x_2)| = |1 + |x_1| - 1 - |x_2|| \leq |x_1 - x_2| < \delta = \varepsilon.$$

Example 8. "If $f(x)$ is continuous on X, and $g(x)$ is uniformly continuous on $f(X)$, then the composite function $g(f(x))$ is uniformly continuous on X."

Solution.

Consider the functions $f(x) = \frac{1}{x}$ defined on $X = (0,+\infty)$ and $g(x) = \sin x$ defined on $f(X) = (0,+\infty)$. The first function is not uniformly continuous on X, since for $x_n = \frac{1}{n}$ and $\tilde{x}_n = \frac{1}{n+1}$, $\forall n \in \mathbb{N}$ one has $|x_n - \tilde{x}_n| = |\frac{1}{n} - \frac{1}{n+1}| \underset{n\to\infty}{\to} 0$, while $|f(x_n) - f(\tilde{x}_n)| = |n - (n+1)| = 1 > 0$. The second function is uniformly continuous on $\mathbb{R}$ (see Example 5 in this section) and, consequently, it is uniformly continuous on $(0,+\infty)$. At the same time, the composite function $h(x) = g(f(x)) = \sin\frac{1}{x}$ is not uniformly continuous on $(0,+\infty)$, because for $x_k = \frac{1}{2k\pi}$ and $\tilde{x}_k = \frac{2}{(4k+1)\pi}$, one gets $|x_k - \tilde{x}_k| = |\frac{1}{2k\pi} - \frac{2}{(4k+1)\pi}| \underset{k\to\infty}{\to} 0$, but $|h(x_k) - h(\tilde{x}_k)| = |1 - 0| > 0$, $\forall k \in \mathbb{N}$.

Remark 1. The statement with the opposite conclusion is also false. Indeed, the function $f(x) = \frac{1}{x}$ defined on $X = (0,1)$ is non-uniformly continuous (just choose the same sequences of points $x_n = \frac{1}{n}$ and $\tilde{x}_n = \frac{1}{n+1}$ as above), and the

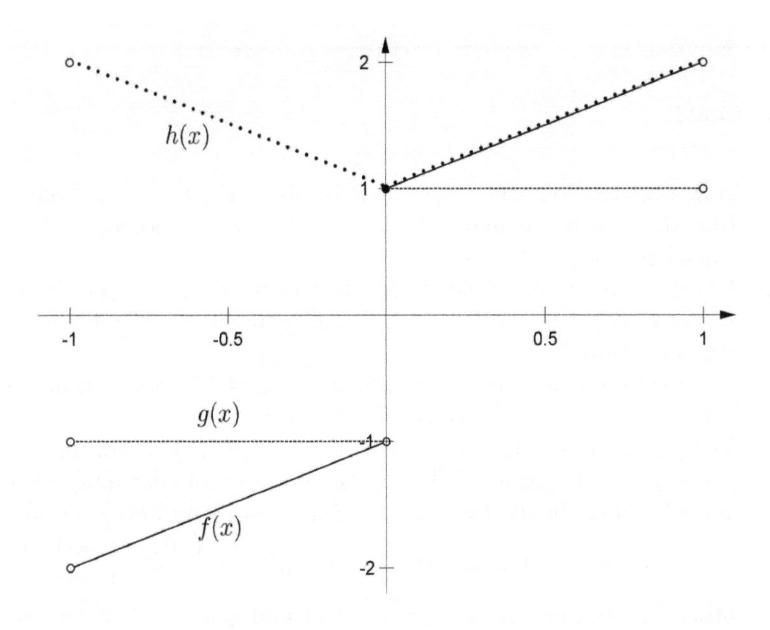

FIGURE 3.5.4: Example 7

function $g(x) = \frac{1}{x}$ defined on $\tilde{X} = (1, +\infty)$ is uniformly continuous, since for $\forall \varepsilon > 0$ and any pair of points x_1, $x_2 \in (1, +\infty)$ such that $|x_1 - x_2| < \delta = \varepsilon$ it follows that

$$|g(x_1) - g(x_2)| = \left| \frac{1}{x_1} - \frac{1}{x_2} \right| = \frac{|x_1 - x_2|}{|x_1| \cdot |x_2|} < \varepsilon.$$

At the same time, the composite function $h(x) = g(f(x)) = x$ is uniformly continuous on $\mathbb{R}$ and, therefore, on $(0, 1)$.

Remark 2. The change of order of a non-uniform continuous and uniformly continuous functions in the composite function leads to the same results.

Remark 3. The well-known result on uniform continuity of composite function states that if $f(x)$ is uniformly continuous on X, and $g(x)$ is uniformly continuous on $f(X)$, then the composite function $g(f(x))$ is uniformly continuous on X. If one tries to weaken the conditions on one of the functions, then the result may be wrong.

Remark 4. Of course, the composite function is continuous on X as composition of continuous functions.

Exercises

1. Show that the following statement is false: "if $f(x)$ is continuous at 0 and $f(0) = 0$, then the inequality $|f(x)| \leq |x|^{\alpha}$ holds for some $\alpha > 0$ at least in some neighborhood of 0".

2. Provide a counterexample to the statement "if $f(x) + g(x)$ is continuous at a, then both $f(x)$ and $g(x)$ are continuous at a". What about other arithmetic operations?

3. Give a counterexample to the statement: "if $\cos f(x)$ is continuous at a, then $f(x)$ is also continuous". What about $\cosh f(x)$?

4. Verify if the following statement is true: "if $g(f(x))$ is continuous at a, then $f(x)$ is also continuous". What if the condition of continuity of $g(y)$ at $f(a)$ is added? (Hint: in the last case try $f(x) = \sin \frac{1}{x}$ and $g(y) = 0$ at $a = 0$; to not involve constant functions one can use $g(y) = \begin{cases} 0, & y \in [-1,1] \\ y^2 - 1, & otherwise \end{cases}$; still another counterexample is $f(x) = \tilde{D}(x)$ and $g(y) = y^2$ at an arbitrary point a.)

5. Verify if the following statement is true: "if $g(f(x))$ and $f(x)$ are continuous at a, then $g(y)$ is continuous at $f(a)$".

6. Suppose $f(x)$ and $g(x)$ are such that both $f(g(x))$ and $g(f(x))$ are defined in a neighborhood of a and $f(g(x))$ is continuous at a. Is $g(f(x))$ also continuous?

7. Give a counterexample to the statement that "a continuous on $[a,b]$ function can not possess infinitely many local extrema on this interval". (This is a weakened version of the statement in Example 2, section 3.3.)

8. Demonstrate that the counterexamples suggested in Remark to Example 2, section 3.3, work.

9. Provide a counterexample to the following statement: "if $f(x)$ is increasing and continuous on the interval $[c,a]$ and on the interval $(a,b]$, then it is increasing on $[c,b]$".

10. Give a counterexample to the statement: "a function defined on $\mathbb{R}$ cannot be continuous at each point $x \in \mathbb{Z}$ and discontinuous at all other points".

11. It is known that the convexity condition on an open interval (a,b) - $f(\alpha x_1 + (1-\alpha)x_2) \leq \alpha f(x_1) + (1-\alpha)f(x_2)$, $\forall x_1, x_2 \in (a,b)$ and $\forall \alpha \in (0,1)$ - implies continuity of $f(x)$ on (a,b). Show that this implication fails in the case of a closed interval. (Hint: consider $f(x) = \begin{cases} 0, & x \in (-1,1) \\ 1, & x = 0, x = 1 \end{cases}$)

12. It is known that any function $f(x)$ that maps continuously a closed interval on itself has a fixed point, that is, such point c in the closed interval that $f(c) = c$. Show that this is not true for an open interval. (Hint: consider $f(x) = x^2$ on $(0,1)$.)

13. Provide a counterexample to the following statement: "if $f(x)$ is continuous and invertible on domain X, then the inverse function $f^{-1}(y)$ is continuous on $Y = f(X)$". (Hint: consider $f(x) = \begin{cases} x\,,\ x \le 0 \\ x-1\,,\ x > 1 \end{cases}$)

14. Provide a counterexample to the following statement: "if $f(x)$ is continuous on $X \subset \mathbb{R}$, $X \neq \mathbb{R}$, then there is a continuous extension of $f(x)$ from X onto $\mathbb{R}$, that is, such function $g(x)$ continuous on $\mathbb{R}$, that $g(x) = f(x), \forall x \in X$". (Hint: use $f(x) = \sin\frac{1}{x}$, $X = (0, +\infty)$; notice that the statement is true for a closed set X.)

15. Demonstrate that the function $f(x) = x\sin x + x + \frac{1}{x}$ proposed in Remark 1 to Example 10, section 3.3 is a counterexample to the corresponding statement.

16. Provide counterexamples for Remark 2 to Example 10 in section 3.3.

17. Give a counterexample to the statement in Example 1 in section 3.4 such that an open interval is transformed onto the closed interval under continuous mapping.

18. Show that the statement "if $f(x)$ is continuous and has a compact image, then its domain is a bounded set" is false. (This is a strengthened version of the statement in Example 3, section 3.4.)

19. Show that the following statement is false: "if $f(x)$ is continuous and bounded on a closed set, then it attains its global minimum and maximum on this set".

20. Provide a counterexample to the following statement: "if $f(x)$ is continuous on $[a, b]$ except for only one point, then $f(x)$ is bounded on $[a, b]$".

21. Give a counterexample to the statement: "if $f(x)$ is defined, but not continuous on $[a, b]$, it cannot attain its global extrema on this interval".

22. Give a counterexample to the statement in Example 11, section 3.4 with interval (a, b) substituted by $[a, b)$.

23. According to the Intermediate Value theorem, any connected set has a connected image under continuous mapping. Verify if the same is true in the "inverse" direction: "if $f(x)$ is continuous on $\mathbb{R}$, then any connected set in $f(X)$ has a connected inverse image". If not, construct a counterexample.

24. Verify if the following "definition" of the uniform continuity is correct: "for every $\varepsilon > 0$ and every $\delta > 0$ whenever $x_1, x_2 \in S$ and $|x_1 - x_2| < \delta$ it follows that $|f(x_1) - f(x_2)| < \varepsilon$". If not, provide a counterexample.

25. Demonstrate that the function $\sin x^2$ in Remark 1 to Example 1 of section 3.4 is a counterexample to the statement in that Example.

26. Show that the statement "if $f(x)$ is uniformly continuous and $g(x)$ is not uniformly continuous on X, then $h(x) = f(x) \cdot g(x)$ is not uniformly continuous on X" is false by providing a counterexample. Even a strengthened version with non-continuous function $g(x)$ is false. (Hint: try $f(x) = |x|$ and $g(x) = \operatorname{sgn} x$ on $X = \mathbb{R}$.)

27. Provide counterexamples to the statements mentioned in Remark 2 to Example 8 in section 3.4, that is, to the following statements: "if $f(x)$ is uniformly continuous on X, and $g(x)$ is continuous on $f(X)$, then the composite

function $g(f(x))$ is uniformly continuous on X" and "if $f(x)$ is uniformly continuous on X, and $g(x)$ is continuous on $f(X)$, then the composite function $g(f(x))$ is not uniformly continuous on X".

28. Show that the following statement is false: "if $f(x)$ is uniformly continuous on any interval $[a,b]$, then it is uniformly continuous on $\mathbb{R}$".

29. Since the violation of uniform continuity for one of the functions in the composition can lead to loss of uniform continuity by the composite function, it may be tempting to state that "if $f(x)$ is non-uniformly continuous on X, and $g(x)$ is non-uniformly continuous on $f(X)$, then the composite function $g(f(x))$ is non-uniformly continuous on X". Show that it is not true by providing a counterexample.

30. A function $f(x)$ is called Lipschitz-continuous on an interval I if there exists a constant $C \geq 0$ such that $|f(x) - f(y)| \leq C|x - y|$ for $\forall x, y \in I$. It is straightforward to see that the Lipschitz-continuity implies the uniform continuity. What about the converse?

Chapter 4

Differentiation

4.1 Elements of theory

Concepts

Differentiability at a point. A function $f(x)$ defined on X is called *differentiable at a point* $a \in X$, if for any point x in a neighborhood of a the following representation holds: $f(x) = f(a) + A \cdot (x-a) + \alpha \cdot (x-a)$, where A is a constant and α is a function of x such that $\alpha \underset{x\to a}{\to} 0$.

Remark 1. The same definition is frequently written in the terms of the increments of the argument $\Delta x = x - a$ and the function $\Delta f = f(x) - f(a)$ as follows: $\Delta f = A \cdot \Delta x + \alpha \cdot \Delta x$, where A is a constant and α is a function of Δx such that $\alpha \underset{\Delta x\to 0}{\to} 0$.

Differential. The expression $A \cdot \Delta x \equiv df$ is called the *differential* of $f(x)$ at a point a, and the difference $\gamma = \Delta f - df$ is called the remainder. Note that the term df is the linear (with respect to Δx) part of Δf, and this is also the main part of Δf in a small neighborhood of a if $A \neq 0$. Thus, the differential represents a linear approximation to the function increment in a neighborhood of a.

Derivative at a point. Given a function $f(x)$ defined on a neighborhood of a point a, the *derivative* of $f(x)$ at a (denoted $f'(a)$ or $\frac{df}{dx}(a)$) is defined by $f'(a) = \lim\limits_{x\to a} \frac{f(x)-f(a)}{x-a}$ provided this limit exists and is finite.

Theorem. A function $f(x)$ is differentiable at a point a if, and only if, there exists $f'(a)$. Moreover, in the definition of differentiability $A = f'(a)$.

Remark 1. Sometimes the equality $A = f'(a)$ is considered to be a part of the definition of differentiability.

Remark 2. Due to the last Theorem, a function possessing a derivative at a is also called differentiable at a.

One-sided derivatives. Given a function $f(x)$ defined at a point a and on a right-hand (left-hand) neighborhood of a, the *right-hand (left-hand) derivative* of $f(x)$ at a is defined by $f'_+(a) = \lim\limits_{x \to a_+} \frac{f(x)-f(a)}{x-a}$ $\left(f'_-(a) = \lim\limits_{x \to a_-} \frac{f(x)-f(a)}{x-a}\right)$ provided this limit exists and is finite.

Differentiation on a set. If $f'(x)$ is defined at every point of a set S, then $f(x)$ is *differentiable on* S.

Remark . It is worth to notice that in the case when endpoints belong to the interval, for example a closed interval $[a, b]$, the differentiability on $[a, b]$ means the differentiability at every point $c \in (a, b)$ plus the existence of the right-hand derivative at a and the left-hand derivative at b.

Continuous differentiability. If the derivative $f'(x)$ is continuous on a set S, then $f(x)$ is *continuously differentiable* on S.

Higher-order derivatives. If the derivative $g(x) = f'(x)$ is defined on a neighborhood of a point a, and the derivative of $g(x)$ exists at a, then it is called the *second derivative* of $f(x)$ at a and the standard notation is $f''(a) = g'(a)$ (or $\frac{d^2 f}{dx^2}(a)$). In a similar way, consecutively the *n-th derivative* of $f(x)$ at a is defined as the derivative of the $(n-1)$-th derivative of $f(x)$ at a, and the standard notation is $f^{(n)}(a)$ or $\frac{d^n f}{dx^n}(a)$.

Tangent line. Let $f(x)$ be continuous in a neighborhood of a point a, and b be a point in such a neighborhood. The *secant line* through the points $P_a = (a, f(a))$ and $P_b = (b, f(b))$ is defined by the formula: $y - f(a) = \frac{f(b)-f(a)}{b-a}(x-a)$. The *tangent line* is the limit position of the secant lines (if it exists) when P_b tends to P_a (in other words, when b approaches a). Since each secant line is characterized but its slope $\frac{f(b)-f(a)}{b-a}$, to find the limit position of secants is equivalent to find the limit of their slopes. If this limit exists and is finite, then $\lim\limits_{b \to a} \frac{f(b)-f(a)}{b-a} = f'(a)$ and the equation of the tangent is $y - f(a) = f'(a)(x-a)$. If $\lim\limits_{b \to a} \frac{f(b)-f(a)}{b-a} = (\pm)\infty$, then the equation of the tangent is $x = a$. Sometimes the case when $\lim\limits_{b \to a_-} \frac{f(b)-f(a)}{b-a} = +\infty$ and $\lim\limits_{b \to a_+} \frac{f(b)-f(a)}{b-a} = -\infty$ (or vice-versa) is considered to be a different situation called cusp or corner.

Basic properties

Theorem. If $f(x)$ is differentiable on a set, then it is continuous on this set.

Arithmetic (algebraic) properties

If $f(x)$ and $g(x)$ are differentiable at a point x, then

1) $f(x)+g(x)$ is differentiable at x and $(f(x)+g(x))' = f'(x)+g'(x)$;

2) $f(x)-g(x)$ is differentiable at x and $(f(x)-g(x))' = f'(x)-g'(x)$;

3) $f(x)\cdot g(x)$ is differentiable at x and $(f(x)\cdot g(x))' = f'(x)\cdot g(x)+f(x)\cdot g'(x)$;

4) $\frac{f(x)}{g(x)}$ is differentiable at x and $\left(\frac{f(x)}{g(x)}\right)' = \frac{f'(x)\cdot g(x)-f(x)\cdot g'(x)}{g^2(x)}$ (under the additional condition $g(x)\neq 0$).

Chain rule. If $f(x)$ is differentiable at a point x, and $g(y)$ is differentiable at the point $y=f(x)$, then the composite function $g(f(x))$ is differentiable at the point x and $(g(f(x)))' = g'(y)\cdot f'(x)$.

Rolle's Theorem. If $f(x)$ is continuous on $[a,b]$ and differentiable on (a,b) with $f(a)=f(b)$, then there exists a point $c\in(a,b)$ where $f'(c)=0$.

The Mean Value Theorem (Lagrange's Theorem). If $f(x)$ is continuous on $[a,b]$ and differentiable on (a,b), then there exists a point $c\in(a,b)$ where $f'(c)=\frac{f(b)-f(a)}{b-a}$.

Darboux's Theorem. If $f(x)$ is differentiable on a connected set S, then $f'(x)$ satisfies the intermediate value property on S. (In particular, $f'(S)$ is also a connected set.)

Theorem. If $f(x)$ is differentiable on a connected set S, then $f'(x)$ cannot have removable or jump discontinuities on S. (However, $f'(x)$ can have discontinuities of the second kind.)

Inverse Function Theorem.

The first formulation. If $y=f(x)$ is strictly monotone and continuous in a neighborhood of x_0, and there exists $f'(x_0)\neq 0$, then in a neighborhood of the point $y_0=f(x_0)$ there exists the inverse function $x=f^{-1}(y)$ differentiable at y_0 and such that $\frac{df^{-1}}{dy}(y_0)=\left(\frac{df}{dx}(x_0)\right)^{-1}$.

The second formulation. If $y=f(x)$ is continuously differentiable in a neighborhood of x_0 and $f'(x_0)\neq 0$, then in a neighborhood of the point $y_0=f(x_0)$ there exists the inverse function $x=f^{-1}(y)$ differentiable at y_0 and such that $\frac{df^{-1}}{dy}(y_0)=\left(\frac{df}{dx}(x_0)\right)^{-1}$.

Applications

Criterion of a constant function. A differentiable on (a, b) function $f(x)$ is constant on (a, b) if, and only if, $f'(x) = 0$, $\forall x \in (a, b)$.

Criterion of monotonicity. A differentiable on (a, b) function $f(x)$ is increasing (decreasing) on (a, b) if, and only if, $f'(x) \geq 0$ $(f'(x) \leq 0)$, $\forall x \in (a, b)$.

Sufficient condition of strict monotonicity. A differentiable on (a, b) function $f(x)$ is strictly increasing (decreasing) on (a, b) if $f'(x) > 0$ $(f'(x) < 0)$, $\forall x \in (a, b)$.

Critical (stationary) point. If $f'(c) = 0$, then c is called a *critical (stationary)* point.

Fermat's Theorem. If $f(x)$ is differentiable at a local extremum point c, then $f'(c) = 0$.

The first derivative test for local extremum. Let $f(x)$ be differentiable in a neighborhood of a point c. Then
1) if $f'(x) > 0$ in a left-hand neighborhood of c and $f'(x) < 0$ in a right-hand neighborhood of c, then c is a strict local maximum;
2) if $f'(x) < 0$ in a left-hand neighborhood of c and $f'(x) > 0$ in a right-hand neighborhood of c, then c is a strict local minimum;
3) if $f'(x) < 0$ $(f'(x) > 0)$ in both one-sided neighborhoods of c, then c is not a local extremum.

The second derivative test for local extremum. Let $f(x)$ be twice differentiable at a point c and $f'(c) = 0$. Then
1) if $f''(c) < 0$, then c is a strict local maximum;
2) if $f''(c) > 0$, then c is a strict local minimum.

Upward concavity. A function $f(x)$ defined on (a, b) is *upward concave (strictly upward concave)* on (a, b) if the inequality $f(\alpha_1 x_1 + \alpha_2 x_2) \leq \alpha_1 f(x_1) + \alpha_2 f(x_2)$ $(f(\alpha_1 x_1 + \alpha_2 x_2) < \alpha_1 f(x_1) + \alpha_2 f(x_2)$) holds for any points $x_1, x_2 \in (a, b)$ and arbitrary parameters $\alpha_1, \alpha_2 \geq 0$, $\alpha_1 + \alpha_2 = 1$. Another common term is convex (strictly convex) function.

Geometrically it means that the line segment joining $(x_1, f(x_1))$ and $(x_2, f(x_2))$ for $\forall x_1, x_2 \in (a, b)$ lies above the graph of $f(x)$.

Downward concavity. A function $f(x)$ defined on (a, b) is *downward concave (strictly downward concave)* on (a, b) if the inequality $f(\alpha_1 x_1 + \alpha_2 x_2) \geq \alpha_1 f(x_1) + \alpha_2 f(x_2)$ $(f(\alpha_1 x_1 + \alpha_2 x_2) > \alpha_1 f(x_1) + \alpha_2 f(x_2))$ holds for any points $x_1, x_2 \in (a, b)$ and arbitrary parameters $\alpha_1, \alpha_2 \geq 0$, $\alpha_1 + \alpha_2 = 1$.

Geometrically it means that the line segment joining $(x_1, f(x_1))$ and $(x_2, f(x_2))$ for $\forall x_1, x_2 \in (a, b)$ lies below the graph of $f(x)$.

Inflection point. Let $f(x)$ be continuous in a neighborhood of a point

a. If $f(x)$ is concave upward (downward) in a left-hand neighborhood of a and concave downward (upward) in a right-hand neighborhood of a, then the point a is called the *inflection* point.

Theorem. If $f(x)$ is concave on (a,b), then it is continuous on (a,b).

Theorem. Let $f(x)$ be differentiable on (a,b). $f(x)$ is strictly upward (downward) concave on (a,b) if, and only if, the graph of $f(x)$ lies above (below) each tangent line on (a,b) except at the point of contact.

Theorem (criterion of concavity). Let $f(x)$ be twice differentiable on (a,b). $f(x)$ is upward (downward) concave on (a,b) if, and only if, $f''(x) \geq 0$ ($f''(x) \leq 0$) on (a,b).

Theorem (sufficient condition of concavity). Let $f(x)$ be twice differentiable on (a,b). If $f''(x) > 0$ ($f''(x) < 0$) on (a,b), then$f(x)$ is strictly upward (downward) concave on (a,b).

Theorem. If $f(x)$ is twice differentiable at an inflection point c, then $f''(c) = 0$.

Theorem. Let $f(x)$ be twice differentiable in a neighborhood of a point c. Then

1) if $f''(x) > 0$ in a left-hand neighborhood of c and $f''(x) < 0$ in a right-hand neighborhood of c (or $f''(x) < 0$ in a left-hand neighborhood of c and $f''(x) > 0$ in a right-hand neighborhood of c), then c is an inflection point;
2) if $f''(x)$ keeps the same sign in both one-sided neighborhoods of c, then c is not an inflection point.

Asymptotes

The line $y = A$ is a *horizontal asymptote* of $f(x)$ if either $\lim\limits_{x\to-\infty} f(x) = A$ or $\lim\limits_{x\to+\infty} f(x) = A$, or both.

The line $y = Ax + B$ is an *oblique (slant) asymptote* of $f(x)$ if either $\lim\limits_{x\to-\infty} (f(x) - Ax - B) = 0$ or $\lim\limits_{x\to+\infty} (f(x) - Ax - B) = 0$, or both.

The line $x = a$ is a *vertical asymptote* if $\lim\limits_{x\to a_-} f(x) = (\pm)\infty$ or $\lim\limits_{x\to a_+} f(x) = (\pm)\infty$, or both.

4.2 Concepts

Example 1. "If $f(x)$ is not differentiable at a point a, then it is not continuous at this point."

Solution.

The function $f(x) = |x|$ is continuous at $a = 0$, but it is not differentiable there. In fact, $\lim\limits_{x\to 0_+} f(x) = \lim\limits_{x\to 0_+} x = 0$ and $\lim\limits_{x\to 0_-} f(x) = \lim\limits_{x\to 0_-} (-x) = 0$, so $\lim\limits_{x\to 0} f(x) = 0 = f(0)$. At the same time, $\lim\limits_{x\to 0_+} \frac{f(x)-f(0)}{x-0} = \lim\limits_{x\to 0_+} \frac{x}{x} = 1$, while $\lim\limits_{x\to 0_-} \frac{f(x)-f(0)}{x-0} = \lim\limits_{x\to 0_-} \frac{-x}{x} = -1$. Since the one-sided derivatives are different, the derivative at $a = 0$ does not exist.

Remark 1. This is the false converse to the theorem stating that if $f(x)$ is differentiable at a then it is continuous at this point.

Remark 2. There is the famous Weierstrass function continuous on $\mathbb{R}$ and nowhere differentiable, as well as other examples of this kind. However, the study of such functions requires the knowledge of the uniform convergence of series of functions, which is out of the scope of this book.

Example 2. "If $f(x)$ is differentiable in both right-hand and left-hand neighborhoods of a point a and $\lim\limits_{x\to a_+} f'(x) = \lim\limits_{x\to a_-} f'(x)$, then $f(x)$ is differentiable at a."

Solution.

For any point $a \neq 0$ the function $f(x) = \operatorname{sgn} x$ has zero derivative: for every $x < 0$, $f(x) = -1$ and so $f'(x) = 0$; for every $x > 0$, $f(x) = 1$ and then $f'(x) = 0$. Therefore, $\lim\limits_{x\to 0_+} f'(x) = \lim\limits_{x\to 0_-} f'(x) = 0$. However, $f(x)$ is not differentiable at $a = 0$ because it is not even continuous at this point.

Remark 1. The same counterexample works for a strengthened (but still false) version of this statement: "if $f(x)$ is differentiable at every $x \neq a$ and $\lim\limits_{x\to a} f'(x) = A$, then $f(x)$ is differentiable at a".

Remark 2. The following counterexample shows another kind of non-differentiability for the original statement: $f(x) = \sqrt[3]{x}$ is differentiable at every point $a \neq 0$, and $\lim\limits_{x\to 0} f'(x) = \lim\limits_{x\to 0} \frac{1}{3\sqrt[3]{x^2}} = +\infty$. Moreover, the function is continuous at $a = 0$, and nevertheless the derivative does not exist at $a = 0$. Notice that this counterexample does not work for the statement in Remark 1, where the finite limit is required.

Remark 3. If we put aside the case of infinite limits in the original statement, then both statements (original and in Remark 1) will be true if we add the condition of continuity of $f(x)$ at a.

Remark 4. A "negative" version is true and applicable in practice: if $f(x)$ is differentiable in both right-hand and left-hand neighborhoods of a point a and $\lim\limits_{x\to a_+} f'(x) \neq \lim\limits_{x\to a_-} f'(x)$, then $f(x)$ is not differentiable at a.

Example 3. "If $f(x)$ is differentiable on a neighborhood of a point a, then its derivative is continuous at this point."

Solution.

The function $f(x) = \begin{cases} x^2 \sin\frac{1}{x}, & x \neq 0 \\ 0, & x = 0 \end{cases}$ is differentiable at any x and its derivative is $f'(x) = \begin{cases} 2x\sin\frac{1}{x} - \cos\frac{1}{x}, & x \neq 0 \\ 0, & x = 0 \end{cases}$. Indeed, for any $x \neq 0$ the derivative is readily found by applying the arithmetic and chain rules, while for $x = 0$ the derivative can be calculated by appealing to the definition where we first evaluate the expression $\left|\frac{f(x)-f(0)}{x-0}\right| = \left|x\sin\frac{1}{x}\right| \leq |x|$ for $x \neq 0$ and then letting $x \to 0$ we get $f'(0) = 0$. However, the function $f'(x)$ is discontinuous at zero: $\lim\limits_{x\to 0} 2x\sin\frac{1}{x} = 0$, but $\lim\limits_{x\to 0}\cos\frac{1}{x}$ does not exist (it can be shown in the same way as for $\sin\frac{1}{x}$ in Example 4, section 2.2).

Remark . Darboux's theorem specifies what types of discontinuities are admissible by a derivative function. It states that if $f(x)$ is differentiable on (a,b), then $f'(x)$ cannot have a removable or jump discontinuity on (a,b). (But $f'(x)$ still may have discontinuities of the second kind like in the shown counterexample.)

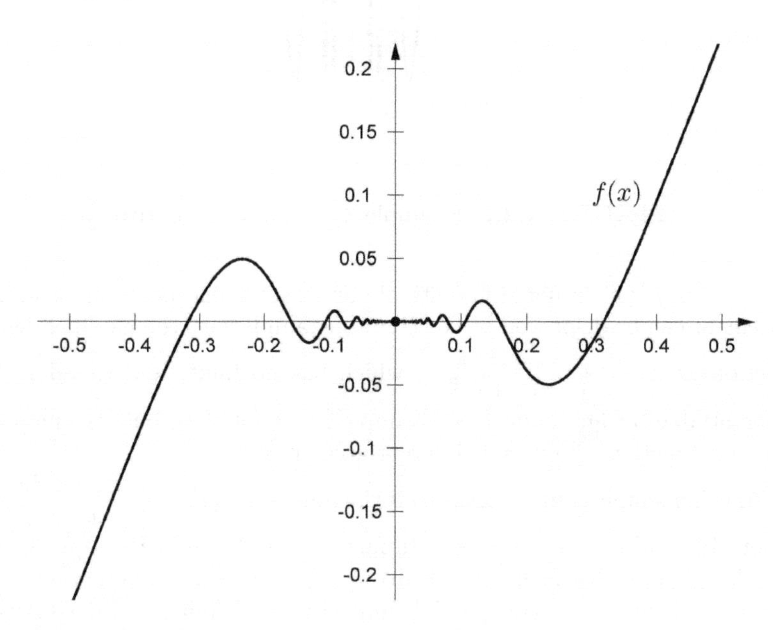

FIGURE 4.2.1: Example 3, graph of the function

Example 4. "If a function $f(x)$ is not differentiable at a point a, then $f^2(x)$ also is not differentiable at a."

Solution.

The function $f(x) = |x|$ is not differentiable at $a = 0$ (see Example 1 in this section), but $f^2(x) = x^2$ is differentiable on $\mathbb{R}$.

Remark 1. The converse is true.

Remark 2. This Example can be easily extended to a set of points in the following form: "if a function $f(x)$ is not differentiable at any point in

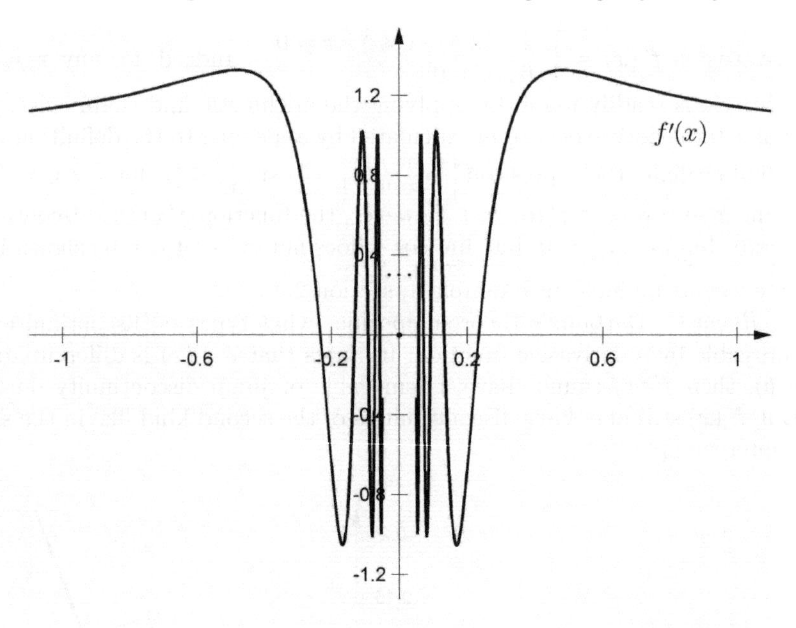

FIGURE 4.2.2: Example 3, graph of the derivative

(a,b), then $f^2(x)$ is not differentiable at least at one point in (a,b)". This statement can be disproved by the counterexample with the modified Dirichlet function $\tilde{D}(x) = \begin{cases} 1, & x \in \mathbb{Q} \\ -1, & x \in \mathbb{I} \end{cases}$, which has no limit (and therefore is not differentiable) at any point (see Example 2, section 2.2), but its square is the constant function $\tilde{D}^2(x) = 1$ differentiable on $\mathbb{R}$.

Another simple counterexample is the function $f(x) = \begin{cases} x, & x \in \mathbb{Q} \\ -x, & x \in \mathbb{I} \end{cases}$ that is not differentiable at any $x \in \mathbb{R}$. In fact, it was shown in Example 3, section 3.3, that $f(x)$ is discontinuous at any $x \neq 0$, so it is not differentiable at any $x \neq 0$, and for $x = 0$ two partial limits in the definition of derivative give different results:

$$\lim_{x \to 0, x \in \mathbb{Q}} \frac{f(x) - f(0)}{x - 0} = \lim_{x \to 0, x \in \mathbb{Q}} \frac{x}{x} = 1$$

and

$$\lim_{x \to 0, x \in \mathbb{I}} \frac{f(x) - f(0)}{x - 0} = \lim_{x \to 0, x \in \mathbb{I}} \frac{-x}{x} = -1,$$

so the derivative does not exist at $x = 0$. However, $f^2(x) = x^2$ is differentiable on $\mathbb{R}$.

Example 5. "If a function $f(x)$ is not differentiable at a point a, then $|f(x)|$ also is not differentiable at this point."

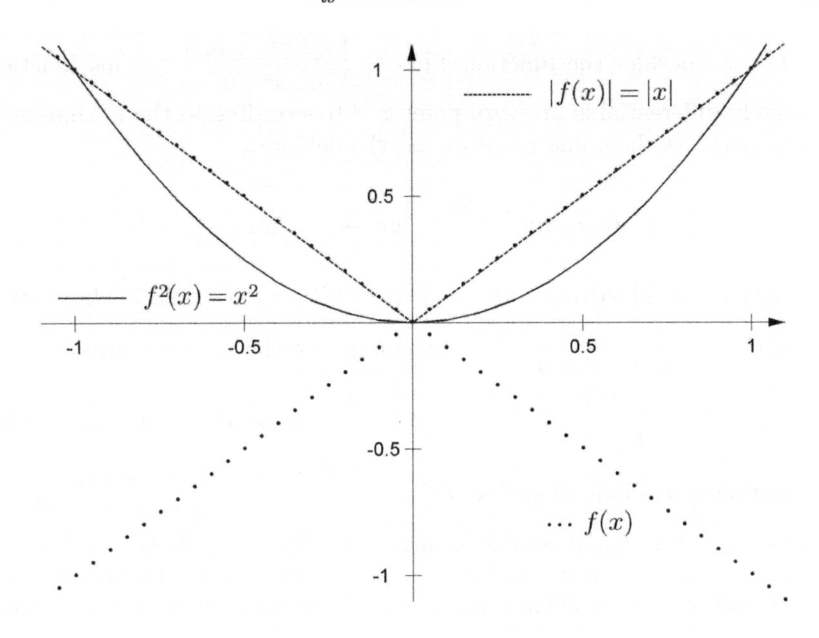

FIGURE 4.2.3: Example 4, Remark 2 and Example 5, the function $f(x)$

Solution.

The same functions in Remark 2 to the previous Example 4 work here: both $\tilde{D}(x) = \begin{cases} 1, & x \in \mathbb{Q} \\ -1, & x \in \mathbb{I} \end{cases}$ and $f(x) = \begin{cases} x, & x \in \mathbb{Q} \\ -x, & x \in \mathbb{I} \end{cases}$ are not differentiable at any $x \in \mathbb{R}$, but $\left|\tilde{D}(x)\right| = 1$ is differentiable on $\mathbb{R}$ and $|f(x)| = |x|$ is differentiable on $\mathbb{R} \backslash \{0\}$.

Remark 1. The following extended statement is also false: "if a function $f(x)$ is not differentiable at any point in (a, b), then $|f(x)|$ is not differentiable at least at one point in (a, b)". The same functions can be used for the counterexamples: the function $\tilde{D}(x)$ on $\mathbb{R}$, and the function $f(x)$ on $(0, +\infty)$.

Remark 2. The converse is also false. The corresponding example is $f(x) = x$, which is differentiable on $\mathbb{R}$, but $|f(x)| = |x|$ is not differentiable at $a = 0$.

Remark 3. It is worth to note that relation between $f(x)$ and $|f(x)|$ in terms of differentiability is different from that for continuous functions. For continuous functions the continuity of $f(x)$ implies the continuity of $|f(x)|$, while the converse implication does not hold. For differentiability both implications are not true.

Example 6. "If a function $f(x)$ is infinitely differentiable on $\mathbb{R}$ and $f(0) = f'(0) = f''(0) = \ldots = 0$, that is, $f^{(n)}(0) = 0$, $\forall n \in \mathbb{N} \cup \{0\}$, then $f(x) \equiv 0$ on $\mathbb{R}$."

Solution.

Let us consider the function $f(x) = \begin{cases} e^{-\frac{1}{x^2}}, & x \neq 0 \\ 0, & x = 0 \end{cases}$. This function is infinitely differentiable at every point $x \neq 0$ according to the arithmetic and chain rules. At the point $x = 0$ we use the definition:

$$f'(0) = \lim_{x \to 0} \frac{e^{-\frac{1}{x^2}}}{x} = \lim_{t \to \infty} \frac{t}{e^{t^2}} = \lim_{t \to \infty} \frac{1}{2te^{t^2}} = 0$$

(applying the substitution $x = \frac{1}{t}$ and L'Hospital's rule). Therefore, we have $f'(x) = \begin{cases} \frac{2}{x^3} e^{-\frac{1}{x^2}}, & x \neq 0 \\ 0, & x = 0 \end{cases}$. Similarly, for the second derivative we get $f''(x) = \begin{cases} -\frac{6}{x^4} e^{-\frac{1}{x^2}} + \left(\frac{2}{x^3}\right)^2 e^{-\frac{1}{x^2}}, & x \neq 0 \\ 0, & x = 0 \end{cases}$. Generalizing, it can be shown by mathematical induction that $f^{(n)}(x) = \begin{cases} P_{3n}\left(\frac{1}{x}\right) e^{-\frac{1}{x^2}}, & x \neq 0 \\ 0, & x = 0 \end{cases}$, $\forall n \in \mathbb{N}$, where $P_{3n}(t)$ is a polynomial of degree $3n$. We have already deduced this formula for $n = 1$ and $n = 2$. Let us suppose now that the formula is true for $n = k$ and prove that it holds for $n = k + 1$. At every point $x \neq 0$ we have

$$f^{(k+1)}(x) = \left(P_{3k}\left(\frac{1}{x}\right) e^{-\frac{1}{x^2}} \right)'$$

$$= P_{3k}\left(\frac{1}{x}\right) e^{-\frac{1}{x^2}} \cdot \frac{2}{x^3} + P_{3k+1}\left(\frac{1}{x}\right) e^{-\frac{1}{x^2}} = P_{3(k+1)}\left(\frac{1}{x}\right) e^{-\frac{1}{x^2}}.$$

For the point $x = 0$, applying the substitution $x = \frac{1}{t}$ and L'Hospital's rule, we obtain:

$$f^{(k+1)}(0) = \lim_{x \to 0} \frac{f^{(k)}(x) - f^{(k)}(0)}{x - 0} = \lim_{x \to 0} \frac{P_{3k}\left(\frac{1}{x}\right) e^{-\frac{1}{x^2}}}{x} = \lim_{t \to \pm\infty} \frac{P_{3k+1}(t)}{e^{t^2}} = 0.$$

Therefore, the formula for the derivatives is proved. Hence, $f(x)$ satisfied the conditions of the statement, but $f(x)$ is equal to zero only at the origin.

Example 7. "If $\lim_{h \to 0} \frac{f(x+h) - f(x-h)}{2h} = A$ at a point x, then $f'(x) = A$."
Solution.
The function $f(x) = |x|$ satisfies the above condition at $x = 0$, but it does not differentiable at this point.

Remark 1. Actually for $x = 0$ any even function non-differentiable at $x = 0$ provides a counterexample to this statement. For instance, one can choose even Dirichlet's function, which does not have a limit at any point.

Remark 2. If the differentiability of $f(x)$ at x is added to the condition, then the statement becomes true.

Remark 3. The converse is true.

Example 8. "If $\lim_{h\to 0} \frac{f(x+h)-2f(x)+f(x-h)}{h^2} = A$ at a point x, then $f''(x) = A$."

Solution.

The function $f(x) = \operatorname{sgn} x$ satisfies the above condition at $x = 0$, but it does not even have a limit at this point.

Remark 1. Actually for $x = 0$ any odd function discontinuous at the origin provides a counterexample to this statement.

Remark 2. If the second derivative $f''(x)$ exists, then the statement becomes true.

Remark 3. The converse is true.

Example 9. "If $f(x)$ is continuous in a neighborhood of a point a and is differentiable at a, then there is a neighborhood of a where $f(x)$ is differentiable."

Solution.

The function $f(x) = \begin{cases} x^2 \left|\sin \frac{\pi}{x}\right|, & x \neq 0 \\ 0, & x = 0 \end{cases}$ is continuous at $x \neq 0$ according to the composition and arithmetic properties of continuous functions, and at $x = 0$ one gets $\lim_{x\to 0} f(x) = 0 = f(0)$. Therefore, $f(x)$ is continuous on $\mathbb{R}$. The definition of this function can be rewritten as follows:

$$f(x) = \begin{cases} x^2 \sin \frac{\pi}{x}, & \frac{1}{2n+1} < x < \frac{1}{2n},\ n \in \mathbb{Z} \\ -x^2 \sin \frac{\pi}{x}, & \frac{1}{2n} < x < \frac{1}{2n-1},\ n \in \mathbb{Z} \\ 0, & x = 0,\ x = \frac{1}{n},\ n \in \mathbb{Z}\setminus\{0\} \end{cases}.$$

Then for $x \neq 0$ and $x \neq \frac{1}{n}$ the derivative can be found by applying the arithmetic and chain rules:

$$f'(x) = \begin{cases} 2x \sin \frac{\pi}{x} - \pi \cos \frac{\pi}{x}, & \frac{1}{2n+1} < x < \frac{1}{2n},\ n \in \mathbb{Z} \\ -2x \sin \frac{\pi}{x} + \pi \cos \frac{\pi}{x}, & \frac{1}{2n} < x < \frac{1}{2n-1},\ n \in \mathbb{Z} \end{cases}.$$

Let us show that the derivative does not exist at any point $x_n = \frac{1}{2n}, n \in \mathbb{Z}\setminus\{0\}$. In fact,

$$\lim_{x\to x_n-} f'(x) = \lim_{x\to x_n-} \left(2x \sin \frac{\pi}{x} - \pi \cos \frac{\pi}{x}\right) = -\pi,$$

while

$$\lim_{x\to x_n+} f'(x) = \lim_{x\to x_n+} \left(-2x \sin \frac{\pi}{x} + \pi \cos \frac{\pi}{x}\right) = \pi.$$

Note that $x_n = \frac{1}{2n} \underset{n\to\pm\infty}{\to} 0$, but, at the same time the function is differentiable at $x = 0$:

$$f'(0) = \lim_{x\to 0} \frac{f(x) - f(0)}{x - 0} = \lim_{x\to 0} x \left|\sin \frac{\pi}{x}\right| = 0.$$

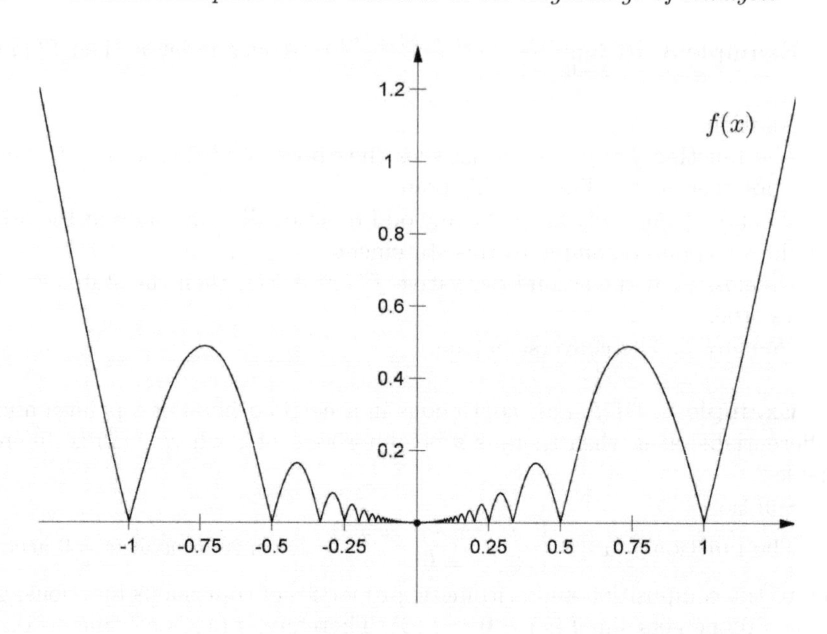

FIGURE 4.2.4: Example 9

4.3 Local properties

Example 1. "If $h(x) = f(x) + g(x)$ is differentiable at a point a, then both $f(x)$ and $g(x)$ are differentiable at a."

Solution.

The functions $f(x) = |x|$ and $g(x) = -|x|$ are not differentiable at the point $a = 0$ (see Example 1, section 4.2), but their sum $h(x) = f(x) + g(x) = 0$ is differentiable at 0.

Remark 1. Similar statements and counterexamples can be given for other arithmetic operations. For example, for the ratio the corresponding counterexample is $f(x) = (|x| + 1)\cos x$ and $g(x) = |x| + 1$. At the point $a = 0$ the one-sided derivatives of $f(x)$ are different:

$$f'_+(0) = \lim_{x \to 0_+} \frac{(x+1)\cos x - 1}{x} = \lim_{x \to 0_+} \frac{x\cos x - 2\sin^2 x/2}{x} = 1$$

and

$$f'_-(0) = \lim_{x \to 0_-} \frac{(-x+1)\cos x - 1}{x} = \lim_{x \to 0_-} \frac{-x\cos x - 2\sin^2 x/2}{x} = -1.$$

The situation is similar for the one-sided derivatives of $g(x)$:

$$g'_+(0) = \lim_{x\to 0_+} \frac{(x+1)-1}{x} = 1 \text{ and } g'_-(0) = \lim_{x\to 0_-} \frac{(-x+1)-1}{x} = -1.$$

Hence, the functions $f(x)$ and $g(x)$ are non-differentiable at $a = 0$. At the same time, the ratio $h(x) = \frac{f(x)}{g(x)} = \cos x$ is differentiable on $\mathbb{R}$.

Remark 2. The reverse is true (it is one of the arithmetic rules of differentiation).

Example 2. "If $h(x) = f(x) \cdot g(x)$ is differentiable at a point a and $f(x)$ is differentiable at the same point, then $g(x)$ is also differentiable at a."

Solution.

The function $f(x) = x$ is differentiable at $a = 0$ and $g(x) = \sqrt[3]{x}$ is not differentiable at $a = 0$ (the corresponding limit is infinite $\lim_{x\to 0} \frac{\sqrt[3]{x}-0}{x} = \lim_{x\to 0} \frac{1}{\sqrt[3]{x^2}} = +\infty$), but their product $h(x) = f(x) \cdot g(x) = \sqrt[3]{x^4}$ is differentiable at 0, because $\lim_{x\to 0} \frac{\sqrt[3]{x^4}-0}{x} = \lim_{x\to 0} \sqrt[3]{x} = 0$.

Remark 1. The above counterexample shows also that the following statement is not correct: "if $f(x)$ is differentiable at a point a, and $g(x)$ is non-differentiable at a, then $f(x) \cdot g(x)$ is non-differentiable at a". On the other hand, the statement: "if $f(x)$ is differentiable at a point a, and $g(x)$ is non-differentiable at a, then $f(x) \cdot g(x)$ is differentiable at a" is not true either. The counterexample is $f(x) = 1$ and $g(x) = |x|$, $a = 0$. It means that the situation under the conditions of the statements is indeterminate.

Remark 2. Similar statement can be given for the ratio and the corresponding counterexample can be $f(x) = (|x|+1)\sin x$ and $g(x) = |x|+1$ at $a = 0$.

Remark 3. For the sum or difference the corresponding statement is true: if $h(x) = f(x) \pm g(x)$ is differentiable at a point a and $f(x)$ is differentiable at the same point, then $g(x)$ is also differentiable at a. Or, equivalently, if $f(x)$ is differentiable at a point a, and $g(x)$ is non-differentiable at a, then $f(x) \pm g(x)$ is non-differentiable at a.

Example 3. "If $f(x)$ is not differentiable at a point a, and $g(x)$ is differentiable at the point $f(a)$, then $g(f(x))$ is not differentiable at a."

Solution.

The function $f(x) = |x|$ is not differentiable at $a = 0$ and $g(x) = x^2$ is differentiable at $f(0) = 0$. However, $g(f(x)) = x^2$ is differentiable at $a = 0$.

Remark 1. If we interchange the definitions of the functions - setting $f(x) = x^2$ and $g(x) = |x|$ - and keep the same $a = 0$, then we obtain the counterexample to a similar statement: "if $f(x)$ is differentiable at a point a, and $g(x)$ is not differentiable at the point $f(a)$, then $g(f(x))$ is not differentiable at a".

Remark 2. Of course, the opposite conclusion is also not correct: if $f(x)$ is not differentiable at a point a, and $g(x)$ is differentiable at the point $f(a)$, it may happen that $g(f(x))$ is not differentiable at a. For example, $f(x) = |x|$ and $g(x) = x$, $a = 0$. It means that the situation under the conditions of the statement is indeterminate.

Remark 3. This is a false "negative" version of the chain rule.

Example 4. "If $f(x)$ is not differentiable at a point a, and $g(x)$ is not differentiable at the point $f(a)$, then $g(f(x))$ is not differentiable at a."

Solution.

The function $f(x) = g(x) = \begin{cases} 1/x, \ x \neq 0 \\ 0, \ x = 0 \end{cases}$ is not differentiable at $a = 0$ (it has a discontinuity of the second kind at 0), but the function $g(f(x)) = x$ is differentiable at 0.

Remark 1. Another interesting example involves two functions non-differentiable at any point: $f(x) = -g(x) = \begin{cases} x, \ x \in \mathbb{Q} \\ -x, \ x \in \mathbb{I} \end{cases}$. Their composition $g(f(x)) = -x$ is differentiable on $\mathbb{R}$.

Remark 2. Of course, it may happen that under the above conditions the function $g(f(x))$ will not be differentiable at a, for example, $f(x) = |x|$ and $g(x) = |x|$, $a = 0$. It means that the situation under the conditions of the statement is indeterminate.

Example 5. "If $f(x)$ is differentiable on $\mathbb{R}$, then $f'(x)$ is bounded in a neighborhood of each point."

Solution.

The function $f(x) = \begin{cases} x^2 \sin 1/x^2, \ x \neq 0 \\ 0, \ x = 0 \end{cases}$ is differentiable on $\mathbb{R}$. Indeed, for any $x \neq 0$, applying the arithmetic and chain rules, we obtain $f'(x) = 2x \sin \frac{1}{x^2} - \frac{2}{x} \cos \frac{1}{x^2}$, and at $x = 0$, appealing to the definition, we get

$$f'(0) = \lim_{x \to 0} \frac{x^2 \sin 1/x^2 - 0}{x - 0} = \lim_{x \to 0} x \sin \frac{1}{x^2} = 0$$

(the last result follows from the evaluation $\left| x \sin \frac{1}{x^2} \right| \leq |x|$, $x \neq 0$). Hence, $f'(x) = \begin{cases} 2x \sin \frac{1}{x^2} - \frac{2}{x} \cos \frac{1}{x^2}, \ x \neq 0 \\ 0, \ x = 0 \end{cases}$. Now, choosing $x_k = \frac{1}{\sqrt{2k\pi}}$, $k \in \mathbb{N}$ we obtain $\lim_{x_k \to 0} f'(x_k) = \lim_{k \to +\infty} \left(-2\sqrt{2k\pi} \cos 2k\pi \right) = -\infty$, that is, $f'(x)$ is unbounded in any neighborhood of 0.

Example 6. "There is no function differentiable only at one point of its domain."

Solution.

The function $f(x) = \begin{cases} x^2, \ x \in \mathbb{Q} \\ 0, \ x \in \mathbb{I} \end{cases}$ is not differentiable at any point $a \neq 0$. In fact, this function is not even continuous at any $a \neq 0$: $\lim_{x \to a, \, x \in \mathbb{Q}} f(x) =$

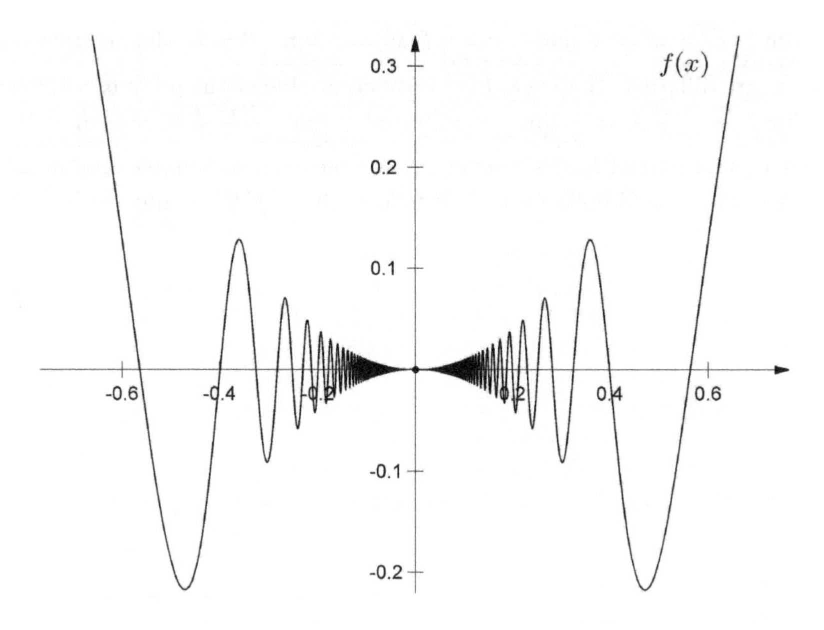

FIGURE 4.3.1: Example 5, graph of the function

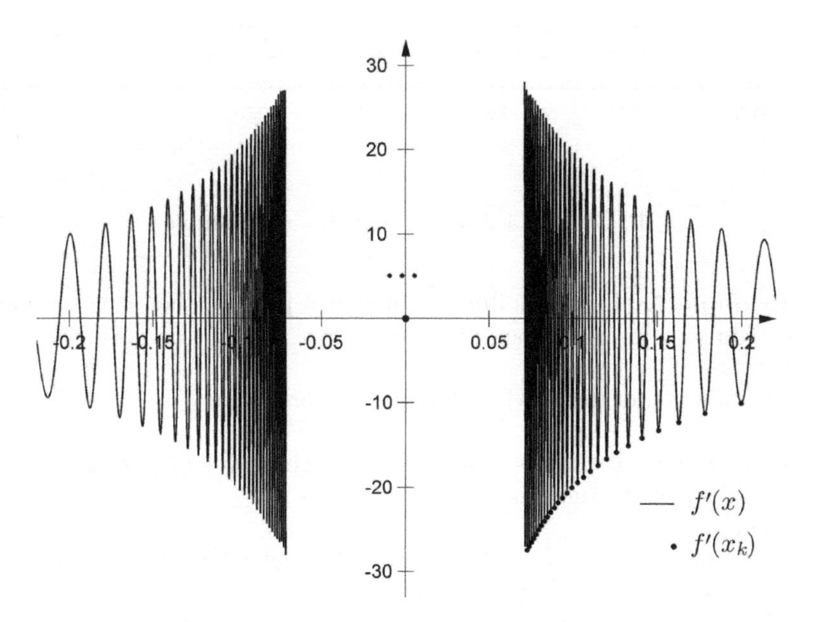

FIGURE 4.3.2: Example 5, graph of the derivative

$\lim\limits_{x\to a,\, x\in\mathbb{Q}} x^2 = a^2 \neq 0$ and $\lim\limits_{x\to a,\, x\in\mathbb{I}} f(x) = \lim\limits_{x\to a, x\in\mathbb{I}} 0 = 0$, that is, two partial limits are different. However, $f(x)$ is differentiable at the point $a = 0$, because $\lim\limits_{x\to 0,\, x\in\mathbb{Q}} \frac{f(x)-f(0)}{x-0} = \lim\limits_{x\to 0,\, x\in\mathbb{Q}} x = 0$ and $\lim\limits_{x\to 0,\, x\in\mathbb{I}} \frac{f(x)-f(0)}{x-0} = \lim\limits_{x\to 0, x\in\mathbb{I}} 0 = 0$, that is, two partial limits coincide. Since these partial limits involve all real numbers in a neighborhood of 0, it follows that $\exists f'(0) = \lim\limits_{x\to 0} \frac{f(x)-f(0)}{x-0} = 0$.

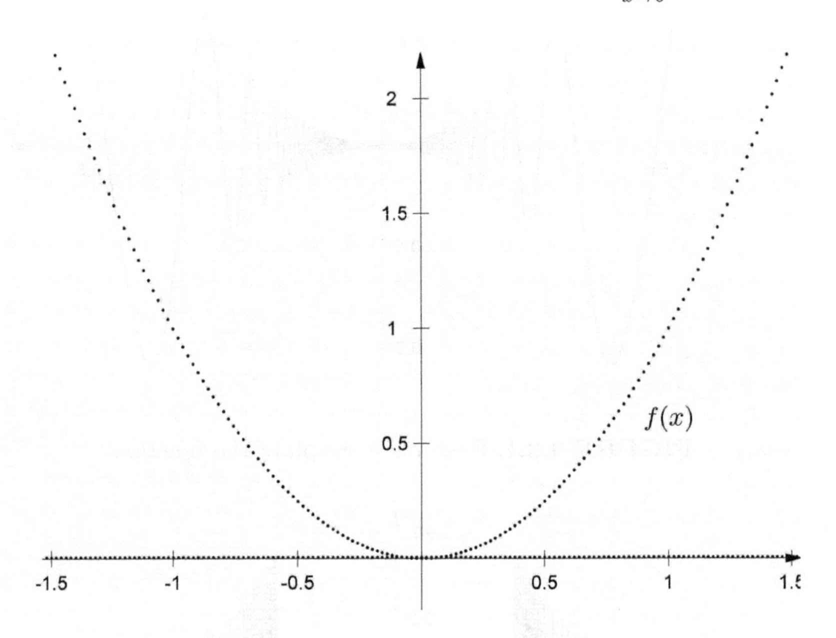

FIGURE 4.3.3: Example 6 and Example 7

Example 7. "If $f(x)$ is differentiable at a point a, then it is continuous at least in some neighborhood of a."

Solution.

The function of the previous Example 6 works here.

Example 8. "If there is a tangent line to the graph of $f(x)$ at a point c, then $f(x)$ is differentiable at c."

Solution.

The function $f(x) = \sqrt[3]{x}$ has a vertical tangent line at the point $c = 0$, but it is not differentiable at this point. In fact, the derivative is infinite: $\lim\limits_{x\to 0} \frac{f(x)-f(0)}{x-0} = \lim\limits_{x\to 0} \frac{\sqrt[3]{x}}{x} = +\infty$, so $f(x)$ is not differentiable at $c = 0$, but $x = 0$ is a vertical tangent line to the graph of $f(x)$.

Example 9. "If $f(x)$ is not differentiable at x_0, then either the inverse function does not exist in a neighborhood of the point $y_0 = f(x_0)$ or (if it exists) it is not differentiable at y_0."

Solution.

The function $y = f(x) = \sqrt[3]{x}$ is strictly monotone and continuous in a neighborhood of $x_0 = 0$, but its derivative does not exists at the origin. However, the inverse function $x = f^{-1}(y) = y^3$ is defined and differentiable on $\mathbb{R}$, including at $y_0 = 0$.

Remark . This example shows that the conditions of the Inverse Function Theorem are just sufficient.

Example 10. "If $f(x)$ is differentiable on $\mathbb{R}$, $f'(x)$ is bounded in a neighborhood of a point x_0 and $f'(x_0) \neq 0$, then in a neighborhood of the point $y_0 = f(x_0)$ there exists the inverse function $x = f^{-1}(y)$."

Solution.

The function $f(x) = \begin{cases} x + 2x^2 \sin\frac{1}{x}, & x \neq 0 \\ 0, & x = 0 \end{cases}$ is differentiable at every real point. In fact, for any $x \neq 0$ the derivative can be calculated by using the arithmetic and chain rules: $f'(x) = 1 + 4x\sin\frac{1}{x} - 2\cos\frac{1}{x}$. For $x = 0$, we should appeal to definition:

$$f'(0) = \lim_{x\to 0} \frac{x + 2x^2 \sin 1/x}{x} = \lim_{x\to 0}\left(1 + 2x\sin\frac{1}{x}\right) = 1.$$

Hence,

$$f'(x) = \begin{cases} 1 + 4x\sin\frac{1}{x} - 2\cos\frac{1}{x}, & x \neq 0 \\ 1, & x = 0 \end{cases} \quad \text{and } f'(0) \neq 0.$$

Besides, $f'(x)$ is bounded on $[-1, 1]$:

$$\left|1 + 4x\sin\frac{1}{x} - 2\cos\frac{1}{x}\right| \leq 1 + 4|x| + 2 \leq 7.$$

Thus, all the conditions of the statement are satisfied. Nevertheless, one can show that there is no neighborhood of $x_0 = 0$ on which $f(x)$ is a one-to-one correspondence. Let us consider, for simplicity, a right-hand neighborhood of 0. On such a neighborhood $\sin\frac{1}{x}$ takes all the values in $[-1, 1]$, and therefore, $f(x)$ oscillates between the parabolas $x - 2x^2$ and $x + 2x^2$. The apex of the parabola $x - 2x^2$ is at the point $x = \frac{1}{4}$. Consider $\forall\delta > 0$, $\delta < \frac{1}{4}$ and $\delta_1 = \frac{\delta}{\pi} < \delta$. Let us find a point x_1 such that $f(x_1) = x_1 - 2x_1^2$ and $x_1 < \delta_1$. From the first condition, $\sin\frac{1}{x_1} = -1$, that gives $x_k = \frac{1}{2k\pi - \pi/2}$, $k \in \mathbb{N}$, and the second condition imposes restriction on k: $2k - \frac{1}{2} > \frac{1}{\delta}$, that is, $k > \frac{1}{4} + \frac{1}{2\delta}$. Let us choose some fixed k_1, which satisfies the last condition, and the corresponding $x_1 = \frac{1}{2k_1\pi - \pi/2} \in (0, \delta_1) \subset (0, \delta)$. Note that the function value at the auxiliary point $x_2 = \frac{1}{2k_1\pi + \pi/2}$ is larger than at x_1:

$$f(x_2) - f(x_1) = x_2 + 2x_2^2 \sin\left(2k_1\pi + \frac{\pi}{2}\right) - x_1 - 2x_1^2 \sin\left(2k_1\pi - \frac{\pi}{2}\right)$$

$$= \frac{1}{2k_1\pi + \pi/2} + \frac{2}{(2k_1\pi + \pi/2)^2} - \frac{1}{2k_1\pi - \pi/2} + \frac{2}{(2k_1\pi - \pi/2)^2}$$

$$= \frac{(4-\pi)\cdot 4k_1\pi^2 + (4+\pi)\cdot \pi^2/4}{(2k_1\pi + \pi/2)^2 (2k_1\pi - \pi/2)^2} > 0.$$

Consider one more point $x_3 = \frac{1}{2(k_1+1)\pi - \pi/2} < x_1 < \frac{1}{4}$. Since $2x - x^2$ is increasing on $\left(0, \frac{1}{4}\right)$, it follows that $f(x_3) < f(x_1)$. Hence, we have the three points $x_3 < x_2 < x_1 < \delta$ such that $f(x_3) < f(x_1) < f(x_2)$. Applying the Intermediate Value Theorem to the continuous function $f(x)$ on the interval $[x_3, x_2]$, we conclude that for $\forall A$ such that $f(x_3) < A < f(x_2)$, in particular, for $A = f(x_1)$, there exists a point $x_A \in [x_3, x_2]$ at which $f(x_A) = A = f(x_1)$. Thus, $f(x)$ takes the same value $f(x_A) = f(x_1)$ at two different points of the interval $(0, \delta)$, that is, $f(x)$ is not injective on $(0, \delta)$. Since it holds for $\forall \delta > 0$, the function $f(x)$ is not one-to-one in any neighborhood of $x_0 = 0$ and an inverse does not exist in a neighborhood of the origin.

Remark. This false statement is a weakened version of the Inverse Function Theorem where the condition of the continuity of the derivative is replaced by the weaker condition of its boundedness. In fact, the derivative of $f(x)$ is discontinuous at 0: for $x_n = \frac{1}{2n\pi + \pi} \underset{n\to\infty}{\to} 0$ $(n \in \mathbb{N})$ one has $f'(x_n) = 1 + 2 = 3 \neq 1 = f'(0)$.

4.4 Global properties

Example 1. "If both functions $f(x)$ and $g(x)$ are differentiable on $\mathbb{R}$ and $f(x) > g(x)$ on $\mathbb{R}$, then there exists at least one point in $\mathbb{R}$ where $f'(x) > g'(x)$."

Solution.

Evidently, the functions $f(x) = e^{-x}$ and $g(x) = -e^{-x}$ are differentiable on $\mathbb{R}$ and $f(x) > g(x)$ on $\mathbb{R}$, but $f'(x) = -e^{-x} < e^{-x} = g'(x)$ on $\mathbb{R}$.

Example 2. "If a function is differentiable and monotone on $\mathbb{R}$, then its derivative is also monotone on $\mathbb{R}$."

Solution.

The function $f(x) = 2x + \sin x$ is differentiable and strictly increasing on $\mathbb{R}$ $(f'(x) = 2 + \cos x > 0,\ \forall x \in \mathbb{R})$, but its derivative is not monotone on $\mathbb{R}$.

Remark . The following variation of the statement is also false: "if a function is differentiable and non-monotone on $\mathbb{R}$ then its derivative is also non-monotone on $\mathbb{R}$". It can be disproved by the counterexample $f(x) = x^2$.

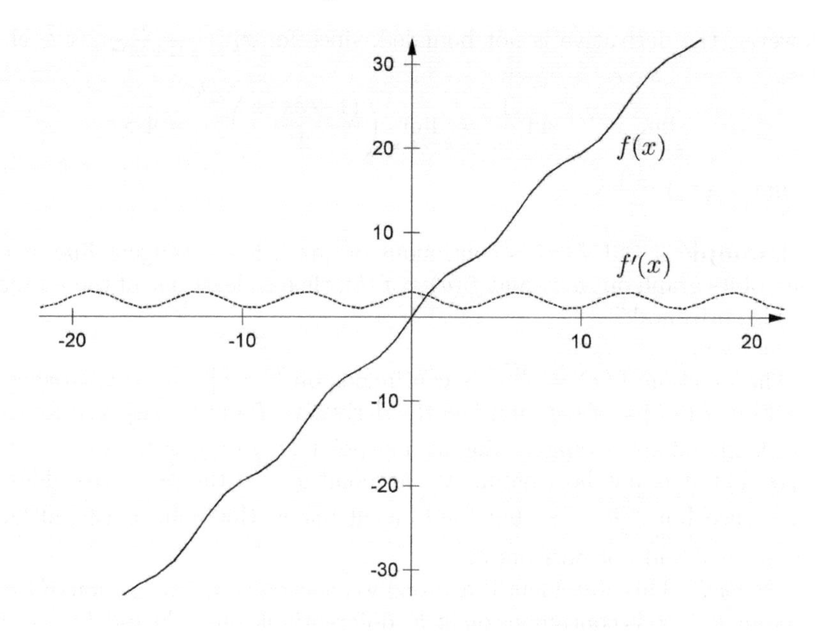

FIGURE 4.4.1: Example 2

Example 3. "If $f(x)$ is differentiable on $[a,b]$ then $f'(x)$ is bounded on $[a,b]$."

Solution.

This is a "global" version of the statement in Example 5, section 4.3, and we can use the same function for a counterexample. The function $f(x) = \begin{cases} x^2 \sin\frac{1}{x^2}, & x \neq 0 \\ 0, & x = 0 \end{cases}$ has the derivative $f'(x) = \begin{cases} 2x\sin\frac{1}{x^2} - \frac{2}{x}\cos\frac{1}{x^2}, & x \neq 0 \\ 0, & x = 0 \end{cases}$ (see Example 5, section 4.3 for details). However, $f'(x)$ is unbounded on $[-1,1]$, because $\lim\limits_{x\to 0} 2x\sin\frac{1}{x^2} = 0$, and for $x_k = \frac{1}{\sqrt{2k\pi}}$, $k \in \mathbb{N}$ it follows

$$\lim_{x_k\to 0} \frac{2}{x_k}\cos\frac{1}{x_k^2} = \lim_{k\to+\infty} 2\sqrt{2k\pi}\cos 2k\pi = +\infty,$$

so $\lim\limits_{x_k\to 0} f'(x_k) = -\infty$.

Another similar counterexample is $f(x) = \begin{cases} x^{4/3}\cos\frac{1}{x}, & x \neq 0 \\ 0, & x = 0 \end{cases}$ considered on $[-1,1]$. This function is differentiable on $[-1,1]$:

$$f'(x) = \begin{cases} \frac{4}{3}x^{1/3}\cos\frac{1}{x} + x^{-2/3}\sin\frac{1}{x}, & x \neq 0 \\ 0, & x = 0 \end{cases}.$$

However, the derivative is not bounded, since for $x_k = \frac{2}{(1+4k)\pi}$, $k \in \mathbb{N}$ one gets

$$\lim_{x_k \to 0} x_k^{-2/3} \sin \frac{1}{x_k} = \lim_{k \to +\infty} \left(\frac{(1+4k)\pi}{2} \right)^{2/3} = +\infty,$$

so $\lim_{x_k \to 0} f'(x_k) = +\infty$.

Example 4. "If $f(x)$ is continuous on $[a,b]$, has a tangent line at each point of its graph on (a,b) and $f(a) = f(b)$, then at least one of these tangent lines is horizontal."

Solution.

The function $f(x) = \sqrt[3]{x^2}$ is continuous on $[-a,a]$, $\forall a > 0$, satisfies the condition $f(-a) = f(a)$, and has the derivative $f'(x) = \frac{2}{3\sqrt[3]{x}} \neq 0$ for every $x \neq 0$. Therefore, a tangent line at any point $(x, f(x))$, $x \in [-a,0) \cup (0,a]$, exists, but it is not horizontal. At the point $x = 0$ the derivative does not exist, since $\lim_{x \to 0} \frac{\sqrt[3]{x^2}}{x} = \infty$, but the tangent line at this point exists, although it is vertical and not horizontal.

Remark . This statement is a wrong weakened geometric version of Rolle's theorem: if $f(x)$ is continuous on $[a,b]$, differentiable on (a,b) and $f(a) = f(b)$ then there is a point $c \in (a,b)$ such that $f'(c) = 0$. (It implies that the tangent line at the point $(c, f(c))$ is horizontal.)

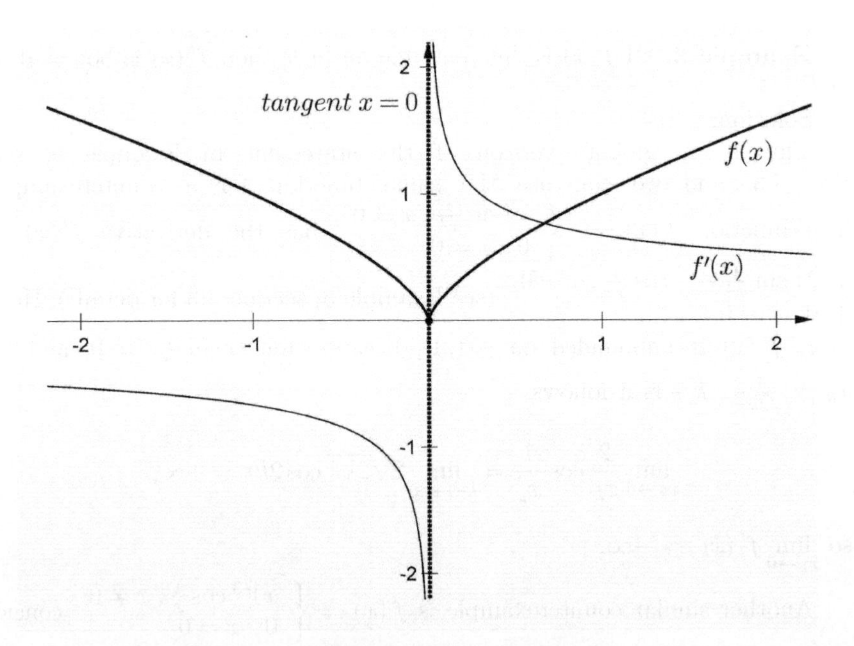

FIGURE 4.4.2: Example 4

Example 5. "If $f(x)$ is bounded on $[a,b]$, differentiable on (a,b) and $f(a)=f(b)=0$, then its derivative is zero at some point in (a,b). "

Solution.

The function $f(x)=x-[x]$ is bounded on $[0,1]$, differentiable on $(0,1)$ ($f(x)=x$ on $(0,1)$) and $f(0)=f(1)=0$, but $f'(x)=1$ for every $x\in(0,1)$.

Remark . This is a wrong version of Rolle's theorem with dropped condition of the continuity of $f(x)$ on $[a,b]$.

Example 6. "If $f(x)$ is continuous on $[a,b]$ and differentiable on (a,b), then for any $c\in(a,b)$ there are two points $x_1,x_2\in(a,b)$ such that $f'(c)=\frac{f(x_2)-f(x_1)}{x_2-x_1}$."

Solution.

The function $f(x)=x^3$ is continuous on $[-1,1]$ and differentiable on $(-1,1)$. However, for $c=0$ there is no pair $x_1,x_2\in(-1,1)$ such that $f'(0)=0=\frac{f(x_2)-f(x_1)}{x_2-x_1}$, because $f(x)=x^3$ is strictly increasing on $[-1,1]$.

Remark 1. This is a wrong reformulation of the Medium Value theorem: if $f(x)$ is continuous on $[a,b]$ and differentiable on (a,b), then for any $x_1,x_2\in[a,b]$ there is a point $c\in(a,b)$ such that $f'(c)=\frac{f(x_2)-f(x_1)}{x_2-x_1}$. (It implies that the tangent line at the point $(c,f(c))$ is parallel to the line passing through the points $(x_1,f(x_1))$ and $(x_2,f(x_2))$.)

Remark 2. The above statement can be also considered as a wrong extension of the following theorem: if $f(x)$ is convex on $[a,b]$ and differentiable on (a,b), then for any $c\in(a,b)$ there are two points $x_1,x_2\in(a,b)$ such that $f'(c)=\frac{f(x_2)-f(x_1)}{x_2-x_1}$.

Example 7. "If a function $f(x)$ is differentiable on (a,b) and $f'(x)$ is unbounded on (a,b), then $f(x)$ is also unbounded on (a,b)."

Solution.

The function $f(x)=\sqrt[3]{x}$ is differentiable on $(0,1)$ and its derivative $f'(x)=\frac{1}{3\sqrt[3]{x^2}}$ is unbounded on $(0,1)$, but the function values are in $(0,1)$.

Remark. This is the false converse to the following correct statement: if a function $f(x)$ is differentiable on a bounded interval (a,b) and $f(x)$ is unbounded on (a,b), then $f'(x)$ is also unbounded on (a,b). If (a,b) is unbounded, then the last statement is false. The corresponding counterexample is: $f(x)=x$ on $\mathbb{R}$.

Example 8. "If a function $f(x)$ is differentiable on a finite interval (a,b) and $\lim\limits_{x\to a_+}f(x)=\infty$, then $\lim\limits_{x\to a_+}f'(x)=\infty$."

Solution.

The function $f(x)=\frac{1}{x}+\cos\frac{1}{x}$ is differentiable on $(0,1)$ and $\lim\limits_{x\to 0_+}f(x)=+\infty$. However, the derivative $f'(x)=-\frac{1}{x^2}+\frac{1}{x^2}\sin\frac{1}{x}$ does not have a limit (finite or infinite) when x approaches 0. Indeed, for $x_n=\frac{2}{(4n+1)\pi}$, $n\in\mathbb{N}$ we have $\lim\limits_{x_n \to 0_+} f'(x_n) = \lim\limits_{n\to\infty} 0 = 0$, but for $x_k = \frac{2}{(4k-1)\pi}$, $k \in \mathbb{N}$ it follows that $\lim\limits_{x_k \to 0_+} f'(x_k) = \lim\limits_{k\to\infty}\left(-\frac{2}{x_k^2}\right) = -\infty$.

Example 9. "If $f(x)$ is differentiable on $(a, +\infty)$ and both the function and its derivative are bounded on $(a, +\infty)$, then the existence of $\lim\limits_{x\to+\infty} f(x)$ implies the existence of $\lim\limits_{x\to+\infty} f'(x)$, and vice-versa."

Solution.

First, consider $f(x) = \frac{\sin x^2}{x}$ on $(1, +\infty)$. Since $\left|\frac{\sin x^2}{x}\right| \le \frac{1}{x}$ for $x > 1$, it follows that $\lim\limits_{x\to+\infty} \frac{\sin x^2}{x} = 0$. However, $f'(x) = 2\cos x^2 - \frac{\sin x^2}{x^2}$ has no limit at infinity. It is sufficient to compare two partial limits: the first corresponding to the sequence $x_n = \sqrt{2n\pi}$, $n \in \mathbb{N}$ gives

$$\lim_{x_n\to+\infty} f'(x_n) = \lim_{n\to+\infty} f'\left(\sqrt{2n\pi}\right) = \lim_{n\to+\infty} 2 = 2,$$

and the second related to $x_k = \sqrt{\pi + 2k\pi}$, $k \in \mathbb{N}$ results in

$$\lim_{x_k\to+\infty} f'(x_k) = \lim_{k\to+\infty} f'\left(\sqrt{\pi + 2k\pi}\right) = \lim_{k\to+\infty} (-2) = -2.$$

Since two partial limits are different, the limit $\lim\limits_{x\to+\infty} f'(x)$ does not exist.

Second, consider $f(x) = \cos(\ln x)$ on $(1, +\infty)$. The derivative $f'(x) = -\frac{\sin(\ln x)}{x}$ has zero limit at infinity $\lim\limits_{x\to+\infty} f'(x) = 0$, because $\left|-\frac{\sin(\ln x)}{x}\right| \le \frac{1}{x}$ for $x > 1$. However, the function has no limit at infinity, since for the partial limit with the points $x_n = e^{2n\pi}$, $n \in \mathbb{N}$ we have

$$\lim_{x_n\to+\infty} f(x_n) = \lim_{n\to+\infty} f\left(e^{2n\pi}\right) = \lim_{n\to+\infty} 1 = 1,$$

while for $x_k = e^{\pi+2k\pi}$, $k \in \mathbb{N}$ the result is different

$$\lim_{x_k\to+\infty} f(x_k) = \lim_{k\to+\infty} f\left(e^{\pi+2k\pi}\right) = \lim_{k\to+\infty} (-1) = -1.$$

Remark 1. Other interesting counterexamples are $f(x) = e^{-x}\cos e^{x}$ for the case when $\lim\limits_{x\to+\infty} f(x)$ exists, but $\lim\limits_{x\to+\infty} f'(x)$ does not, and $f(x) = \sin\sqrt[3]{x}$ for the opposite case.

Remark 2. It is simple to modify the above examples in such a way that the function or its derivative will be unbounded. For example, the function $f(x) = e^{-x}\cos e^{2x}$ is bounded on $(0, +\infty)$ and $\lim\limits_{x\to+\infty} f(x) = 0$, but its derivative $f'(x) = -e^{-x}\cos e^{2x} - 2e^{x}\sin e^{2x}$ has no limit at infinity and is unbounded: for $x_n = \frac{1}{2}\ln(2n\pi)$, $n \in \mathbb{N}$ one gets

$$\lim_{x_n\to+\infty} f'(x_n) = \lim_{n\to+\infty}\left(-e^{-x_n}\cos e^{2x_n}\right) = -\lim_{n\to+\infty}\frac{1}{\sqrt{2n\pi}} = 0,$$

while for $x_k = \frac{1}{2}\ln\left(\frac{\pi}{2} + 2k\pi\right)$, $k \in \mathbb{N}$ one obtains

$$\lim_{x_k \to +\infty} f'(x_k) = \lim_{k \to +\infty} \left(-2e^{x_k} \sin e^{2x_k}\right) = -\lim_{k \to +\infty} 2\sqrt{\frac{\pi}{2} + 2k\pi} = -\infty.$$

In the opposite direction, the function $f(x) = \sqrt[3]{x} \sin \sqrt[3]{x}$ has no limit at infinity and is unbounded: choosing $x_n = \left(\frac{\pi}{2} + 2n\pi\right)^3$, $n \in \mathbb{N}$ one gets

$$\lim_{x_n \to +\infty} f(x_n) = \lim_{n \to +\infty} \left(\frac{\pi}{2} + 2n\pi\right) = +\infty.$$

However, its derivative $f'(x) = \frac{1}{3\sqrt[3]{x^2}} \sin \sqrt[3]{x} + \frac{1}{3\sqrt[3]{x}} \cos \sqrt[3]{x}$ is bounded and has limit at infinity equal to zero.

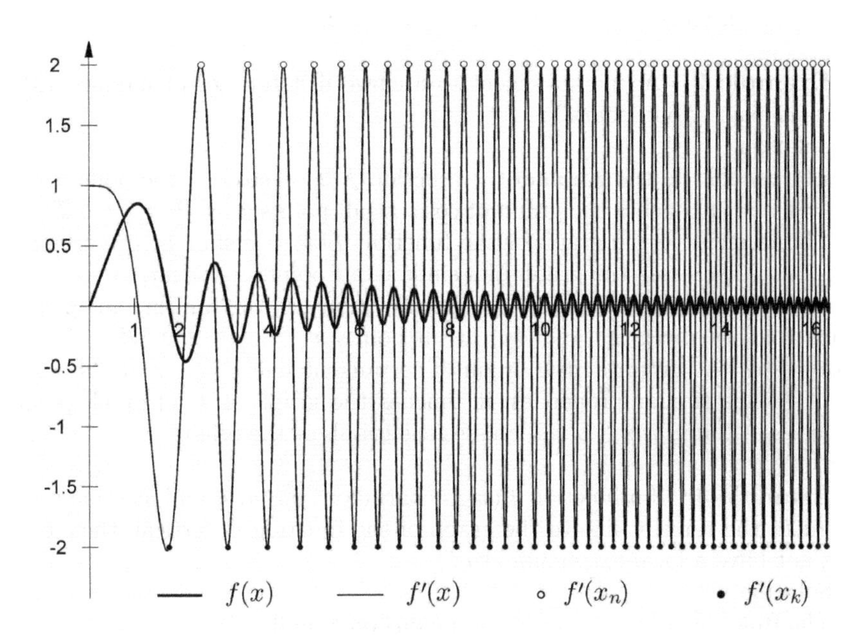

FIGURE 4.4.3: Example 9, graph of the first function

Example 10. "If $f'(x) = 0$ on a set S then $f(x) = const$ on this set."
Solution.
Let $f(x) = \begin{cases} 1, & x \in (0,1) \\ 2, & x \in (1,2) \end{cases}$ with the set $S = (0,1) \cup (1,2)$. Evidently, $f'(x) = 0$ at every point $x \in S$, but $f(x)$ is not a constant function.

Remark . The statement will be correct if the set S is connected (that is, S is an interval).

4.5 Applications

4.5.1 Tangent line

Example 1. "A line is tangent to a curve at a point if it touches, but does not cross the curve at that point."

Solution.

The tangent to the graph of the function $f(x) = x^3$ at the origin is the x-axis, which crosses the graph at the same point.

Example 2. "A line is tangent to a curve if it touches or intersects the curve at exactly one point."

Solution.

The tangent to the graph of the function $f(x) = \cos x$ at the point $x = 0$ is $y = 1$ and it touches the graph of $\cos x$ at all points $x_n = 2n\pi$, $\forall n \in \mathbb{Z}$.

The tangent to the graph of the function $f(x) = x \sin x$ at the origin is $y = 0$ and it crosses the graph of $x \sin x$ at all points $x_n = n\pi$, $\forall n \in \mathbb{Z}$. At the same time, at the point $x = \frac{\pi}{2}$ this function has the tangent line $y = x$, which touches the graph of $f(x)$ at all points $x_k = \frac{\pi}{2} + 2k\pi$, $\forall k \in \mathbb{Z}$ and also crosses the graph at the origin. The tangent line $y = -x$ at the point $x = -\frac{\pi}{2}$ has similar properties: it touches the graph of $f(x)$ at all points $x_m = -\frac{\pi}{2} + 2m\pi$, $\forall m \in \mathbb{Z}$ and crosses the graph at the origin.

Example 3. "If a function $f(x)$ is continuous on (a, b) and at some point $c \in (a, b)$ the tangent line to the graph of the function is vertical, then $f(x)$ does not have a local extremum at c."

Solution.

The function $f(x) = \sqrt[3]{x^2}$ is continuous on $\mathbb{R}$ and

$$\lim_{x \to 0_+} \frac{f(x) - f(0)}{x - 0} = \lim_{x \to 0_+} \frac{1}{\sqrt[3]{x}} = +\infty,$$

$$\lim_{x \to 0_-} \frac{f(x) - f(0)}{x - 0} = \lim_{x \to 0_-} \frac{1}{\sqrt[3]{x}} = -\infty,$$

which means that $x = 0$ is the vertical tangent to the graph of $f(x)$. At the same time $c = 0$ is a strict local (and global) minimum of $f(x)$.

Remark. Let us consider the statement with the opposite conclusion: "if $f(x)$ is continuous on (a, b) and at some point $c \in (a, b)$ the tangent line to the graph of the function is vertical, then $f(x)$ has a local extremum at c". This statement is also false and a simple counterexample is $f(x) = \sqrt[3]{x}$ at $c = 0$.

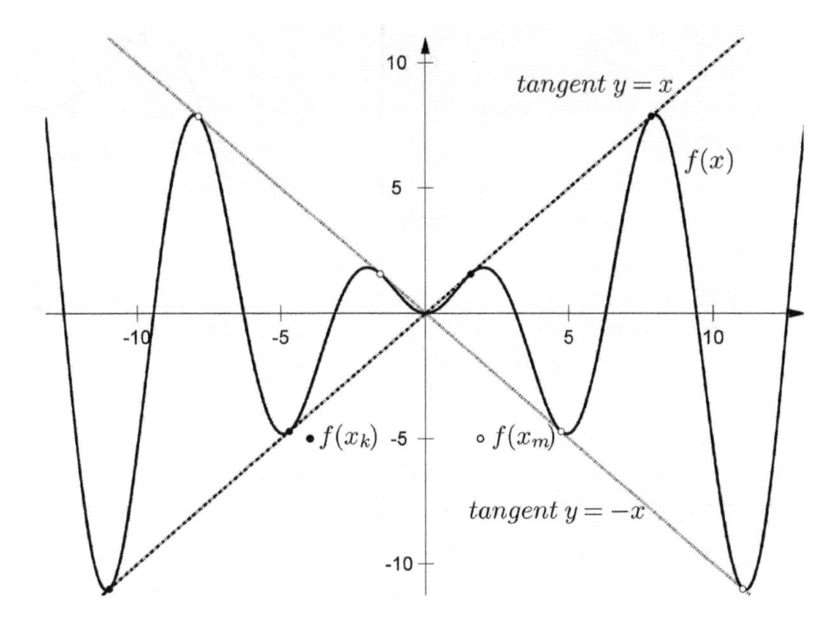

FIGURE 4.5.1: Example 2, graph of the second function

Example 4. "If $f(x)$ is continuous on (a, b) and at some point $c \in (a, b)$ a tangent line does not exist, then $f(x)$ does not have a local extremum at c."

Solution.

The function $f(x) = |x|$ does not have a tangent at $c = 0$, but this point is a strict local (and global) minimum of this function.

Remark . Let us consider the statement with the opposite conclusion: "if $f(x)$ is continuous on (a, b) and at some point $c \in (a, b)$ a tangent line does not exist, then $f(x)$ has a local extremum at c". This statement is also false and a simple counterexample is $f(x) = \begin{cases} x, x < 0 \\ 3x, x \geq 0 \end{cases}$ at $c = 0$.

4.5.2 Monotonicity and local extrema

Example 5. "If $f(x)$ is strictly increasing and differentiable on (a, b), then $f'(x) > 0$ on (a, b)."

Solution.

The function $f(x) = x^3$ is strictly increasing and differentiable on $(-1, 1)$, but $f'(0) = 0$.

Remark 1. This is the wrong converse to the following statement: if

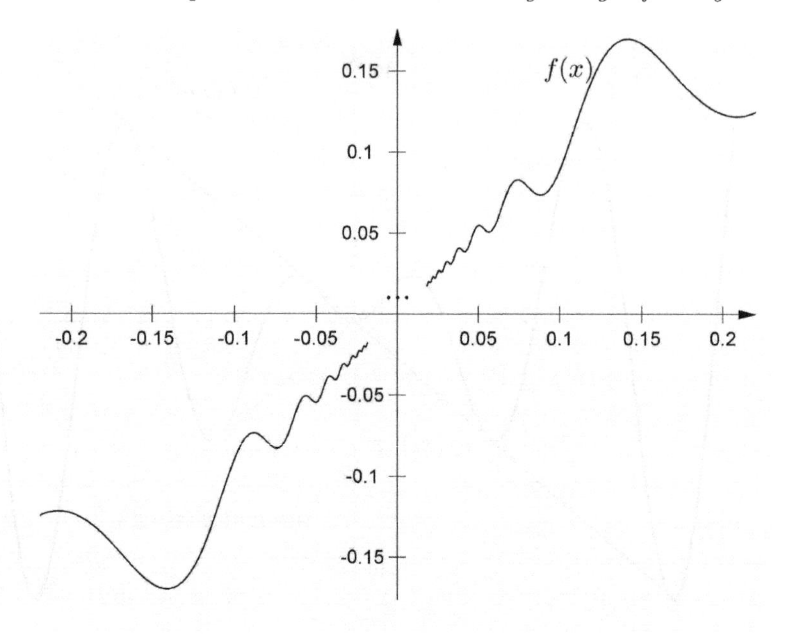

FIGURE 4.5.2: Example 6, graph of the function

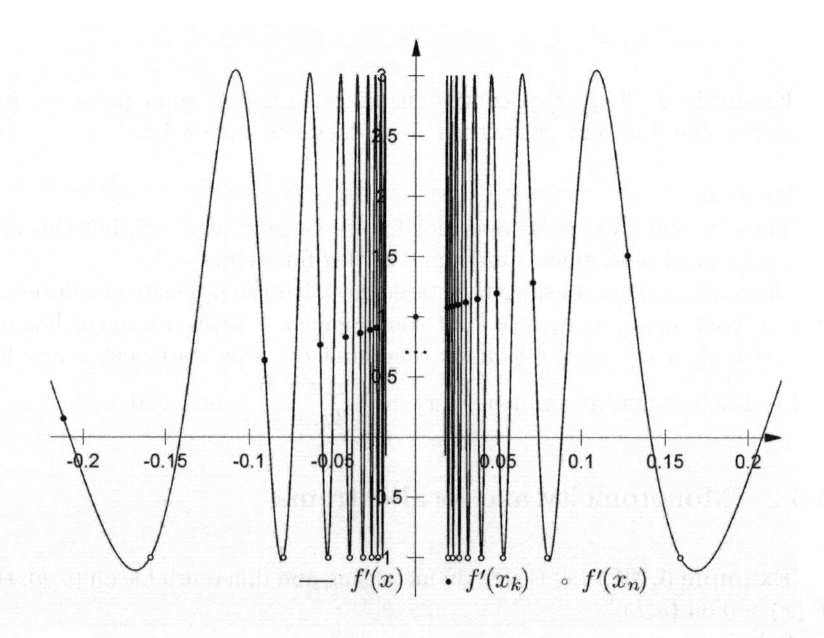

FIGURE 4.5.3: Example 6, graph of the derivative

$f'(x) > 0$ on (a,b), then $f(x)$ is strictly increasing on (a,b). This is also the wrong extension of the statement: if $f(x)$ is increasing and differentiable on (a,b), then $f'(x) \geq 0$ on (a,b).

Remark 2. Of course, the same is true for decreasing functions.

Example 6. "If $f'(c) > 0$, then $f(x)$ is increasing in a neighborhood of the point c."

Solution.

The function $f(x) = \begin{cases} x + 2x^2 \sin\frac{1}{x}, & x \neq 0 \\ 0, & x = 0 \end{cases}$ is differentiable at every real point, and its derivative is $f'(x) = \begin{cases} 1 + 4x\sin\frac{1}{x} - 2\cos\frac{1}{x}, & x \neq 0 \\ 1, & x = 0 \end{cases}$ and $f'(0) > 0$ (see Example 10, section 4.3 for details). However, in every neighborhood of zero $f'(x)$ has both positive and negative values: for the points $x_k = \frac{2}{\pi(4k+1)}$, $k \in \mathbb{Z}$, we obtain $f'(x_k) = 1 + \frac{8}{\pi(4k+1)} > 0$; while for the points $x_n = \frac{1}{2n\pi}$, $n \in \mathbb{Z}\setminus\{0\}$, we obtain $f'(x_n) = 1 - 2 < 0$. It means that $f(x)$ cannot be increasing in any neighborhood of the point $c = 0$.

Remark 1. Of course, the same is true for the relation between the condition $f'(c) < 0$ and decreasing of the function.

Remark 2. If one adds the condition of continuity of $f'(x)$ at the point c, then the statement becomes true.

Example 7. "If $f'(c) = 0$, then the point c is a local extremum of the function $f(x)$."

Solution.

The function $f(x) = x^3$ is differentiable at any real point and $f'(0) = 0$, but $c = 0$ is not a local extremum, because $f(x) = x^3$ is a strictly increasing function on $\mathbb{R}$.

Example 8. "If c is a local extremum of $f(x)$, then $f'(c) = 0$."

Solution.

The function $f(x) = |x|$ is not differentiable at $c = 0$, but this point is a local (and global) minimum of this function.

Remark . Another good example is the function $f(x) = \sqrt{|x|}$ at $c = 0$.

Remark to Examples 7 and 8. Each of these Examples is a wrong modification of the necessary condition for a local extremum: if $f(x)$ is differentiable at a point c and c is a local extremum, then $f'(c) = 0$.

Example 9. "If c is a local extremum of $f(x)$ and $f(x)$ is twice differentiable at c, then $f''(c) \neq 0$."

Solution.

The function $f(x) = x^4$ is twice differentiable at $c = 0$ and the origin is a local (and global) minimum of this function, but $f''(0) = 0$.

Remark . This is the false converse to the Second Derivative test: if $f(x)$

is twice differentiable at a stationary point c and $f''(c) \neq 0$, then c is a local extremum of $f(x)$.

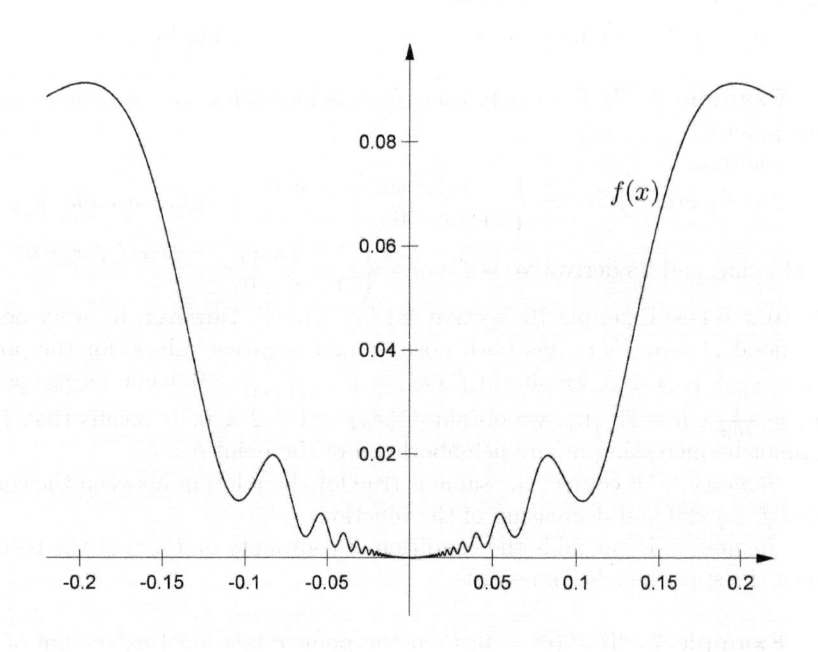

FIGURE 4.5.4: Example 10, graph of the function

Example 10. "If the derivative $f'(x)$ is defined in a neighborhood of a point c, and it does not change the sign passing through this point, then c is not a local extremum."

Solution.

The function $f(x) = \begin{cases} 2x^2 + x^2 \cos \frac{1}{x}, & x \neq 0 \\ 0, & x = 0 \end{cases}$ is differentiable at every real point. In fact, for any $x \neq 0$ the derivative can be calculated by using the arithmetic and chain rules: $f'(x) = 4x + 2x \cos \frac{1}{x} + \sin \frac{1}{x}$. For $x = 0$ applying the definition we obtain

$$f'(0) = \lim_{x \to 0} \frac{2x^2 + x^2 \cos 1/x}{x} = \lim_{x \to 0} \left(2x + x \cos \frac{1}{x} \right) = 0.$$

Hence, $f'(x) = \begin{cases} 4x + 2x \cos \frac{1}{x} + \sin \frac{1}{x}, & x \neq 0 \\ 0, & x = 0 \end{cases}$. Now, using the points $x_k = \frac{2}{\pi(4k+1)}$, $k \in \mathbb{Z}$, we obtain $f'(x_k) = \frac{8}{\pi(4k+1)} + 1 > 0$, $k \in \mathbb{Z}$. At the same time, for the points $x_n = \frac{2}{\pi(4n-1)}$, $n \in \mathbb{Z}$, we get $f'(x_n) = \frac{8}{\pi(4n-1)} - 1 < 0$, $n \in \mathbb{Z}$. Note that in any neighborhood of the point $c = 0$ there exist both points x_k and x_n (for k and n sufficiently large), that is, in any neighborhood of zero

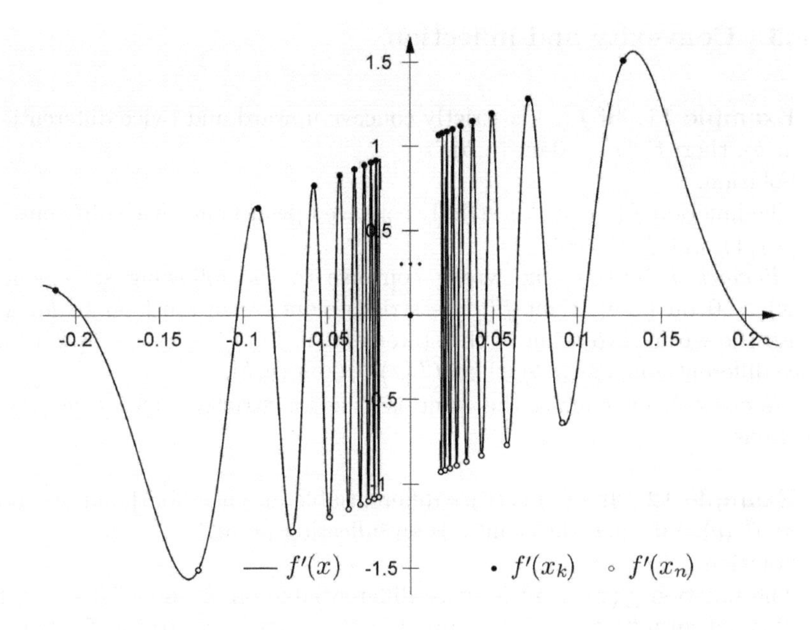

FIGURE 4.5.5: Example 10, graph of the derivative

there are both positive and negative values of $f'(x)$. Therefore, the derivative does not preserve the sign in one-sided neighborhoods of zero, so $f'(x)$ does not change its sign passing through the point $c = 0$. On the other hand, the following evaluation is satisfied: $f(x) = x^2 \left(2 + \cos \frac{1}{x}\right) > 0 = f(0)$, $\forall x \neq 0$. Therefore, by the definition, $c = 0$ is a strict local (and global) minimum of $f(x)$.

Remark 1. This is a warning to use with caution the following popular "rule": "if $f'(x)$ exists in a neighborhood of a point c, but it does not change sign passing through this point, then c is not a local extremum". This "rule" is a non precise shortened version of a part of the so-called First Derivative test: if $f'(x)$ exists in a neighborhood of a point c, and $f'(x)$ keeps the same sign on both sides of c, then c is not a local extremum.

Remark 2. An analogous false "rule", also disproved by this counterexample, is the following: "if $f(x)$ is continuous in a neighborhood of a local minimum point c, then $f(x)$ is decreasing in a left-hand neighborhood of c and increasing in a right-hand neighborhood of c" (and a similar "rule" for a local maximum). This "rule" is a non precise shortened version of another part of the First Derivative test: if $f'(x)$ exists in a neighborhood of a point c, and $f'(x)$ changes sign passing through the point c, then c is a local extremum.

4.5.3 Convexity and inflection

Example 11. "If $f(x)$ is strictly concave upward and twice differentiable on (a,b), then $f''(x)>0$ on (a,b)."

Solution.

The function $f(x)=x^4$ is strictly concave upward and twice differentiable on $(-1,1)$, but $f''(0)=0$.

Remark 1. This is the wrong converse to the following statement: if $f''(x)>0$ on (a,b), then $f(x)$ is strictly concave upward on (a,b). This is also the wrong extension of the statement: if $f(x)$ is concave upward and twice differentiable on (a,b), then $f''(x)\geq 0$ on (a,b).

Remark 2. Of course, the same is true for strictly concave downward functions.

Example 12. "If $f(x)$ is twice differentiable in a neighborhood of a point c and $f''(c)=0$, then the point c is an inflection point."

Solution.

The function $f(x)=x^4$ is twice differentiable on $\mathbb{R}$ and $f''(0)=0$, but $c=0$ is not an inflection point, since $f(x)$ is concave upward on $\mathbb{R}$ ($f''(x)=12x^2>0$ for $\forall x\neq 0$).

Example 13. "If c is an inflection point of $f(x)$, then $f''(c)=0$."

Solution.

The function $f(x)=\begin{cases}-x^2,\ x<0\\ x^4,\ x\geq 0\end{cases}$ is not twice differentiable at $c=0$ ($f''(x)=-2$ for $\forall x<0$, $f''(x)=12x^2$ for $\forall x>0$, so $\lim\limits_{x\to 0_-} f''(x)=-2\neq 0=\lim\limits_{x\to 0_+} f''(x)$), but this point is an inflection point for $f(x)$, because the function has a downward concavity at negative points ($f''(x)=-2<0$ for $\forall x<0$) and an upward concavity at positive points ($f''(x)=12x^2>0$ for $\forall x>0$).

Remark to Examples 12 and 13. Each of these Examples is the wrong converse to the necessary condition of an inflection point: if $f(x)$ is twice differentiable at a point c and the point c is an inflection point, then $f''(c)=0$.

Example 14. "If a function $f(x)$ is twice differentiable in a deleted neighborhood of a point c, and a tangent line at the point c lies above the graph of $f(x)$ in a left-hand neighborhood of c and below the graph of $f(x)$ in a right-hand neighborhood of c, then c is an inflection point."

Solution.

The function $f(x)=\begin{cases}2x^3+x^3\sin\frac{1}{x},\ x\neq 0\\ 0\ ,\ x=0\end{cases}$ is differentiable on $\mathbb{R}$ and twice differentiable on $\mathbb{R}\backslash\{0\}$. In fact, for any $x\neq 0$ the first and second

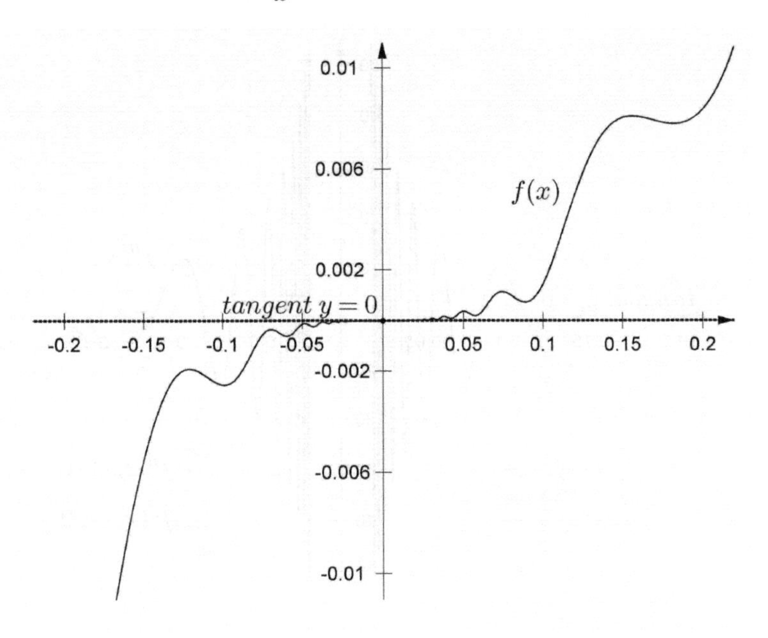

FIGURE 4.5.6: Example 14, graph of the function

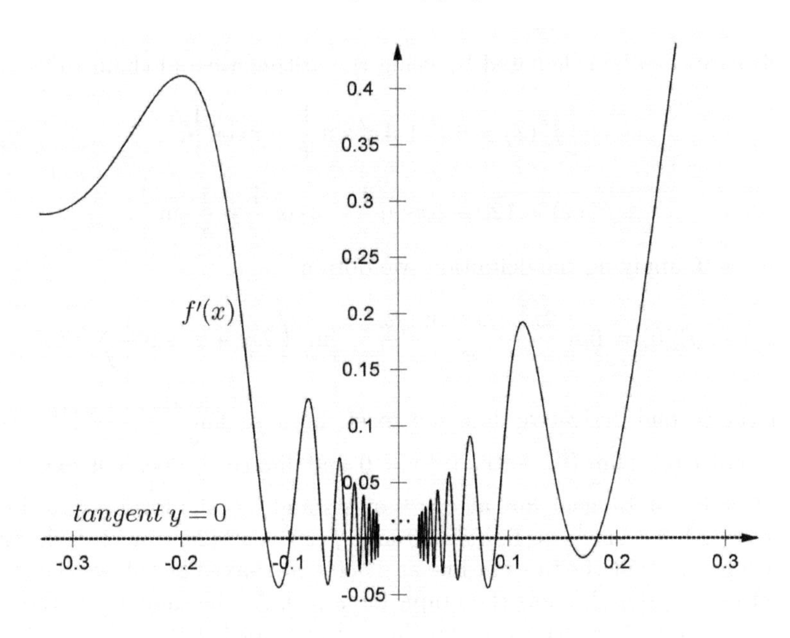

FIGURE 4.5.7: Example 14, graph of the derivative

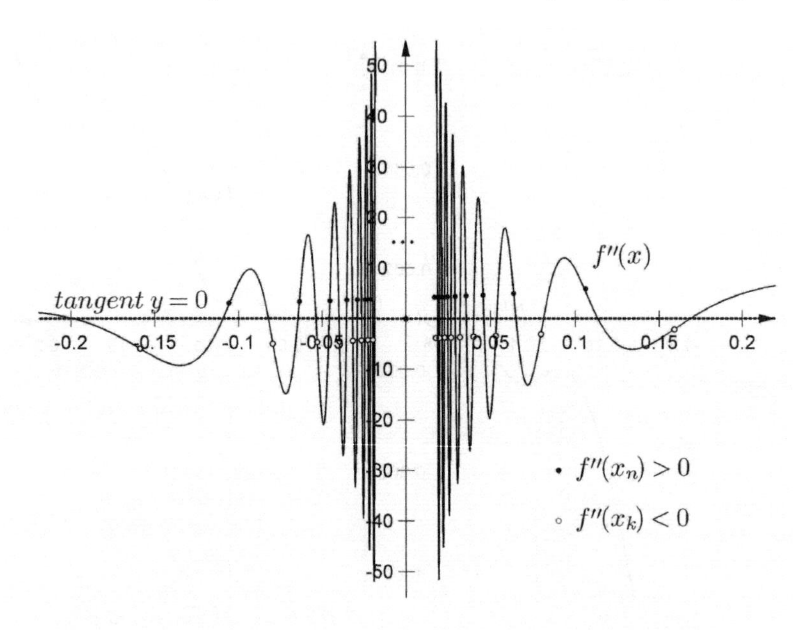

FIGURE 4.5.8: Example 14, graph of the second derivative

derivatives can be calculated by using the arithmetic and chain rules:

$$f'(x) = 6x^2 + 3x^2 \sin\frac{1}{x} - x\cos\frac{1}{x},$$

$$f''(x) = 12x + 6x\sin\frac{1}{x} - 4\cos\frac{1}{x} - \frac{1}{x}\sin\frac{1}{x}.$$

For $x = 0$, applying the definition, we obtain

$$f'(0) = \lim_{x\to 0}\frac{2x^3 + x^3\sin 1/x}{x} = \lim_{x\to 0}\left(2x^2 + x^2\sin\frac{1}{x}\right) = 0,$$

but the second derivative does not exist, because $\lim\limits_{x\to 0}\frac{6x^2+3x^2\sin 1/x - x\cos 1/x}{x}$ does not exist ($\lim\limits_{x\to 0}\left(6x + 3x\sin\frac{1}{x}\right) = 0$ and $\lim\limits_{x\to 0}\cos\frac{1}{x}$ does not exist). Since $f'(0) = 0$, the tangent line at $c = 0$ exists and has the form $y = 0$. Besides, $f(x) = x^3\left(2 + \sin\frac{1}{x}\right) < 0$, $\forall x < 0$ and $f(x) = x^3\left(2 + \sin\frac{1}{x}\right) > 0$, $\forall x > 0$, which means that the tangent line at $c = 0$ (the x-axis) lies above the graph of $f(x)$ for $x < 0$ and below the graph for $x > 0$. At the same time, the second derivative does not keep the sign in any one-sided neighborhood of $c = 0$. Indeed, for $x_k = \frac{1}{2k\pi}$, $k \in \mathbb{Z}\backslash\{0\}$ we obtain $f''(x_k) = \frac{12}{2k\pi} - 4 < 0$, while for $x_n = \frac{1}{2n\pi+\pi}$, $n \in \mathbb{Z}$ it follows $f''(x_n) = \frac{12}{2n\pi+\pi} + 4 > 0$. Therefore, $f(x)$ does not have any kind of concavity in any one-sided neighborhood of $c = 0$. Hence, the point $c = 0$ is not an inflection point.

Remark . This is a warning to use with caution the following too simplified "rule": "if the tangent line at the point c lies above (below) the graph of $f(x)$ in a left-hand neighborhood of c and below (above) the graph of $f(x)$ in a right-hand neighborhood of c, then c is an inflection point". Or even more simplified "definition": "c is called an inflection point of $f(x)$ if the tangent line to the graph of $f(x)$ at $(c, f(c))$ crosses the graph". These "rules" are non precise shortened versions of the following theorem: if $f(x)$ is twice differentiable in a deleted neighborhood of a point c, and $f''(x)$ changes its sign passing through the point c, then c is an inflection point.

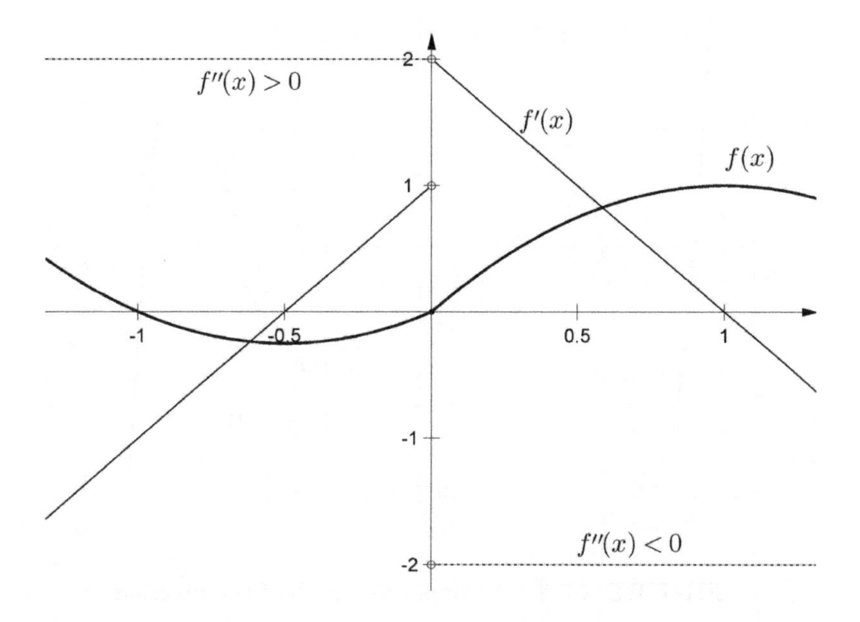

FIGURE 4.5.9: Example 15

Example 15. "If $f(x)$ is continuous on (a, b) and is not differentiable at a point $c \in (a, b)$, then c is not an inflection point of the graph of $f(x)$."

Solution.

The function $f(x) = \begin{cases} x^2 + x, x < 0 \\ 2x - x^2, x \geq 0 \end{cases}$ is continuous on $\mathbb{R}$ and does not have a derivative at $c = 0$, since the one-sided derivatives are different:

$$f'_-(0) = \lim_{x \to 0_-} \frac{f(x) - f(0)}{x - 0} = \lim_{x \to 0_-} (x + 1) = 1,$$

$$f'_+(0) = \lim_{x \to 0+} \frac{f(x) - f(0)}{x - 0} = \lim_{x \to 0_+} (2 - x) = 2.$$

At the same time, $f''(x) = 2$ for $x < 0$, and $f''(x) = -2$ for $x > 0$, that is, the function is concave upward on $(-\infty, 0)$ and downward on $(0, +\infty)$. Therefore, $c = 0$ is an inflection point.

Remark . Let us consider the statement with the opposite conclusion: "if $f(x)$ is continuous on (a, b) and is not differentiable at a point $c \in (a, b)$, then c is an inflection point of the graph of $f(x)$". This statement is also false and a simple counterexample is $f(x) = \begin{cases} x^2 + x, x < 0 \\ x^2 - x, x \geq 0 \end{cases}$ at $c = 0$.

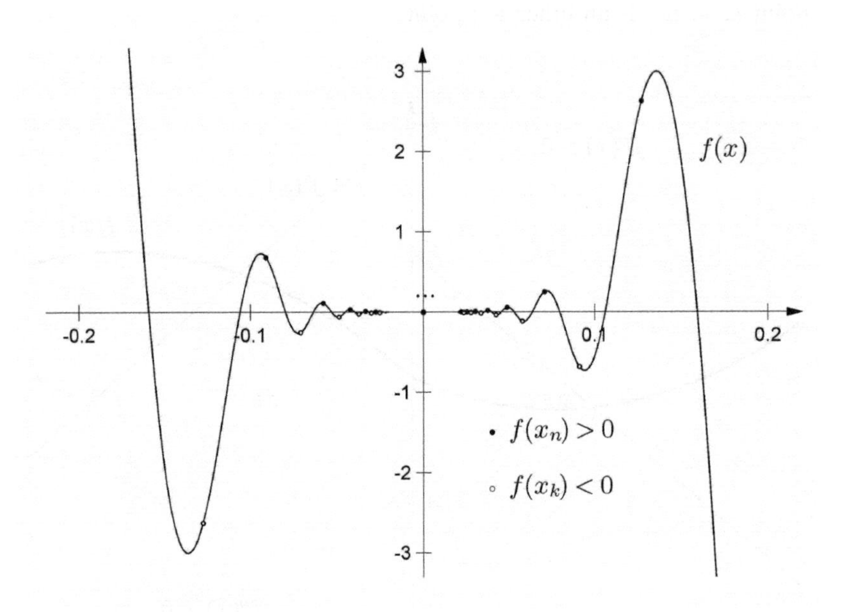

FIGURE 4.5.10: Example 16, graph of the function

Example 16. "If $f(x)$ is continuous in a neighborhood of a point c and $f'(c) = f''(c) = 0$, then the point c is either a local extremum or an inflection point of $f(x)$."

Solution.

Let $f(x) = \begin{cases} x^4 \sin \frac{1}{x}, \ x \neq 0 \\ 0, \ x = 0 \end{cases}$. Notice that in any deleted neighborhood of $x = 0$ there exist points $x_n = \frac{2}{(4n+1)\pi}$, $\forall n \in \mathbb{Z}$ where $f(x_n) = x_n^4 > 0 = f(0)$, and also there exist points $x_k = \frac{2}{(4k-1)\pi}$, $\forall k \in \mathbb{Z}$ such that $f(x_k) = -x_k^4 < 0 = f(0)$, that is, $x = 0$ is not a local extremum of $f(x)$. On the other hand, the derivative can be easily found using the arithmetic and chain rules for $x \neq 0$ and the definition of the derivative for $x = 0$ (see Example 5, section 4.3 and Example 14 in this section for analogous calculations): $f'(x) = \begin{cases} 4x^3 \sin \frac{1}{x} - x^2 \cos \frac{1}{x}, \ x \neq 0 \\ 0, \ x = 0 \end{cases}$. Furthermore, the

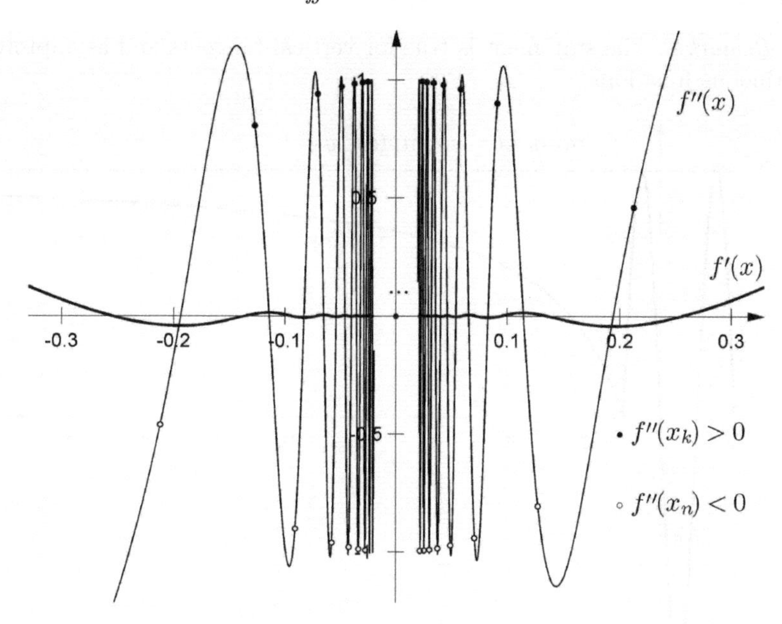

FIGURE 4.5.11: Example 16, graphs of the first and second derivatives

second derivative also exists and can be found by applying similar calculations: $f''(x) = \begin{cases} 12x^2 \sin\frac{1}{x} - 6x\cos\frac{1}{x} - \sin\frac{1}{x}, & x \neq 0 \\ 0, & x = 0 \end{cases}$. At the points $x_n = \frac{2}{(4n+1)\pi}, \forall n \in \mathbb{Z}\setminus\{0\}$ the second derivative is negative $f''(x_n) = 12x_n^2 - 1 < 0$, and at the points $x_k = \frac{2}{(4k-1)\pi}, \forall k \in \mathbb{Z}\setminus\{0\}$ the second derivative is positive $f''(x_k) = -12x_k^2 + 1 > 0$. So the function has no concavity in any one-sided neighborhood of $x = 0$. Therefore, $x = 0$ is not an inflection point of $f(x)$.

4.5.4 Asymptotes

Example 17. "If $y = c$ is a tangent line to the graph of $f(x)$, then $y = c$ cannot be an asymptote to the same graph."

Solution.

The function $f(x) = \begin{cases} \sin \pi x, & x < 1 \\ \frac{x-1}{x+1}, & x \geq 1 \end{cases}$ has the tangent line $y = 1$ at the point $x = \frac{1}{2}$, since $f'\left(\frac{1}{2}\right) = \pi \cos\frac{\pi}{2} = 0$, and the same line is the horizontal asymptote, because $\lim\limits_{x\to+\infty} f(x) = \lim\limits_{x\to+\infty} \frac{x-1}{x+1} = 1$.

Remark 1. The statement is also false for vertical tangents and asymptotes. A simple counterexample is $f(x) = \begin{cases} \frac{1}{x}, & x < 0 \\ \sqrt[3]{x}, & x \geq 0 \end{cases}$ at the point $x = 0$.

Remark 2. The statement is true for vertical tangents and asymptotes of continuous functions.

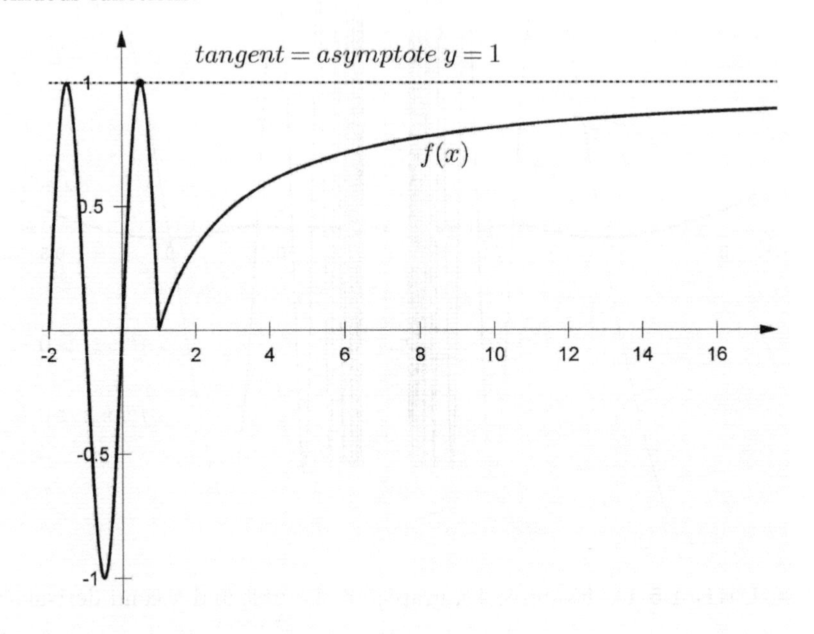

FIGURE 4.5.12: Example 17

Example 18. "The graph of a function cannot cross or touch its asymptote."

Solution.

The function $f(x) = \frac{\sin x}{x}$ has the horizontal asymptote $y = 0$ (since $\lim\limits_{x\to+\infty} f(x) = \lim\limits_{x\to+\infty} \frac{\sin x}{x} = 0$) and the graph of $f(x)$ crosses this asymptote infinitely many times at the points $x_n = n\pi$, $\forall n \in \mathbb{N}$.

Another function $f(x) = e^{-x}(1 - \cos x)$ has the horizontal asymptote $y = 0$ (since $\lim\limits_{x\to+\infty} f(x) = \lim\limits_{x\to+\infty} e^{-x}(1 - \cos x) = 0$) and the graph of $f(x)$ touches this asymptote infinitely many times at the points $x_n = 2n\pi$, $\forall n \in \mathbb{N}$ (at all these points, $y = 0$ is the tangent line to the graph of $f(x)$).

Example 19. "If $f(x)$ and $g(x)$ are continuous in a neighborhood of a point a, and $g(a) = 0$, then the function $\frac{f(x)}{g(x)}$ has a vertical asymptote at the point a."

Solution.

Although the functions $f(x) = \sin x$ and $g(x) = x$ are continuous on $\mathbb{R}$, and $g(x) = x$ equals 0 at the point $x = 0$, the ratio $h(x) = \frac{f(x)}{g(x)} = \frac{\sin x}{x}$ has a finite limit at 0: $\lim\limits_{x\to 0} \frac{\sin x}{x} = 1$. Hence, $h(x)$ has no asymptote at $x = 0$.

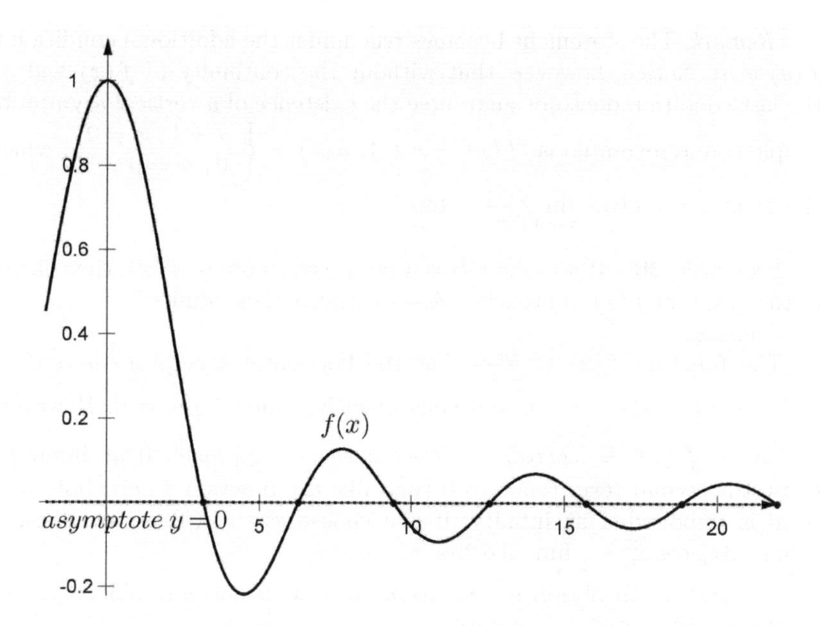

FIGURE 4.5.13: Example 18, the first counterexample

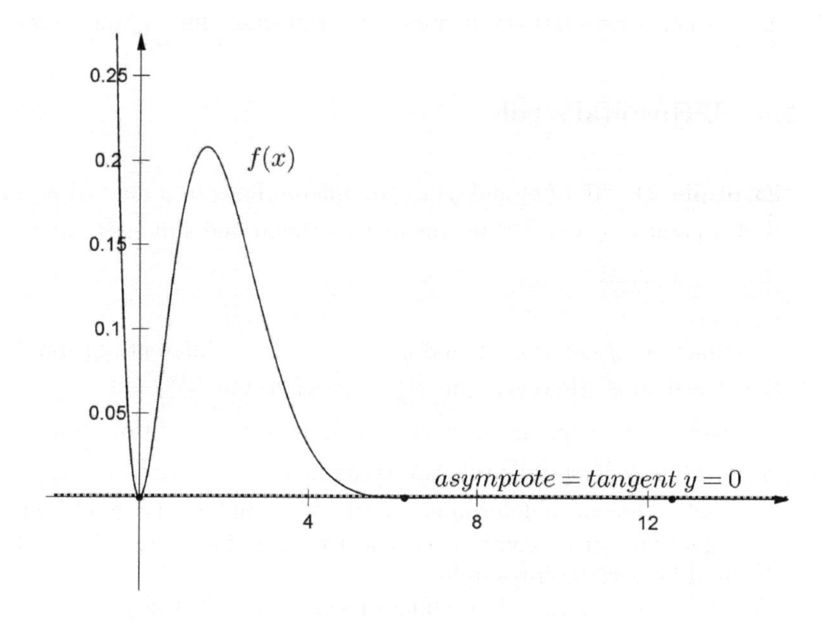

FIGURE 4.5.14: Example 18, the second counterexample

Remark. The statement becomes true under the additional condition that $f(a) \neq 0$. Notice, however, that without the continuity of $f(x)$ and $g(x)$, the last condition does not guarantee the existence of a vertical asymptote. A simple counterexample is: $f(x) = x + 1$, $g(x) = \begin{cases} x+1, & x \neq 0 \\ 0, & x = 0 \end{cases}$, which at the point $x = 0$ gives $\lim\limits_{x\to 0} \frac{f(x)}{g(x)} = \lim\limits_{x\to 0} \frac{x+1}{x+1} = 1$.

Example 20. "If $y = Ax + B$ is a slant asymptote to $f(x)$, then the slope of the graph of $f(x)$ approaches A, as x approaches infinity."

Solution.

The function $f(x) = \frac{\sin x^3}{x}$ has the horizontal asymptote $y = 0$, since $0 \leq \left|\frac{\sin x^3}{x}\right| \leq \frac{1}{|x|} \underset{x\to+\infty}{\to} 0$, and consequently, $\lim\limits_{x\to+\infty} \frac{\sin x^3}{x} = 0$. However, its derivative $f'(x) = 3x\cos x^3 - \frac{\sin x^3}{x^2}$ does not approach 0 at infinity, because the second term tends to 0 (just like the function $f(x)$), but the first term is unbounded at infinity: if one choose $x_n = \sqrt[3]{2n\pi}$, $\forall n \in \mathbb{N}$, then $\lim\limits_{x_n\to+\infty} 3x_n \cos x_n^3 = \lim\limits_{n\to+\infty} 3\sqrt[3]{2n\pi} = +\infty$.

Remark 1. To obtain a slant asymptote with non-zero slope, just use a similar function $f(x) = x + \frac{\sin x^3}{x}$.

Remark 2. This example is related to Example 9 in section 4.4, and it shows that a tempting try to apply derivative inside the limit is not correct: if $\lim\limits_{x\to+\infty} (f(x) - Ax - B) = 0$ it does not mean that $\lim\limits_{x\to+\infty} (f'(x) - A) = 0$.

4.5.5 L'Hospital's rule

Example 21. "If $f(x)$ and $g(x)$ are differentiable in a deleted neighborhood of a point c, $g'(x) \neq 0$ in this neighborhood and $\lim\limits_{x\to c} \frac{f'(x)}{g'(x)}$ exists, then $\lim\limits_{x\to c} \frac{f(x)}{g(x)} = \lim\limits_{x\to c} \frac{f'(x)}{g'(x)}$."

Solution.

The functions $f(x) = x + 2$ and $g(x) = x + 1$ are differentiable on $\mathbb{R}$, and $g'(x) = 1 \neq 0$ on $\mathbb{R}$. However, $\lim\limits_{x\to 0} \frac{f(x)}{g(x)} = 2$, while $\lim\limits_{x\to 0} \frac{f'(x)}{g'(x)} = 1$.

Remark 1. It happens because the important condition $\lim\limits_{x\to c} f(x) = \lim\limits_{x\to c} g(x) = 0$ in L'Hospital's rule is missing here.

Remark 2. For an indeterminate form $\frac{\infty}{\infty}$ a similar false modification of L'Hospital's rule can be constructed and the same functions $f(x)$ and $g(x)$ can be used for a counterexample.

Remark 3. For both indeterminate forms $\frac{0}{0}$ and $\frac{\infty}{\infty}$ the point c can also be infinite. For instance, for the latter form the following counterexample can be given: if $f(x) = 3 - \frac{1}{x^2}$ and $g(x) = 1 - \frac{1}{x}$ on $(0, +\infty)$, then $f'(x) = \frac{2}{x^3}$, $g'(x) = \frac{1}{x^2} \neq 0$ and $\lim\limits_{x\to+\infty} \frac{f(x)}{g(x)} = 3 \neq 0 = \lim\limits_{x\to+\infty} \frac{f'(x)}{g'(x)}$. It happens because the important condition $\lim\limits_{x\to+\infty} f(x) = \lim\limits_{x\to+\infty} g(x) = \infty$ in L'Hospital's rule is missing here.

Example 22. "If $f(x)$ and $g(x)$ are differentiable in a deleted neighborhood of a point c, $g'(x) \neq 0$ in this neighborhood and $\lim\limits_{x\to c} f(x) = \lim\limits_{x\to c} g(x) = 0$, then $\lim\limits_{x\to c} \frac{f(x)}{g(x)} = \lim\limits_{x\to c} \frac{f'(x)}{g'(x)}$."

Solution.

The functions $f(x) = x^2 \cos\frac{1}{x}$ and $g(x) = \sin x$ are differentiable in $\mathbb{R}\setminus\{0\}$, $g'(x) = \cos x \neq 0$ in $(-1,1)$ and $\lim\limits_{x\to 0} x^2 \cos\frac{1}{x} = \lim\limits_{x\to 0} \sin x = 0$ (for the first limit we can apply the evaluation $\left|x^2\cos\frac{1}{x}\right| \leq x^2$). Nevertheless, $\lim\limits_{x\to 0} \frac{f(x)}{g(x)} \neq \lim\limits_{x\to 0} \frac{f'(x)}{g'(x)}$. In fact, since $\lim\limits_{x\to 0} \frac{x}{\sin x} = 1$ and $\lim\limits_{x\to 0} x\cos\frac{1}{x} = 0$ (applying a similar evaluation $\left|x\cos\frac{1}{x}\right| \leq |x|$), we can calculate the original limit just applying the arithmetic rules:

$$\lim_{x\to 0} \frac{f(x)}{g(x)} = \lim_{x\to 0} \frac{x^2\cos\frac{1}{x}}{\sin x} = \lim_{x\to 0} \frac{x}{\sin x} \cdot \lim_{x\to 0} x\cos\frac{1}{x} = 1\cdot 0 = 0.$$

At the same time,

$$\lim_{x\to 0} \frac{f'(x)}{g'(x)} = \lim_{x\to 0} \frac{2x\cos\frac{1}{x} + \sin\frac{1}{x}}{\cos x}$$

does not exist, because $\lim\limits_{x\to 0} 2x\cos\frac{1}{x} = 0$ and $\lim\limits_{x\to 0} \cos x = 1$, but the limit $\lim\limits_{x\to 0} \sin\frac{1}{x}$ does not exist.

Remark 1. This statement is a wrongly "weakened" version of L'Hospital's rule for an indeterminate form $\frac{0}{0}$ where the condition of the existence of $\lim\limits_{x\to c} \frac{f'(x)}{g'(x)}$ (finite or infinite) is omitted.

Remark 2. For an indeterminate form $\frac{\infty}{\infty}$ a similar false modification of L'Hospital's rule can be constructed: "if $f(x)$ and $g(x)$ are differentiable in a deleted neighborhood of a point c, $g'(x) \neq 0$ in some neighborhood of c, and $\lim\limits_{x\to c} f(x) = \lim\limits_{x\to c} g(x) = \infty$, then $\lim\limits_{x\to c} \frac{f(x)}{g(x)} = \lim\limits_{x\to c} \frac{f'(x)}{g'(x)}$". The corresponding counterexample can be given in the following form: for $f(x) = 2x - \sin x$ and $g(x) = 2x + \sin x$, the limits at infinity are infinite $\lim\limits_{x\to+\infty} f(x) = \lim\limits_{x\to+\infty} g(x) = +\infty$, both functions are differentiable on $\mathbb{R}$ (with $g'(x) \neq 0$) and $\lim\limits_{x\to+\infty} \frac{f(x)}{g(x)} = 1$, but $\lim\limits_{x\to+\infty} \frac{f'(x)}{g'(x)} = \lim\limits_{x\to+\infty} \frac{2-\cos x}{2+\cos x}$ does not exist, since partial limits are different

$$\lim_{x\to+\infty, x=2n\pi} \frac{2-\cos x}{2+\cos x} = \lim_{n\to+\infty} \frac{1}{3} = \frac{1}{3} \neq 3 = \lim_{k\to+\infty} 3 = \lim_{x\to+\infty, x=\pi+2k\pi} \frac{2-\cos x}{2+\cos x}.$$

Example 23. "If $f(x)$ and $g(x)$ are differentiable on $\mathbb{R}$, $\lim_{x\to+\infty} f(x) = \lim_{x\to+\infty} g(x) = +\infty$ and $\lim_{x\to+\infty} \frac{f'(x)}{g'(x)}$ exists, then $\lim_{x\to+\infty} \frac{f(x)}{g(x)} = \lim_{x\to+\infty} \frac{f'(x)}{g'(x)}$."

Solution.

The functions $f(x) = 2x + \sin 2x$ and $g(x) = (2x + \sin 2x) e^{\sin x}$ are differentiable on $\mathbb{R}$ and approaches infinity, because $|\sin 2x| \le 1$ and $e^{\sin x} \ge e^{-1}$ on $\mathbb{R}$, while $2x$ approaches infinity. The limit of the ratio of the derivatives exists and can be calculated as follows:

$$\lim_{x\to+\infty} \frac{f'(x)}{g'(x)} = \lim_{x\to+\infty} \frac{2 + 2\cos 2x}{(2 + 2\cos 2x) e^{\sin x} + (2x + \sin 2x)\cos x e^{\sin x}}$$

$$= \lim_{x\to+\infty} \frac{4\cos x}{(4\cos x + 2x + \sin 2x) e^{\sin x}} = 0$$

(again $4\cos x$ and $\sin 2x$ are bounded on $\mathbb{R}$, $e^{\sin x} \ge e^{-1}$ on $\mathbb{R}$, and $2x$ approaches infinity). Nevertheless, the limit of the ratio of the functions $\lim_{x\to+\infty} \frac{f(x)}{g(x)} = \lim_{x\to+\infty} \frac{1}{e^{\sin x}}$ does not exist, since partial limits of $e^{-\sin x}$, as x approaches infinity, assume all the values in $\left[e^{-1}, e\right]$.

Remark 1. This statement is a wrongly "weakened" version of L'Hospital's rule for an indeterminate form $\frac{\infty}{\infty}$ where the condition that $g'(x) \ne 0$ in some neighborhood of infinity is omitted.

Remark 2. A similar example can be provided for an indeterminate form $\frac{\infty}{\infty}$ at a finite point c.

Exercises

1. Show that the following "definition" of differentiability at a point is not correct: "a function $f(x)$ defined on X is called differentiable at a point $a \in X$, if for any point x in a neighborhood of a the following representation holds: $f(x) = f(a) + A \cdot (x-a) + \alpha \cdot (x-a)$, where A is a constant and $|\alpha| \le C|x|$ for some $C > 0$". (Hint: this "definition" is too restrictive, try the differentiable function $f(x) = x^{4/3}$ at $a = 0$.)

2. Provide a counterexample to the statement "if $f(x)$ is continuous at a, then there exists at least one of the one-sided derivatives at a".

3. Show that $f(x) = \begin{cases} x-1, \, x<0 \\ x, \, x \ge 0 \end{cases}$ is another counterexample to the statement in Example 2, section 4.2.

4. Show that $f(x) = \sin x$ is a counterexample to the statement "if $f(x)$ is differentiable then $|f(x)|$ is also differentiable" where the differentiability of $|f(x)|$ does not hold at infinitely many points (compare with Remark 2 to Example 5, section 4.2).

5. Analyze if the following statement is true: "if $f(x)$ is infinitely differentiable on $\mathbb{R}$ and $f^{(n)}(x) = 0$ for $x \in (-\infty, 0)$, then $f(x) \equiv 0$ on $\mathbb{R}$." (Hint: compare this statement with that of Example 6, section 4.2 and modify the function of that Example.)

6. Verify if the following statement is true: if at some x fixed $f'(x) = A$, then $\lim\limits_{h\to 0} \frac{f(x+2h)+f(x)-2f(x-h)}{4h} = A$. What about the converse statement? Provide a counterexample to the false statement.

Formulate similar statements for $\lim\limits_{h\to 0} \frac{4f(x+h)-3f(x)-f(x+2h)}{2h} = A$ in the form of counterexample?

7. Verify if the following statement is true: if at some x fixed $f''(x) = A$, then $\lim\limits_{h\to 0} \frac{f(x+2h)-2f(x+h)+f(x)}{h^2} = A$. What about the converse statement? Provide a counterexample to the false statement.

Formulate similar statements for $\lim\limits_{h\to 0} \frac{f(x+2h)-f(x+h)-f(x)+f(x-h)}{2h^2} = A$ in the form of counterexample?

8. Give a counterexample to the statement in Example 1, section 4.3, with the sum substituted by the product of two functions.

9. Provide a counterexample to the following statement: "if $f'(x)$ exists and is bounded on (a,b), then $f'(x)$ cannot have infinitely many points of discontinuity". (Hint: consider on $(-1,1)$ the function $f(x) = \begin{cases} x^2 g_n(-x), \, x \in \left(-\frac{1}{n}, -\frac{1}{n+1}\right] \\ 0, \, x=0 \\ x^2 g_n(x), \, x \in \left[\frac{1}{n+1}, \frac{1}{n}\right) \end{cases}$, where $g_n(x) = g\left(\frac{x-1/(n+1)}{n(n+1)}\right)$, $g(x) = x(x-1)^2 \sin\frac{\pi}{(x-1)^2}$.)

10. It is well-known that the derivative of a differentiable periodic function is also periodic with the same period. Show that the converse is false. Formulate the result as a counterexample.

11. Provide a counterexample to the statement "if $f(x)$ is differentiable and periodic with a fundamental period T, then $f'(x)$ is also periodic with the fundamental period T".

12. Give a counterexample to the statement: "if $f(x)$ is differentiable on an interval, then it is uniformly continuous on the same interval". What about the converse? Consider separately open and closed intervals.

13. It is well-known that a function $f(x)$ that has a bounded derivative on a (finite or infinite) interval is uniformly continuous on that interval. Show that the uniform continuity and differentiability on an interval do not guarantee boundedness of derivative. (Hint: try $f(x) = \sqrt{x}$ on $(0,1)$.)

14. Recall that a function $f(x)$ is Lipschitz-continuous on an interval I if there exists a constant $C \geq 0$ such that $|f(x) - f(y)| \leq C|x-y|$ for $\forall x, y \in I$. Show that Lipschitz-continuity does not imply differentibility. What about the converse? What if $f(x)$ is infinitely differentiable on I?

15. Any critical point of $f(x)$ is also a critical point of $f^2(x)$. Is the converse correct?

16. There are two apparently natural and similar definitions of a increasing at a point a function. The first says that $f(x)$ is increasing at a point a if there exists a neighborhood of a where $f(x)$ is increasing; and the second determines that $f(x)$ is increasing at a point a if there exists a neighborhood of a such that $f(x) \leq f(a)$ for $\forall x < a$ and $f(x) \geq f(a)$ for $\forall x > a$. (Analogous definitions can be specified for a strictly increasing function.) Show that these definitions are not equivalent. (Hint: use the function of Example 6, section 4.5.)

17. Verify that the function $f(x) = \begin{cases} x^4 \sin^2 \frac{1}{x}, & x \neq 0 \\ 0, & x = 0 \end{cases}$ provides a counterexample to the following intuitive "rule": "if $f(x)$ is a smooth function, say twice differentiable, in a neighborhood of a minimum point a, then $f(x)$ should be decreasing in some interval to the left of a and increasing in some interval to the right". Compare the result with Example 10 in section 4.5.

18. Explain why the following "definition" of upward concavity is wrong: "a function $f(x)$ is upward concave on (a,b) if the inequality $f(\alpha_1 a + \alpha_2 b) \leq \alpha_1 f(a) + \alpha_2 f(b)$ holds for arbitrary parameters $\alpha_1, \alpha_2 \geq 0$, $\alpha_1 + \alpha_2 = 1$". Give an example of function satisfying this property, but not upward concave.

19. Provide a counterexample to the statement: "if $f(x)$ and $g(x)$ are downward concave on an interval, then their product is also a downward concave function on this interval". Show that the conclusion that the product is an upward concave function is also wrong. What if $f(x)$ and $g(x)$ are upward concave?

20. If $f(x)$ is strictly upward concave and $g(x)$ is strictly downward concave on $\mathbb{R}$, then both $f(x) + g(x)$ and $f(x) \cdot g(x)$ may maintain the same type of concavity over $\mathbb{R}$. Show this by example and formulate this statement in the form suitable for counterexamples.

21. It is well-known that the convexity on an interval (a, b) implies the continuity on the same interval. Does it imply differentiability on (a, b) ? Give examples for both strict and non-strict cases.

22. A function $f(x)$ is continuous on $[a, b]$ and strictly convex on $[a, c]$ and $[c, b]$, for some $c \in (a, b)$. Does it convex on $[a, b]$?

23. The theorem states that if $\lim_{x\to+\infty} f'(x) = A$ and $\lim_{x\to+\infty} (f(x) - xf'(x)) = B$, then $y = Ax + B$ is a slant asymptote of $f(x)$. Show that the converse is not true. (Hint: compare with Example 20 in section 4.5.)

24. Explain why a direct application of L'Hospital's rule does not work in the case $f(x) = (2x + 1)^{2/3}$, $g(x) = (x + 2)^{2/3}$, $x \to +\infty$. Show that $\lim_{x\to+\infty} \frac{(2x+1)^{2/3}}{(x+2)^{2/3}} = 2^{2/3}$ and formulate this as a counterexample.

25. Show that $f(x) = x^2 \sin \frac{1}{x}$ and $g(x) = x$ provide another counterexample to the statement in Example 22, section 4.5.

26. Show that $f(x) = \frac{x}{2} - \frac{1}{4} \sin 2x$ and $g(x) = \left(\frac{x}{2} - \frac{1}{4} \sin 2x\right)(\cos x + 2)^2$ represent another counterexample to the statement in Example 23, section 4.5.

Chapter 5

Integrals

5.1 Elements of theory

Indefinite integral

Antiderivative and Indefinite Integral. A function $f(x)$ is *integrable* on a set S if there exists a differentiable on S function $F(x)$ such that $F'(x) = f(x)$ for $\forall x \in S$. The function $F(x)$ is called an *antiderivative (or a primitive)* of the function $f(x)$ on S. If $F(x)$ is an antiderivative of $f(x)$ on S, then the set of all antiderivatives is called an *indefinite integral* of $f(x)$ on S and it has the form $F(x) + C$, where C is an arbitrary real constant. The standard notation is $F(x) + C = \int f(x)\, dx$.

Basic properties:
1) if $f(x)$ is differentiable on S, then $\int f'(x)\, dx = f(x) + C$ on S.
2) if $f(x)$ is integrable on S, then $\left(\int f(x)\, dx\right)' = f(x)$ on S.
3) if $f(x)$ and $g(x)$ are integrable on S, then $f(x) + g(x)$ is also integrable on S and

$$\int f(x) + g(x)\, dx = \int f(x)\, dx + \int g(x)\, dx.$$

4) if $f(x)$ is integrable on S and c is non-zero constant, then $cf(x)$ is also integrable on S and

$$\int cf(x)\, dx = c \int f(x)\, dx.$$

Change of variable.
If $\varphi(t)$ is differentiable on T and $f(x)$ is integrable on $S = \varphi(T)$, then the

function $f(\varphi(t))\varphi'(t)$ is integrable on T and

$$\int f(x)\,dx = \int f(\varphi(t))\,\varphi'(t)\,dt.$$

Integration by parts.

If $f(x)$ and $g(x)$ are differentiable on S, and $f(x)g'(x)$ is integrable on S, then $f'(x)g(x)$ is also integrable on S and

$$\int f'(x)\,g(x)\,dx = f(x)\,g(x) - \int f(x)\,g'(x)\,dx.$$

Definite (Riemann) integral

Riemann Integral. Let $f(x)$ be defined on $[a,b]$. A partition P is a finite ordered set of points $x_0, x_1, \ldots, x_{n-1}, x_n$ in $[a,b]$ such that

$$P = \{x_i : \ a = x_0 < x_1 < \ldots < x_{n-1} < x_n = b\}.$$

We denote the length of a subinterval $[x_{i-1}, x_i]$ by $\Delta x_i = x_i - x_{i-1}$ and the partition diameter by $\Delta = \max\limits_{1\le i\le n} \Delta x_i$. The *Riemann sum* of $f(x)$ corresponding to the specific partition P and the specific choice of points $c_i \in [x_{i-1}, x_i]$ is

$$S(f; P, c) = \sum_{i=1}^{n} f(c_i)\cdot \Delta x_i,$$

in the abbreviated form $S(f)$. If there exists a finite limit of Riemann sums of $f(x)$, as Δ approaches 0, which does not depend on a choice of partitions and points c_i, then the function $f(x)$ is *Riemann integrable* on $[a,b]$ and the value of the limit is called the *Riemann (definite) integral* of $f(x)$ on $[a,b]$:

$$\int_a^b f(x)\,dx = \lim_{\Delta\to 0} S(f).$$

Additionally, if $a = b$, then $\int_a^b f(x)\,dx = 0$, and if $a > b$, then $\int_a^b f(x)\,dx = -\int_b^a f(x)\,dx$.

Remark 1. If there is no confusion with an indefinite integral, the Riemann integrable function can be also called an integrable function and the Riemann integral just an integral. We will use these abbreviations in this section.

Remark 2. All results below are formulated for $a < b$.

Theorem. If $f(x)$ is integrable on $[a,b]$, then it is bounded there.

Theorem. If $f(x)$ is continuous on $[a,b]$, then it is integrable there.

Theorem. If $f(x)$ is bounded on $[a,b]$ and is continuous on $[a,b]$ except at a finite number of points, then $f(x)$ is integrable on $[a,b]$.

Theorem. If $f(x)$ is bounded on $[a,b]$ and is continuous on $[a,b]$ except at a countable set of points, then $f(x)$ is integrable on $[a,b]$.

Lebesgue's Criterion. $f(x)$ is integrable on $[a,b]$ if, and only if, $f(x)$ is bounded on $[a,b]$ and the set of its discontinuity points has (Lebesgue) measure zero.

Remark. Any finite or countable set has Lebesgue measure zero.

Comparative Properties

1) If $f(x)$ and $g(x)$ are integrable on $[a,b]$ and $f(x) \le g(x)$ on $[a,b]$, then

$$\int_a^b f(x)\,dx \le \int_a^b g(x)\,dx.$$

2) If $f(x)$ is integrable on $[a,b]$, $f_{\text{inf}} = \inf_{x\in[a,b]} f(x)$, $f_{\text{sup}} = \sup_{x\in[a,b]} f(x)$, then

$$f_{\text{inf}} \cdot (b-a) \le \int_a^b f(x)\,dx \le f_{\text{sup}} \cdot (b-a).$$

In particular, if $f(x)$ is continuous on $[a,b]$, $f_{\text{min}} = \min_{x\in[a,b]} f(x)$, $f_{\text{max}} = \max_{x\in[a,b]} f(x)$, then

$$f_{\text{min}} \cdot (b-a) \le \int_a^b f(x)\,dx \le f_{\text{max}} \cdot (b-a).$$

Arithmetic (algebraic) properties

1) If $f(x)$ and $g(x)$ are integrable on $[a,b]$, then $f(x)+g(x)$ is also integrable on $[a,b]$ and

$$\int_a^b f(x) + g(x)\,dx = \int_a^b f(x)\,dx + \int_a^b g(x)\,dx.$$

2) If $f(x)$ is integrable on $[a,b]$ and c is an arbitrary constant, then $cf(x)$ is also integrable on $[a,b]$ and

$$\int_a^b cf(x)\,dx = c\int_a^b f(x)\,dx.$$

3) If $f(x)$ is integrable on $[a,b]$, then $|f(x)|$ is also integrable on $[a,b]$ and

$$\left|\int_a^b f(x)\,dx\right| \le \int_a^b |f(x)|\,dx.$$

4) If $f(x)$ and $g(x)$ are integrable on $[a,b]$, then $f(x)g(x)$ is also integrable on $[a,b]$.
5) $f(x)$ is integrable on $[a,b]$, if and only if, $f(x)$ is integrable on $[a,c]$ and $[c,b]$, where c is some point in $[a,b]$. Furthermore,

$$\int_a^b f(x)\,dx = \int_a^c f(x)\,dx + \int_c^b f(x)\,dx.$$

The Mean Value Theorem for integral. If $f(x)$ is continuous on $[a,b]$, there exists a point $c \in [a,b]$ such that $\int_a^b f(x)\,dx = f(c)\cdot(b-a)$.

The Fundamental Theorem of Calculus.

The first formulation. If $f(x)$ is Riemann integrable on $[a,b]$ and has an antiderivative on $[a,b]$, that is there exists $F(x)$ such that $F'(x) = f(x)$ on $[a,b]$, then $\int_a^b f(x)\,dx = F(b) - F(a)$.

The second formulation (a weaker version). If $f(x)$ is continuous on $[a,b]$, then $\int_a^b f(x)\,dx = F(b) - F(a)$, where $F(x)$ is one of antiderivatives of $f(x)$ on $[a,b]$.

Remark. If $f(x)$ is continuous on $[a,b]$, then it has an antiderivative $F(x)$ on this interval.

Change of variable.

The first formulation. If $f(x)$ is integrable on $[a,b]$, and $\varphi(t)$ is continuously differentiable and strictly increasing (decreasing) on $[\alpha,\beta]$ with image $[a,b]$, then

$$\int_a^b f(x)\,dx = \int_\alpha^\beta f(\varphi(t))\,\varphi'(t)\,dt \quad \left(\int_b^a f(x)\,dx = \int_\alpha^\beta f(\varphi(t))\,\varphi'(t)\,dt\right).$$

The second formulation. If $f(x)$ is continuous on $[a,b]$, and $\varphi(t)$ is continuously differentiable on $[\alpha,\beta]$ with image $[a,b]$ and $\varphi(\alpha) = a$, $\varphi(\beta) = b$ ($\varphi(\alpha) = b$, $\varphi(\beta) = a$), then

$$\int_a^b f(x)\,dx = \int_\alpha^\beta f(\varphi(t))\,\varphi'(t)\,dt \quad \left(\int_b^a f(x)\,dx = \int_\alpha^\beta f(\varphi(t))\,\varphi'(t)\,dt\right).$$

Integration by parts.

The first formulation. If $f(x)$ and $g(x)$ are differentiable on $[a,b]$ and $f(x)g'(x)$ is integrable on $[a,b]$, then $f'(x)g(x)$ is also integrable on $[a,b]$ and

$$\int_a^b f'(x)g(x)\,dx = f(b)g(b) - f(a)g(a) - \int_a^b f(x)g'(x)\,dx.$$

The second formulation (a weaker version). If $f(x)$ and $g(x)$ are continuously differentiable on $[a, b]$, then

$$\int_a^b f'(x)\, g(x)\, dx = f(b)\, g(b) - f(a)\, g(a) - \int_a^b f(x)\, g'(x)\, dx.$$

Improper integrals

Improper integral of the first kind. Let $f(x)$ be integrable on every interval $[a, b]$, where a is fixed and b is arbitrary, $b > a$. The following limit is called an *improper integral of the first kind* on $[a, +\infty)$:

$$\int_a^{+\infty} f(x)\, dx = \lim_{b \to +\infty} \int_a^b f(x)\, dx.$$

In the same way one can define an *improper integral of the first kind* on $(-\infty, b]$:

$$\int_{-\infty}^b f(x)\, dx = \lim_{a \to -\infty} \int_a^b f(x)\, dx.$$

If the limit exists and is finite, then the improper integral is called *convergent.* Otherwise it is called *divergent.*

Additionally, an *improper integral of the first kind* on $(-\infty, +\infty)$ is defined as follows:

$$\int_{-\infty}^{+\infty} f(x)\, dx = \int_{-\infty}^{a} f(x)\, dx + \int_a^{+\infty} f(x)\, dx$$

(here a is an arbitrary real number). If both integrals in the right-hand side are convergent, then the integral on $(-\infty, +\infty)$ is called convergent.

Remark 1. We do not consider here the principal value of the last integral.

Remark 2. We present below some results for the improper integral $\int_a^{+\infty} f(x)\, dx$. Similar results are true for other improper integrals of the first kind.

Cauchy criterion. An improper integral $\int_a^{+\infty} f(x)\, dx$ is convergent if, and only if, for $\forall \varepsilon > 0$ there is $B > a$ such that for any pair $b_2 > b_1 > B$ it holds $\left| \int_{b_1}^{b_2} f(x)\, dx \right| < \varepsilon$.

General Comparison theorem. Let $0 \le f(x) \le g(x)$ on $[a, +\infty)$. If $\int_a^{+\infty} g(x)\, dx$ converges, then $\int_a^{+\infty} f(x)\, dx$ also converges (equivalently, if $\int_a^{+\infty} f(x)\, dx$ diverges, then $\int_a^{+\infty} g(x)\, dx$ also diverges).

Specific Comparison theorem. If $0 \le f(x) \le \frac{c}{x^p}$ on $[a, +\infty)$, where $c > 0$ is a constant and $p > 1$, then $\int_a^{+\infty} f(x)\, dx$ converges. If $f(x) \ge \frac{c}{x^p}$ on $[a, +\infty)$, where $c > 0$ is a constant and $p \le 1$, then $\int_a^{+\infty} f(x)\, dx$ diverges.

Change of variable.

Let $f(x)$ be continuous on $[a,+\infty)$, and $\varphi(t)$ be continuously differentiable and strictly increasing on $[\alpha,+\infty]$ with image $[a,+\infty)$. Then the improper integrals $\int_a^{+\infty} f(x)\,dx$ and $\int_\alpha^{+\infty} f(\varphi(t))\,\varphi'(t)\,dt$ are both convergent or both divergent and

$$\int_a^{+\infty} f(x)\,dx = \int_\alpha^{+\infty} f(\varphi(t))\,\varphi'(t)\,dt.$$

Integration by parts.

Let $f(x)$ and $g(x)$ be continuously differentiable on $[a,+\infty)$, and $\lim_{x\to+\infty} f(x)\,g(x) = B$. Then the improper integrals $\int_a^{+\infty} f(x)\,g'(x)\,dx$ and $\int_a^{+\infty} f'(x)\,g(x)\,dx$ are both convergent or both divergent and

$$\int_a^{+\infty} f'(x)\,g(x)\,dx = B - f(a)\,g(a) - \int_a^{+\infty} f(x)\,g'(x)\,dx.$$

Improper integral of the second kind. Let $f(x)$ be defined on $[a,b)$, where a and b are finite points, integrable on every interval $[a,c]$, where c is arbitrary between a and b, $a < c < b$, and unbounded in any left-hand neighborhood of b (that is b is a singular point of $f(x)$). The following limit is called an *improper integral of the second kind* at b:

$$\int_a^b f(x)\,dx = \lim_{c\to b_-} \int_a^c f(x)\,dx.$$

In the same way one can define another *improper integral of the second kind* at a singular point a:

$$\int_a^b f(x)\,dx = \lim_{c\to a_+} \int_c^b f(x)\,dx.$$

If the limit exists and is finite, then the improper integral is called *convergent*. Otherwise it is called *divergent*.

Additionally, if $d \in (a,b)$ is an internal singular point, then an *improper integral of the second kind* is defined as follows:

$$\int_a^b f(x)\,dx = \int_a^d f(x)\,dx + \int_d^b f(x)\,dx,\ a < d < b,$$

where the integrals in the right-hand side are the improper integrals at d. If both integrals in the right-hand side are convergent, then the integral in the left-hand side is called convergent.

Remark 1. We do not consider here the principal value of the last integral.

Remark 2. If $f(x)$ is integrable on $[a, b]$, then an improper integral of the second kind coincides with the Riemann integral.

Remark 3. Improper integrals of the second kind can be transformed to improper integrals of the first kind by a change of variable.

Applications

Rectifiable curve. Let $f(x)$ be continuous on $[a, b]$. For an arbitrary partition

$$P = \{x_i : \ a = x_0 < x_1 < \ldots < x_{n-1} < x_n = b\}$$

of $[a, b]$ one can define the length of a polygonal curve inscribed in the graph of $f(x)$:

$$L(f, P) = \sum_{i=1}^{n} \sqrt{(x_i - x_{i-1})^2 + (f(x_i) - f(x_{i-1}))^2}.$$

If a finite limit of these lengths, as the partition diameter Δ approaches 0, exists and does not depend on a choice of partitions, then it is called the *length* of $f(x)$ on $[a, b]$: $L(f) = \lim\limits_{\Delta \to 0} L(f, P)$. In this case the curve of $f(x)$ on $[a, b]$ is called *rectifiable.*

Theorem. If $f(x)$ is continuously differentiable on $[a, b]$, then the curve of $f(x)$ on $[a, b]$ is rectifiable and its length can be calculated by the formula $L(f) = \int_a^b \sqrt{1 + (f'(x))^2} dx$.

Area of a region. Quadrable figure. Let $f(x)$ and $g(x)$ be defined on $[a, b]$. Consider the plane figure that lies between two curves $f(x)$ and $g(x)$ and between two vertical lines $x = a$ and $x = b$. For an arbitrary partition

$$P = \{x_i : \ a = x_0 < x_1 < \ldots < x_{n-1} < x_n = b\}$$

of $[a, b]$ and an arbitrary choice of points $c_i \in [x_{i-1}, x_i]$ one can define the sum of the areas of approximating rectangles as follows:

$$A(f, g, P) = \sum_{i=1}^{n} |f(c_i) - g(c_i)| \cdot (x_i - x_{i-1}).$$

If there exists a finite limit of these sums as the partition diameter Δ approaches 0, and this limit does not depend on a choice of partitions and points c_i, then it is called the *area* of the considered plane figure: $A(f, g) = \lim\limits_{\Delta \to 0} A(f, g, P)$. In this case the plane figure is called *quadrable or squarable.*

Remark . Actually, this is a simplified definition of a quadrable plane figure. The complete definition is much more complex, involving sets of inscribed/circumscribed polygons and the concepts of infimum/supremum, and it is out of scope of this text.

Theorem. If $f(x)$ and $g(x)$ are continuous on $[a,b]$, then the corresponding plane figure is quadrable and its area can be calculated by the formula $A(f,g) = \int_a^b |f(x) - g(x)|\, dx$.

Solid of revolution. A *solid of revolution* is a space figure obtained by rotating a plane region around a straight line (the axis of rotation), which lies on the same plane and does not contain interior points of the region.

Theorem. If $f(x)$ is continuous on $[a,b]$, the plane region lies between $f(x)$ and the x-axis and between vertical sides $x = a$ and $x = b$, and the x-axis is the axis of rotation, then the surface area and the *volume* of the solid of revolution are well-defined and can be calculated by the formulas

$$S(f) = 2\pi \int_a^b |f(x)| \sqrt{1 + (f'(x))^2} dx \quad \text{and} \quad V(f) = \pi \int_a^b f^2(x)\, dx,$$

respectively.

Remark . Variants of these formulas in the cases when a plane figure is defined by two curvilinear boundaries or when the axis of rotation is different of the x-axis, etc., can be easily deduced from these basic formulas (or consulted in textbooks).

5.2 Indefinite integral

Remark . In this section we will say that a function $f(x)$ is integrable when $f(x)$ has an antiderivative, i.e. when its indefinite integral exists.

Example 1. "If $f(x)$ is integrable on an interval, then it is bounded there."

Solution.

For an open interval (a,b) a counterexample is very simple. The function $f(x) = \frac{1}{\cos^2 x}$ is integrable on $\left(-\frac{\pi}{2}, \frac{\pi}{2}\right)$, with one of the antiderivatives being $F(x) = \tan x$, but $f(x)$ is not bounded on this interval.

For a closed interval $[a,b]$ the situation is not so simple, but the function $f(x) = \begin{cases} 2x \sin\frac{1}{x^2} - \frac{2}{x}\cos\frac{1}{x^2}, & x \neq 0 \\ 0, & x = 0 \end{cases}$ considered in Example 5, section 4.3 can provide a counterexample. For convenience we reproduce here similar considerations of that Example using the integral terminology. One of antiderivatives of $f(x)$ on $\mathbb{R}$ is $F(x) = \begin{cases} x^2 \sin\frac{1}{x^2}, & x \neq 0 \\ 0, & x = 0 \end{cases}$. This is easy to check by applying the arithmetic and chain rules of differentiation for any $x \neq 0$ and appealing to the definition for $x = 0$: $\lim\limits_{x\to 0} \frac{F(x)-F(0)}{x-0} = \lim\limits_{x\to 0} x \sin\frac{1}{x^2} = 0$, since

$\left|x \sin \frac{1}{x^2}\right| \leq |x|$. However, $f(x)$ is not bounded in any neighborhood of the origin since for $x_k = \frac{1}{\sqrt{2k\pi}}$, $\forall k \in \mathbb{N}$ one gets $\lim\limits_{x_k \to 0} f(x_k) = \lim\limits_{k \to +\infty} \left(-2\sqrt{2k\pi}\right) = -\infty$. Therefore, the function $f(x)$ is integrable, but unbounded on $[-1, 1]$ (or on any closed interval containing 0).

Example 2. "If $f(x)$ is bounded on an interval, then it is integrable there."

Solution.

The function $f(x) = \operatorname{sgn} x$ is bounded on $\mathbb{R}$, but it is not integrable on any interval containing the origin, because the derivative cannot contain a jump discontinuity (according to Darboux's theorem).

Another (more radical) example is Dirichlet's function $D(x) = \begin{cases} 1, & x \in \mathbb{Q} \\ 0, & x \in \mathbb{I} \end{cases}$, which is bounded on an arbitrary interval, but it is not integrable on any interval.

Example 3. "If $f(x)$ and $g(x)$ are not integrable, then $f(x) + g(x)$ is not integrable too."

Solution.

The functions $f(x) = x - \operatorname{sgn} x$ and $g(x) = \operatorname{sgn} x$ are not integrable on any interval containing the origin, because they have jump discontinuities at zero, but their sum $f(x) + g(x) = x$ is integrable on $\mathbb{R}$.

Remark . Similar false statements can be formulated for other arithmetic operations. For instance, the statement for the product: "if $f(x)$ and $g(x)$ are not integrable, then $f(x) \cdot g(x)$ is not integrable too". It can be disproved by the counterexample with the functions $f(x) = \begin{cases} x - 1, & x < 0 \\ x + 1, & x \geq 0 \end{cases}$ and $g(x) = \begin{cases} -1, & x < 0 \\ 1, & x \geq 0 \end{cases}$. They have no antiderivative on $(-1, 1)$, but their product $f(x) \cdot g(x) = 1 + |x|$ is integrable on $\mathbb{R}$, with one of the antiderivatives being $x + \frac{1}{2}x|x|$.

Example 4. "If $f(x)$ is integrable and $g(x)$ is not integrable on an interval, then $f(x) \cdot g(x)$ is not integrable on this interval."

Solution.

The function $f(x) = x$ is integrable on $\mathbb{R}$, and $g(x) = \operatorname{sgn} x$ is not integrable on any interval containing the origin, but their product $f(x) \cdot g(x) = |x|$ is integrable on $\mathbb{R}$: $\int |x|\, dx = \frac{1}{2}x|x| + C$, $C = const$.

Remark . The statement would be true if instead of the product one considers the sum or difference of an integrable and non-integrable functions.

Example 5. "If $f(x)$ is not integrable on an interval, then $|f(x)|$ is also not integrable on this interval."

Solution.

The function $f(x) = \begin{cases} -1, \ x < 0 \\ 1, \ x \geq 0 \end{cases}$ is not integrable on any interval containing the origin, but the function $|f(x)| = 1$ is integrable on $\mathbb{R}$.

Another (more radical) example is the modified Dirichlet function $\tilde{D}(x) = \begin{cases} 1, \ x \in \mathbb{Q} \\ -1, \ x \in \mathbb{I} \end{cases}$, which is not integrable on any interval, but $\left|\tilde{D}(x)\right| = 1$ is integrable on $\mathbb{R}$.

Remark 1. The converse is true.

Remark 2. Similarly, the following statement on the square is false: "if $f(x)$ is not integrable on an interval, then $f^2(x)$ is also not integrable on this interval". The same functions can be used for counterexample.

Example 6. "If $f(x)$ satisfies the intermediate value property on an interval, then it is integrable there."

Solution.

The function $f(x) = \begin{cases} 2x \sin \frac{1}{x} - \cos \frac{1}{x}, \ x \neq 0 \\ 0, \ x = 0 \end{cases}$ is integrable on $[-1,1]$ with an antiderivative $F(x) = \begin{cases} x^2 \sin \frac{1}{x}, \ x \neq 0 \\ 0, \ x = 0 \end{cases}$ (see Example 3, section 4.2 for details), and $g(x) = \begin{cases} 2x \sin \frac{1}{x}, \ x \neq 0 \\ 0, \ x = 0 \end{cases}$ is integrable on $[-1,1]$ since it is continuous. Therefore, their difference $h(x) = g(x) - f(x) = \begin{cases} \cos \frac{1}{x}, \ x \neq 0 \\ 0, \ x = 0 \end{cases}$ is also integrable on $[-1,1]$. Using the same reasoning as in Example 14, section 3.4, it can be shown that the more general function $\tilde{h}(x) = \begin{cases} \cos \frac{1}{x}, \ x \neq 0 \\ \alpha, \ x = 0 \end{cases}$ has the intermediate value property on $[-1,1]$ for any $\alpha \in [-1,1]$ (for $\alpha = 0$, that is, for $h(x)$ it also follows from Darboux's theorem). However, $\tilde{h}(x)$ is not integrable for any $\alpha \neq 0$, since otherwise the function $\tilde{h}(x) - h(x) = \begin{cases} 0, \ x \neq 0 \\ \alpha, \ x = 0 \end{cases}$ would be also integrable on $[-1,1]$, which is not possible because a derivative cannot contain a removable discontinuity.

Remark 1. In the above counterexample the interval $[-1,1]$ can be substituted by any interval containing 0.

Remark 2. Another similar example can be provided with the function $\tilde{h}(x) = \begin{cases} \frac{2}{x} \cos \frac{1}{x^2}, \ x \neq 0 \\ \alpha, \ x = 0 \end{cases}$ considered on any interval containing 0. This function has the intermediate value property for an arbitrary α, but it is not integrable for $\alpha \neq 0$, because the corresponding function $h(x) = \begin{cases} \frac{2}{x} \cos \frac{1}{x^2}, \ x \neq 0 \\ 0, \ x = 0 \end{cases}$ is integrable as the difference between the two integrable functions $f(x) = \begin{cases} 2x \sin \frac{1}{x^2} - \frac{2}{x} \cos \frac{1}{x^2}, & x \neq 0 \\ 0, & x = 0 \end{cases}$ (see Example 1 in this section) and $g(x) = \begin{cases} 2x \sin \frac{1}{x^2}, & x \neq 0 \\ 0, & x = 0 \end{cases}$ (the latter is continuous).

Remark 3. This is a wrong converse to Darboux's theorem: if $f(x)$ is integrable on a connected set S, then it satisfies the intermediate value property on S.

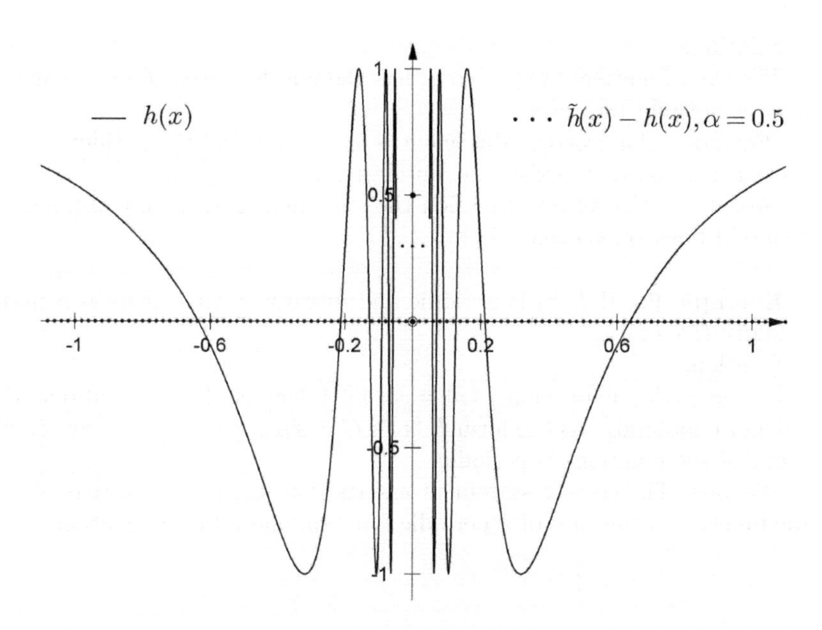

FIGURE 5.2.1: Example 6

Example 7. "The product of two integrable functions is also integrable."
Solution.

Let us consider the three integrable functions of the previous Example 6: $f(x) = \begin{cases} 2x \sin \frac{1}{x} - \cos \frac{1}{x}, & x \neq 0 \\ 0, & x = 0 \end{cases}$, $g(x) = \begin{cases} 2x \sin \frac{1}{x}, & x \neq 0 \\ 0, & x = 0 \end{cases}$ and $h(x) = g(x) - f(x) = \begin{cases} \cos \frac{1}{x}, & x \neq 0 \\ 0, & x = 0 \end{cases}$. It can be shown, that the function $h^2(x) = \begin{cases} \frac{1}{2}\left(1 + \cos \frac{2}{x}\right), & x \neq 0 \\ 0, & x = 0 \end{cases}$ is not integrable on any interval containing 0. In fact, let us suppose, per absurd, that $h^2(x)$ is integrable and denote its primitive by $\tilde{H}(x)$. If $H(x)$ is a primitive of $h(x)$, then $\left(H\left(\frac{x}{2}\right)\right)' = \begin{cases} \frac{1}{2} \cos \frac{2}{x}, & x \neq 0 \\ 0, & x = 0 \end{cases}$ (applying the chain rule for $x \neq 0$ and the definition for $x = 0$). Then $\left(\tilde{H}(x) - H\left(\frac{x}{2}\right)\right)' \equiv \hat{h}(x) = \begin{cases} \frac{1}{2}, & x \neq 0 \\ 0, & x = 0 \end{cases}$, but this is not possible since (according to Darboux's theorem) a derivative cannot have a removable discontinuity.

Remark. A similar statement for Riemann integrable functions is true.

Example 8. "If $f(x)$ is even and integrable, then its antiderivative is odd."

Solution.

The even function $f(x) = \cos x$ has the antiderivative $F(x) = \sin x + 1$, which is not odd (nor even).

Remark 1. The correct statement is that among all the antiderivatives of an even function there exists one antiderivative that is odd.

Remark 2. For an odd function the statement is true: any antiderivative of an odd function is even.

Example 9. "If $f(x)$ is periodic and integrable, then there is a periodic antiderivative of $f(x)$."

Solution.

For the periodic function $f(x) = \cos x + 1$, the set of all the antiderivatives (indefinite integral) has the form $F(x) + C = \sin x + x + C$, $C = const$, where no one of the functions is periodic.

Remark. The correct statement asserts that any antiderivative of a periodic function is the sum of a periodic function and a linear function.

5.3 Definite (Riemann) integral

Remark. In this section we will say that a function $f(x)$ is integrable when there exists an indefinite integral of $f(x)$, and say that $f(x)$ is Riemann integrable when there exists a definite (Riemann) integral of $f(x)$.

Example 1. "If $f(x)$ is bounded on $[a, b]$, then $f(x)$ is Riemann integrable on $[a, b]$."

Solution.

Dirichlet's function $D(x) = \begin{cases} 1, & x \in \mathbb{Q} \\ 0, & x \in \mathbb{I} \end{cases}$ is bounded but not Riemann integrable on any interval $[a, b]$. Indeed, for an arbitrary partition

$$P = \{x_i : a = x_0 < x_1 < \ldots < x_{n-1} < x_n = b;\ \Delta x_i = x_i - x_{i-1},\ i = 1, \ldots, n\}$$

of $[a, b]$ in each of subintervals $[x_{i-1}, x_i]$ there are both rational and irrational points. Choosing rational points $c_i \in [x_{i-1}, x_i]$ in the definition of Riemann sums, we obtain

$$S_{rat}(f) = \sum_{i=1}^{n} f(c_i) \cdot \Delta x_i = \sum_{i=1}^{n} 1 \cdot \Delta x_i = b - a,$$

while for irrational points c_i , we have

$$S_{irr}(f) = \sum_{i=1}^{n} f(c_i) \cdot \Delta x_i = \sum_{i=1}^{n} 0 \cdot \Delta x_i = 0.$$

Therefore, the limits for these sums, as the partition diameter $\Delta = \max_{1 \le i \le n} \Delta x_i$ approaches 0, are different

$$\lim_{\Delta \to 0} S_{rat}(f) = b - a \neq 0 = \lim_{\Delta \to 0} S_{irr}(f),$$

which means that the Riemann integral does not exist.

Remark 1. For the reader accustomed to the upper and lower Riemann integrals, the proof can be made in the following way. For an arbitrary partition P of $[a, b]$, the upper Riemann sum is

$$\bar{S}(f) = \sum_{i=1}^{n} \sup_{x \in [x_{i-1}, x_i]} f(x) \cdot \Delta x_i = \sum_{i=1}^{n} 1 \cdot \Delta x_i = b - a,$$

while the lower sum is

$$\underline{S}(f) = \sum_{i=1}^{n} \inf_{x \in [x_{i-1}, x_i]} f(x) \cdot \Delta x_i = \sum_{i=1}^{n} 0 \cdot \Delta x_i = 0.$$

Therefore, the upper and lower Riemann integrals are different

$$\lim_{\Delta \to 0} \bar{S}(f) = b - a \neq 0 = \lim_{\Delta \to 0} \underline{S}(f),$$

which means that the Riemann integral does not exist.

Remark 2. The converse is true.

Example 2. "If $f(x)$ is Riemann integrable on $[a, b]$, then $f(x)$ is continuous on $[a, b]$."

Solution.

The function $f(x) = \operatorname{sgn} x$ is Riemann integrable but non-continuous on $[0, 1]$. In fact, $f(x)$ has a removable discontinuity at 0 (as the function is considered on $[0, 1]$). Let us show that $f(x)$ is Riemann integrable by the definition. For an arbitrary partition of $[0, 1]$ and an arbitrary choice of the points $c_i \in [x_{i-1}, x_i]$, the following evaluation of the Riemann sums $S(f)$ holds:

$$1 - \Delta x_1 = 0 \cdot \Delta x_1 + \sum_{i=2}^{n} 1 \cdot \Delta x_i \le S(f) = \sum_{i=1}^{n} f(c_i) \cdot \Delta x_i \le \sum_{i=1}^{n} 1 \cdot \Delta x_i = 1.$$

Since $\Delta x_1 \to 0$ when $\Delta \to 0$, it follows that the limit of the Riemann sums, as Δ approaches 0, exists and does not depend on the choice of partition and points c_i: $\lim_{\Delta \to 0} S(f) = 1$, that is $\int_0^1 \operatorname{sgn} x dx = 1$.

Remark 1. For those who use to apply the upper and lower Riemann integrals, the proof can be done as follows. For an arbitrary partition of $[0, 1]$ the upper Riemann sum is

$$\bar{S}(f) = \sum_{i=1}^{n} \sup_{x \in [x_{i-1}, x_i]} f(x) \cdot \Delta x_i = \sum_{i=1}^{n} 1 \cdot \Delta x_i = 1,$$

while the lower sum is

$$\underline{S}(f) = \sum_{i=1}^{n} \inf_{x \in [x_{i-1}, x_i]} f(x) \cdot \Delta x_i = 0 \cdot \Delta x_1 + \sum_{i=2}^{n} 1 \cdot \Delta x_i = 1 - \Delta x_1.$$

Therefore, $\lim_{\Delta \to 0} \bar{S}(f) = 1 = \lim_{\Delta \to 0} \underline{S}(f)$, that is $\int_0^1 \operatorname{sgn} x dx = 1$.

Remark 2. The same considerations can be applied to show that the Heaviside function $H(x) = \begin{cases} 0, x < 0 \\ 1, \ x \geq 0 \end{cases}$ is Riemann integrable on $[-1, 0]$ and $\int_{-1}^{0} H(x)\, dx = 0$. It is just a bit more difficult to show that $f(x) = \operatorname{sgn} x$ and $H(x)$ are Riemann integrable but discontinuous on any interval containing 0.

Remark 3. Another interesting counterexample is the Riemann function $R(x) = \begin{cases} 1/n, \ x \in \mathbb{Q}, x = m/n \\ 0, \ x \in \mathbb{I} \\ 1, \ x = 0 \end{cases}$ (recall that $m \in \mathbb{Z}$, $n \in \mathbb{N}$ and the ratio $\frac{m}{n}$ is in lowest terms), defined in Example 4, section 3.3 where it was shown that $R(x)$ is discontinuous at all rational points (and continuous at all irrational points). It happens that this function is also Riemann integrable. This can be shown by applying the definition, but the proof is somewhat involved. To avoid such complications, the reader familiarized with the Lebesgue measure can appeal to the Lebesgue criterion that asserts that a bounded function is Riemann integrable if the set of the points of its discontinuity is of Lebesgue measure zero, which is the case of the Riemann function that has countably many discontinuities on any interval $[a, b]$. Further, we can find the value of the integral by noting that all the lower Riemann sums are equal to 0:

$$\underline{S}(f) = \sum_{i=1}^{n} \inf_{x \in [x_{i-1}, x_i]} f(x) \cdot \Delta x_i = \sum_{i=1}^{n} 0 \cdot \Delta x_i = 0$$

(since any subinterval of an arbitrary partition contains irrational points). Since the Riemann integral exists, its value is equal to the lower Riemann integral, hence $\int_a^b R(x)\, dx = 0$.

Example 3. "If a function $f(x)$ is integrable on $[a, b]$, then it is Riemann integrable on this interval."

Solution.

The function $f(x) = \begin{cases} 2x\sin\frac{1}{x^2} - \frac{2}{x}\cos\frac{1}{x^2}, & x \neq 0 \\ 0, & x = 0 \end{cases}$ is integrable and unbounded on $[-1,1]$ (see Example 1, section 5.2 for details). So it is not Riemann integrable on this interval.

Example 4. "If a function $f(x)$ is Riemann integrable on $[a,b]$, then it is integrable on $[a,b]$."

Solution.

The function $f(x) = \operatorname{sgn} x$ is Riemann integrable on $[0,1]$ (see Example 2 in this section), but it has no antiderivative on $[0,1]$, because the derivative of any function cannot have removable or jump discontinuities.

Example 5. "If $|f(x)|$ is Riemann integrable on $[a,b]$, then $f(x)$ is also Riemann integrable on $[a,b]$."

Solution.

The modified Dirichlet function $\tilde{D}(x) = \begin{cases} 1, & x \in \mathbb{Q} \\ -1, & x \in \mathbb{I} \end{cases}$ is not Riemann integrable on any interval (it can be shown like in Example 1), but $\left|\tilde{D}(x)\right| = 1$ is Riemann integrable on any interval.

Remark 1. The converse is true.

Remark 2. The corresponding statement for the square of a function is also false: "if $f^2(x)$ is Riemann integrable on $[a,b]$, then $f(x)$ is also Riemann integrable on $[a,b]$". The same modified Dirichlet function $\tilde{D}(x)$ gives a counterexample. Again the converse is true.

Example 6. "If $f(x) \cdot g(x)$ is Riemann integrable on $[a,b]$, then $f(x)$ and $g(x)$ are Riemann integrable on $[a,b]$."

Solution.

If $f(x) = g(x)$, then the problem is reduced to Remark 2 of the previous Example 5.

Of course, the functions may be different. For example, if $f(x) = \begin{cases} \tan x, & x \in (0,\pi/2) \\ 1, & x = 0,\ x = \pi/2 \end{cases}$ and $g(x) = \begin{cases} \cot x, & x \in (0,\pi/2) \\ 1, & x = 0,\ x = \pi/2 \end{cases}$, both functions defined on $\left[0,\frac{\pi}{2}\right]$, then $f(x)$ and $g(x)$ are unbounded on $\left[0,\frac{\pi}{2}\right]$, and consequently, are not Riemann integrable. However, $h(x) = f(x) \cdot g(x) = 1$ is Riemann integrable on $\left[0,\frac{\pi}{2}\right]$.

Remark. The converse is true.

Example 7. "If $f(x)$ is Riemann integrable on $[a,b]$ and $\int_a^b f(x)\,dx = 0$, then $f(x) \equiv 0$ on $[a,b]$."

Solution.

Consider non-zero function $f(x) = x$ on $[-1,1]$. Evidently $f(x)$ is Riemann integrable and $\int_{-1}^{1} x dx = \left.\frac{x^2}{2}\right|_{-1}^{1} = 0$.

Remark 1. The converse is true.

Remark 2. The following generalized version of the statement is also false: "if $f(x)$ and $g(x)$ are Riemann integrable on $[a,b]$ and $\int_a^b f(x)\,dx = \int_a^b g(x)\,dx$, then $f(x) = g(x)$ on $[a,b]$". The corresponding counterexample can be constructed with the functions $f(x) = x$ and $g(x) = \sin x$ on $[-a,a]$, $\forall a > 0$. Again, the converse is true.

Example 8. "If $f(x) \geq 0$ on $[a,b]$ and $\int_a^b f(x)\,dx = 0$, then $f(x) \equiv 0$ on $[a,b]$."

Solution.

The Riemann function $R(x)$ is non-negative on $[0,1]$ and $\int_0^1 R(x)\,dx = 0$ (see details in Remark 3 to Example 2), but it is positive at every rational point.

Remark. The result is correct for continuous on $[a,b]$ functions.

Example 9. "If $f(x)$ is Riemann integrable on $[a,b]$ and $\int_a^b f^2(x)\,dx = 0$, then $f(x) \equiv 0$ on $[a,b]$."

Solution.

For the non-zero integrable function $f(x) = 1 - \operatorname{sgn} x$ considered on $[0,1]$ the integral of $f^2(x) = (1 - \operatorname{sgn} x)^2 = 1 - \operatorname{sgn} x$ equals 0: $\int_0^1 1 - \operatorname{sgn} x dx = \int_0^1 0 dx = 0$. Here we have applied the property that the value of the Riemann integral does not depend on the definition of a function at a finite number of points.

Remark 1. To construct a counterexample with a function assuming positive and negative values at a countable set of points we can use the modified Riemann function:

$$f(x) = \tilde{R}(x) = \begin{cases} 1/\sqrt{n}\,, & x \in \mathbb{Q},\, x = m/n\,,\, n = 2k \\ -1/\sqrt{n}\,, & x \in \mathbb{Q},\, x = m/n\,,\, n = 2k+1 \\ 0\,, & x \in \mathbb{I}\,;\; 1,\; x = 0 \end{cases}$$

(here $m \in \mathbb{Z}$, $n \in \mathbb{N}$, $k \in \mathbb{N}$ and the ratio $\frac{m}{n}$ is in lowest terms). Since $\tilde{R}^2(x) = R(x)$ and $\int_a^b R(x)\,dx = 0$ on any interval $[a,b]$ (see Remark 3 to Example 2), the conditions of the statement are satisfied.

Remark 2. Under the conditions of the statement, it is true that $f(x) = 0$ at all points of $[a,b]$ where $f(x)$ is continuous.

Remark 3. The converse is true.

Example 10. "If $\int_a^b f(x)\,dx \geq 0$, then $f(x) \geq 0$ on $[a,b]$."

Solution.

The function $f(x) = x^2 - 1$ is negative on $[0,1)$ and positive on $(1,2]$, and $\int_0^2 (x^2 - 1)\,dx = \frac{2}{3} > 0$.

Remark 1. The converse is true if $f(x)$ is Riemann integrable.

Remark 2. The statement can be generalized to the following form (also

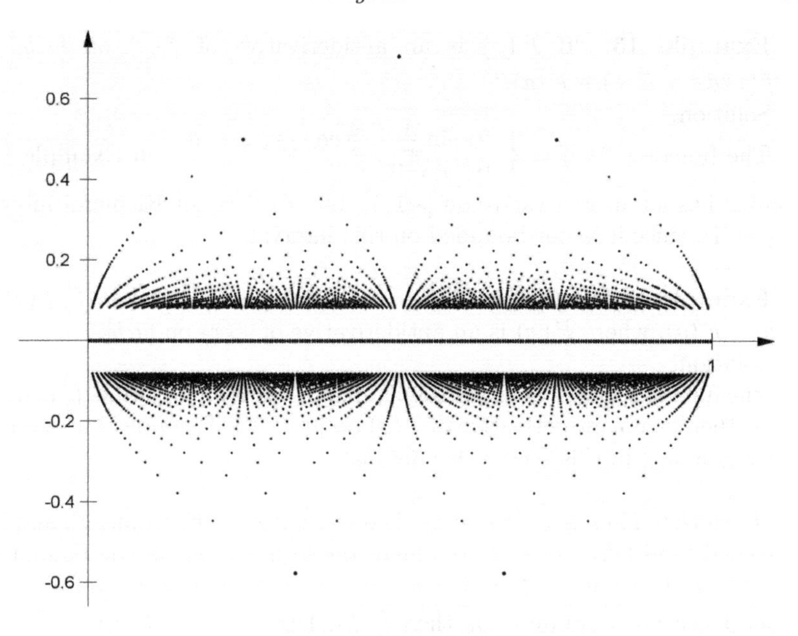

FIGURE 5.3.1: Example 9, Remark 1

false): "if $\int_a^b f(x)\,dx \geq \int_a^b g(x)\,dx$, then $f(x) \geq g(x)$ on $[a,b]$ ". The corresponding counterexample can be provided with the functions $f(x) = x^2$ and $g(x) = 1$ on $[0,2]$. Again, the converse is true.

Example 11. "If $f(x) + g(x)$ is Riemann integrable on $[a,b]$, then both $f(x)$ and $g(x)$ are Riemann integrable on $[a,b]$."

Solution.

$f(x) = D(x)$ and $g(x) = 1 - D(x)$ is a simple counterexample for this case.

Remark. Similar statements for other arithmetic operations are also false. For example, for the ratio the corresponding counterexample can be provided using the modified Dirichlet function: $f(x) = g(x) = \tilde{D}(x)$.

Example 12. "The composition of two Riemann integrable functions is also Riemann integrable."

Solution.

Let $f(x)$ be the Riemann function $f(x) = R(x)$ on $[0,1]$ and $g(x) = \operatorname{sgn} x$ on $\mathbb{R}$. Both functions are Riemann integrable on any interval (see Remark 3 to Example 2). However, their composition is the Dirichlet function $g(f(x)) = \left\{ \begin{array}{l} 1,\ x \in \mathbb{Q} \\ 0,\ x \in \mathbb{I} \end{array} \right\} = D(x)$ on $[0,1]$, that is not Riemann integrable on $[0,1]$.

Example 13. "If $F(x)$ is an antiderivative of $f(x)$ on $[a,b]$, then $\int_a^b f(x)\,dx = F(b) - F(a)$."

Solution.

The function $f(x) = \begin{cases} 2x\sin\frac{1}{x^2} - \frac{2}{x}\cos\frac{1}{x^2}, & x \neq 0 \\ 0, & x = 0 \end{cases}$ in Example 1, section 5.2 has an antiderivative on $[-1,1]$, but $f(x)$ is not Riemann integrable on $[-1,1]$, since it is not bounded on this interval.

Example 14. "If $f(x)$ is Riemann integrable on $[a,b]$, then $\int_a^b f(x)\,dx = F(b) - F(a)$, where $F(x)$ is an antiderivative of $f(x)$ on $[a,b]$."

Solution.

The function $f(x) = \operatorname{sgn} x$ is Riemann integrable on $[0,1]$ and $\int_0^1 \operatorname{sgn} x dx = 1$, but there is no antiderivative of $f(x)$ on $[0,1]$ (see Example 2 in section 5.2 and Example 4 in this section for details).

Remark to Examples 13 and 14. The statements in Examples 13 and 14 are weakened (and false) versions of one of the formulations of the Fundamental Theorem of Calculus: if $f(x)$ is Riemann integrable on $[a,b]$ and $F(x)$ is an antiderivative of $f(x)$ on $[a,b]$, then $\int_a^b f(x)\,dx = F(b) - F(a)$.

Example 15. "If $f(x)$ is Riemann integrable on $[a,b]$, then there exists $c \in [a,b]$ such that $\int_a^b f(x)\,dx = f(c)\cdot(b-a)$."

Solution.

The Heaviside function $H(x) = \begin{cases} 0, & x < 0 \\ 1, & x \geq 0 \end{cases}$ is Riemann integrable (see Remark 2 to Example 2) on $[-1,1]$ and $\int_{-1}^1 H(x)\,dx = 1$, but there is no point where $H(x)$ is equal to $\frac{1}{2}$.

Remark. The additional condition of the continuity of $f(x)$ on $[a,b]$ will turn this false statement in one of the versions of the Mean Value Theorem for integrals.

Example 16. "If $f(x)$ is continuous on $[a,b]$, and $\varphi(t)$ is continuously differentiable on $[\alpha,\beta]$, then $\int_\alpha^\beta f(\varphi(t))\,\varphi'(t)\,dt = \int_a^b f(x)\,dx$, where $a = \varphi(\alpha)$ and $b = \varphi(\beta)$."

Solution.

Let us consider the following calculations applying this "rule". First, calculate $\int_{-1}^2 t^2 dt$ directly: $\int_{-1}^2 t^2 dt = \left.\frac{t^3}{3}\right|_{-1}^2 = 3$. Next, consider the change of the variable using the continuously differentiable function $x = \varphi(t) = t^2$ with $\varphi'(t) = 2t$. The lower limit $\alpha = -1$ corresponds to $a = \varphi(-1) = 1$ and the upper $\beta = 2$ to $b = \varphi(2) = 4$. Now expressing all elements in the terms of variable x we get the right-hand side integral in the form $\int_1^4 \frac{1}{2}\sqrt{x}dx$ with the integrand $f(x) = \frac{1}{2}\sqrt{x}$ continuous on $[1,4]$. Calculating the last integral we obtain: $\int_1^4 \frac{1}{2}\sqrt{x}dx = \frac{1}{3}x^{3/2}\Big|_1^4 = \frac{7}{3}$. Notice, that the conditions of the statement hold, but two results are different.

Remark. The correct formulation of the change of variable rule is: if $f(x)$ is continuous on $[a,b]$, and $\varphi(t)$ is continuously differentiable on $[\alpha,\beta]$ with image $[a,b]$ and $\varphi(\alpha)=a$, $\varphi(\beta)=b$, then $\int_\alpha^\beta f(\varphi(t))\varphi'(t)\,dt = \int_a^b f(x)\,dx$. In the false statement of this Example, the condition of the connection between the domain of $f(x)$ and the image of $\varphi(t)$ was omitted, and it gives a wrong result illustrated by the presented counterexample, where the image of $\varphi(t)=t^2$ when $t\in[-1,2]$ is $[0,4]$, while the interval $[\varphi(-1),\varphi(2)]=[1,4]$ is only a part of the image. Of course, the correct rule provides sufficient conditions, and in some cases when these conditions are not satisfied the formula may still be true, but it may also happen that the formula will not work, as in the counterexample above.

Example 17. "If $f(x)$ and $g(x)$ are continuous on $[a,b]$ and continuously differentiable on (a,b), then $\int_a^b f(x)\,g'(x)\,dx = f(x)\,g(x)|_a^b - \int_a^b f'(x)\,g(x)\,dx$."

Solution.

The functions $f(x)=\sqrt{x}$ and $g(x)=-\cos x$ are continuous on $[0,1]$ and continuously differentiable on $(0,1)$ (with the derivatives $f'(x)=\frac{1}{2\sqrt{x}}$, $g'(x)=\sin x$), but the formula

$$\int_0^1 \sqrt{x}\sin x dx = -\sqrt{x}\cos x\Big|_0^1 + \frac{1}{2}\int_0^1 \frac{\cos x}{\sqrt{x}}dx$$

is not valid, because the integral in the right-hand side does not exist (the function $\frac{\cos x}{\sqrt{x}}$ is not bounded on $[0,1]$, and so it is not Riemann integrable).

Remark 1. One of the correct statements of the integration by parts rule is the following: if $f(x)$ and $g(x)$ are continuously differentiable on $[a,b]$, then

$$\int_a^b f(x)\,g'(x)\,dx = f(x)\,g(x)|_a^b - \int_a^b f'(x)\,g(x)\,dx.$$

While these are the sufficient conditions, we can see that a weak violation of them at one of the endpoints can lead to a wrong result. Of course, since the conditions of the rule are sufficient, it may happen that they are violated, but the rule is still working. For instance, using the same $f(x)=\sqrt{x}$ and changing the second function to $g(x)=\sin x$, we keep the same properties of two functions on $[0,1]$, but the formula of the integration by parts holds in this case:

$$\int_0^1 \sqrt{x}\cos x dx = \sqrt{x}\sin x\Big|_0^1 - \frac{1}{2}\int_0^1 \frac{\sin x}{\sqrt{x}}dx.$$

Although $\frac{\sin x}{\sqrt{x}}$ is not defined at 0, but it is Riemann integrable, because this function is continuous on $(0,1]$ and $\lim\limits_{x\to 0}\frac{\sin x}{\sqrt{x}}=0$, which indicates just a removable discontinuity at 0.

Remark 2. The integral $\int_0^1 \frac{\cos x}{\sqrt{x}} dx$ still exists if it is considered as an improper integral. In this generalized context of the integrability, the formula

$$\int_0^1 \sqrt{x} \sin x dx = -\sqrt{x} \cos x \Big|_0^1 + \int_0^1 \frac{\cos x}{2\sqrt{x}} dx$$

is correct. (Improper integrals will be considered in section 5.4.)

Example 18. "If $f(x)$ is Riemann integrable on $[a, b]$, then the value of the Riemann integral does not change if one alters the values of $f(x)$ at a countable number of points."

Solution.

The functions $f(x) \equiv 0$ and $D(x) = \begin{cases} 1, & x \in \mathbb{Q} \\ 0, & x \in \mathbb{I} \end{cases}$ differ only at rational points, that is on a countable set, however $f(x)$ is Riemann integrable on any $[a, b]$ and $D(x)$ is not.

Remark. The statement becomes true if both functions are Riemann integrable.

Example 19. "If $f(x)$ is continuous on $[a, b]$ except at only one point, then $f(x)$ is Riemann integrable on $[a, b]$."

Solution.

The function $f(x) = \tan x$ has the only discontinuity point $x = \frac{\pi}{2}$ on $[0, \pi]$, but $f(x)$ is not Riemann integrable because it is not bounded on $[0, \pi]$.

Remark. The statement becomes true if $f(x)$ is additionally bounded on $[a, b]$.

5.4 Improper integrals

Example 1. "If $\int_a^{+\infty} f(x) dx$ converges, then $\lim_{x \to +\infty} f(x) = 0$."

Solution.

Consider $f(x) = \cos x^2$ on $[1, +\infty)$. By choosing two sequences of points - $x_k = \sqrt{2k\pi}$, $k \in \mathbb{N}$ and $x_n = \sqrt{2n\pi + \pi}$, $n \in \mathbb{N}$ - we obtain two different partial limits at infinity:

$$\lim_{x_k \to +\infty} f(x_k) = \lim_{k \to +\infty} \cos 2k\pi = 1$$

and

$$\lim_{x_n \to +\infty} f(x_n) = \lim_{n \to +\infty} \cos(2n\pi + \pi) = -1.$$

Therefore, $\lim_{x \to +\infty} f(x)$ does not exist. However, $\int_1^{+\infty} f(x)\, dx$ converges. Indeed, representing

$$f(x) = \cos x^2 = \left(\frac{\sin x^2}{2x}\right)' + \frac{\sin x^2}{2x^2},$$

we obtain for any $b > 1$:

$$\int_1^b \cos x^2 dx = \int_1^b \left(\frac{\sin x^2}{2x}\right)' + \frac{\sin x^2}{2x^2} dx = \frac{\sin b^2}{2b} - \frac{\sin 1}{2} + \frac{1}{2}\int_1^b \frac{\sin x^2}{x^2} dx.$$

The last integral converges, because $\left|\frac{\sin x^2}{x^2}\right| \le \frac{1}{x^2}$ and the integral $\int_1^{+\infty} \frac{1}{x^2} dx$ converges. Also $\lim_{b \to +\infty} \frac{\sin b^2}{2b} = 0$, because $\left|\frac{\sin b^2}{2b}\right| \le \frac{1}{2b} \underset{b \to +\infty}{\to} 0$. Therefore, the integral $\int_1^{+\infty} \cos x^2 dx$ converges.

Remark 1. The statement: "if $\int_a^{+\infty} f(x)\, dx$ converges, then $f(x)$ is bounded at infinity (that is $f(x)$ is bounded on $[c, +\infty)$ for some $c > a$) ", is also false. Evidently, if $\lim_{x \to +\infty} f(x) = 0$ then $f(x)$ is bounded at infinity, and the converse is not true. So this is a strengthened (with the same conditions, but a weakened conclusion) version of the original statement. The function $f(x) = x \cos x^4$ on $[1, +\infty)$ provides a counterexample to the last statement. Indeed, by changing the variable $t = x^2$ the improper integral of $x \cos x^4$ can be reduced to the improper integral of $\cos x^2$: $\int_1^{+\infty} x \cos x^4 dx = \frac{1}{2}\int_1^{+\infty} \cos t^2 dt$. Hence, the integral converges, but for the sequence $x_k = \sqrt[4]{2k\pi}$, $\forall k \in \mathbb{N}$ it follows $\lim_{x_k \to +\infty} f(x_k) = \lim_{k \to +\infty} \sqrt[4]{2k\pi} \cos 2k\pi = +\infty$, that is, $f(x) = x \cos x^4$ is unbounded at infinity.

Remark 2. The readers familiar with infinite series can note the contrast between this statement and the known fact that a general term of a convergent infinite series must tend to zero. (Infinite series will be considered in section 6.3.)

Remark 3. The converse statement: "if $\lim_{x \to +\infty} f(x) = 0$, then $\int_a^{+\infty} f(x)\, dx$ converges", is also false. This is a more elementary result and a simple counterexample is $f(x) = \frac{1}{x}$ on $[1, +\infty)$: $\lim_{x \to +\infty} \frac{1}{x} = 0$, but $\int_1^{+\infty} \frac{1}{x} dx = \lim_{b \to +\infty} \ln b = +\infty$ diverges.

Remark 4. The related correct result states that if $\lim_{x \to +\infty} f(x) = A \neq 0$, then $\int_a^{+\infty} f(x)\, dx$ diverges.

Example 2. "If $f(x)$ is non-negative, continuous and unbounded at infinity (that is $f(x)$ is unbounded on $[c, +\infty)$ for any $c > a$), then $\int_a^{+\infty} f(x)\, dx$ diverges."

Solution.

First let us construct an auxiliary function with a small non-zero support.

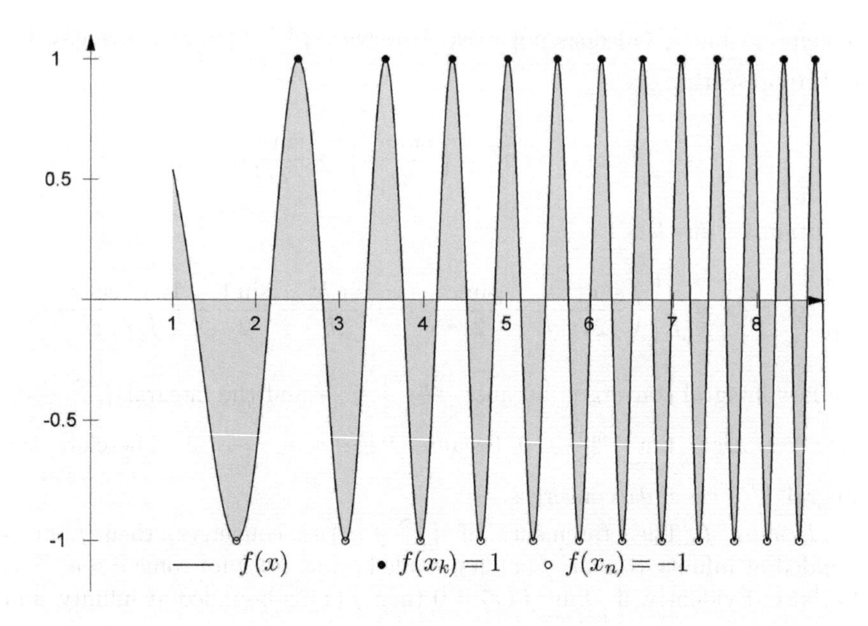

FIGURE 5.4.1: Example 1

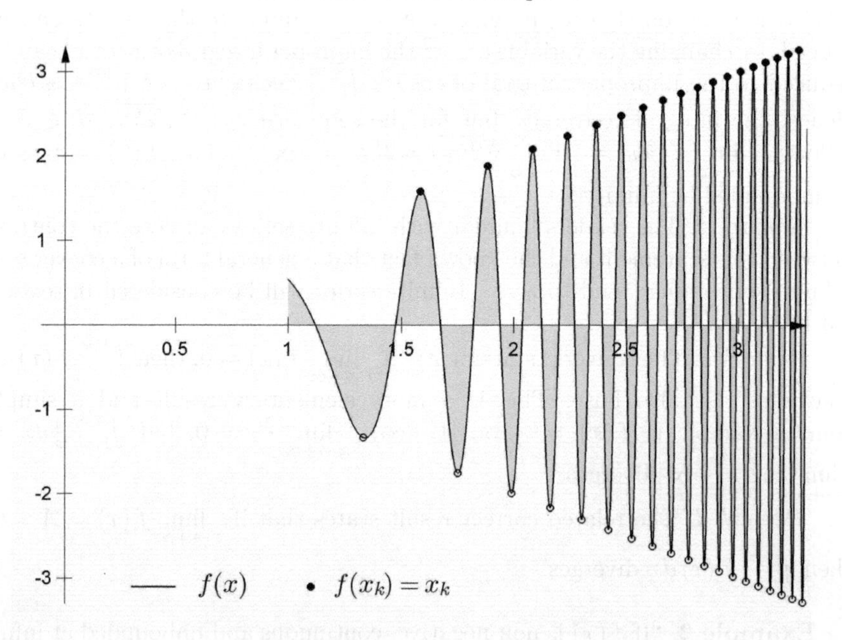

FIGURE 5.4.2: Example 1, Remark 1

Consider an arbitrary $n \in \mathbb{N}$ and choose the interval $A_n = [n - \varepsilon_n, n + \varepsilon_n]$ centered at the point n, where $\varepsilon_n > 0$ is the radius of A_n. The auxiliary function with the support A_n is defined as

$$g_n = \begin{cases} (x - n + \varepsilon_n) \cdot \frac{n}{\varepsilon_n},\ x \in [n - \varepsilon_n, n] \\ -(x - n - \varepsilon_n) \cdot \frac{n}{\varepsilon_n},\ x \in [n, n + \varepsilon_n] \\ 0,\ x \notin A_n \end{cases},$$

that is g_n has the form of a tooth of a saw, with the support A_n and height n, located above the x-axis. It is easy to evaluate the area of this tooth (triangle):

$$\int_{A_n} g_n(x)\, dx = \int_{n-\varepsilon_n}^{n+\varepsilon_n} g_n(x)\, dx = \varepsilon_n n.$$

In particular, if we choose $\varepsilon_n = \frac{1}{n \cdot 10^n}$, that is, the support radius fast decrease along the x-axis, then $\int_{n-\varepsilon_n}^{n+\varepsilon_n} g_n(x)\, dx = \frac{1}{10^n}$. Now we construct the function defined on $[0, +\infty)$: $f(x) = \begin{cases} g_n,\ x \in A_n,\ \forall n \in \mathbb{N} \\ 0,\ x \notin A \end{cases}$, where $A = \bigcup\limits_{n=1}^{\infty} A_n$. Evidently, $f(x)$ is non-negative, continuous (as gluing together the line segments) and unbounded at infinity (since $f(n) = n$). However, the improper integral $\int_0^{+\infty} f(x)\, dx$ converges. In fact, due to the evaluation of the tooth areas we have:

$$\int_0^{+\infty} f(x)\, dx = \sum_{n=1}^{\infty} \frac{1}{10^n} = \frac{1}{9}$$

(the sum of the infinite geometric progression), that is, the improper integral converges.

Remark 1. If the conditions of the statement are weakened, then of course the statement will remain false and the constructed counterexample will work, but at the same time a simpler counterexample can arise. For instance, for the false statement: "if $f(x)$ is a positive and unbounded at infinity, then $\int_a^{+\infty} f(x)\, dx$ diverges" (the condition of continuity is dropped here), there is a simpler counterexample: $f(x) = \begin{cases} e^{-x},\ x \notin \mathbb{N} \\ n,\ x \in \mathbb{N} \end{cases}$.

Remark 2. It is not difficult to construct a counterexample to a strengthened (but still false) version of the above statement: "if $f(x)$ is positive, continuous and unbounded at infinity, then $\int_a^{+\infty} f(x)\, dx$ diverges". For example, taking the advantage of the constructed counterexample, we can sum the constructed function $f(x) = \begin{cases} g_n,\ x \in A_n, \forall n \in \mathbb{N} \\ 0,\ x \notin A = \cup_{n \in \mathbb{N}} A_n \end{cases}$ with any positive function, say $g(x) = \frac{1}{x^2+1}$, continuous on $[0, +\infty)$ and whose improper integral converges. According to the arithmetic properties of improper integrals, $\int_0^{+\infty} h(x)\, dx$, $h(x) = f(x) + g(x)$, is also convergent and $h(x)$ is a positive, continuous and unbounded at infinity function.

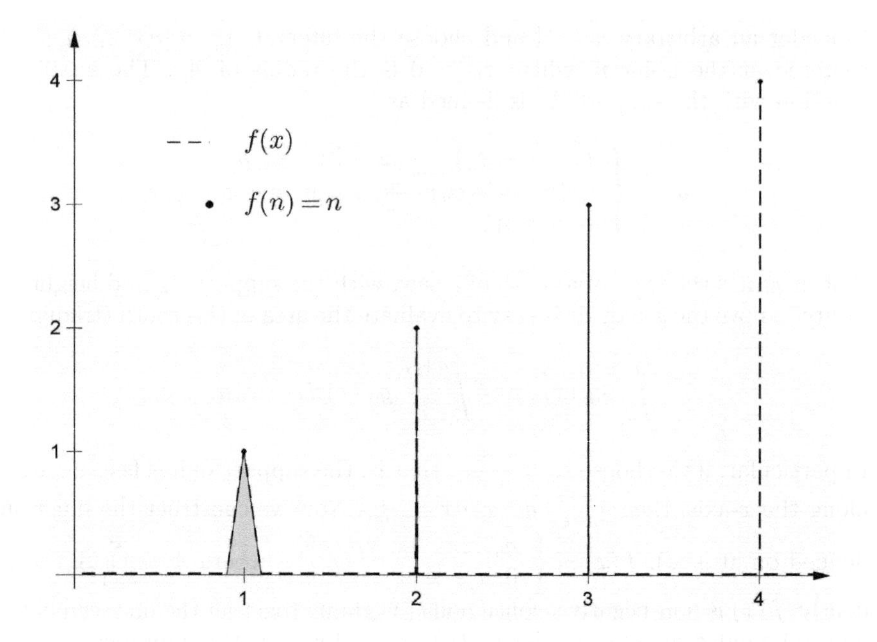

FIGURE 5.4.3: Example 2

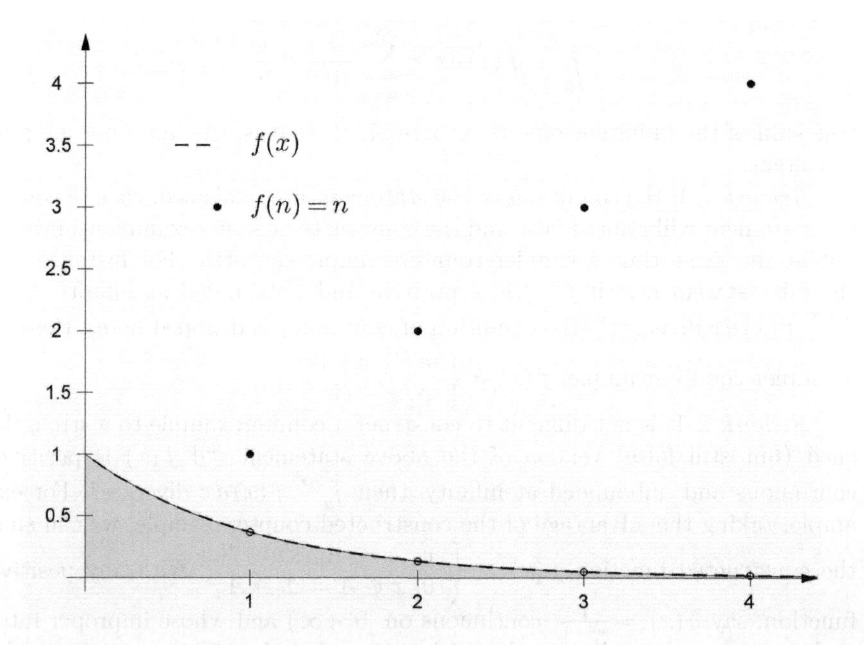

FIGURE 5.4.4: Example 2, Remark 1

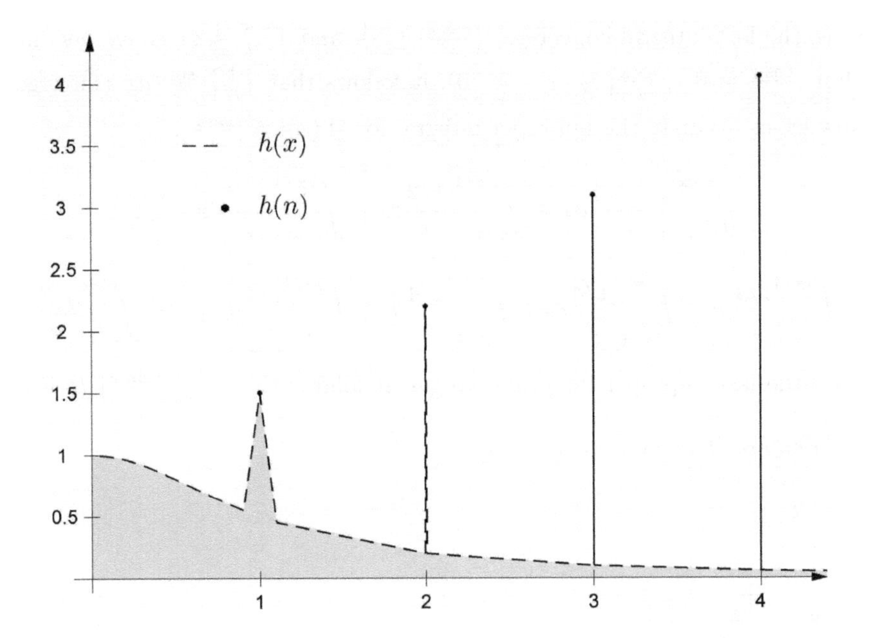

FIGURE 5.4.5: Example 2, Remark 2 and Example 3

Example 3. "If $f(x)$ is positive and continuous on $[a,+\infty)$, and $\int_a^{+\infty} f(x)\,dx$ converges, then $f(x)$ is bounded at infinity."

Solution.

This statement is closely related to that of Example 2, and the function

$$h(x) = f(x) + g(x),\ f(x) = \begin{cases} g_n,\ x \in A_n, \forall n \in \mathbb{N} \\ 0,\ x \notin A = \cup_{n\in\mathbb{N}} A_n \end{cases},\ g(x) = \frac{1}{x^2+1}$$

on $[0,+\infty)$ constructed in Remark 2 of Example 2 provides a counterexample to the current statement also (the reasoning is the same).

Remark. Of course, any weakened version of this statement is also false. For instance, for the statement: "if $f(x)$ is positive and continuous on $[a,+\infty)$, and $\int_a^{+\infty} f(x)\,dx$ converges, then $\lim\limits_{x\to+\infty} f(x) = 0$", the same counterexample can be applied (evidently, if $\lim\limits_{x\to+\infty} f(x) = 0$ then $f(x)$ is bounded at infinity, so the last statement is weaker than the original).

Example 4. "If $\int_a^{+\infty} f(x)\,dx$ converges, then $\int_a^{+\infty} |f(x)|\,dx$ also converges."

Solution.

Let us consider $f(x) = \frac{\sin x}{x}$ on $\left[\frac{\pi}{4},+\infty\right)$. Integrating by parts we obtain

$$\int_{\pi/4}^{+\infty} \frac{\sin x}{x} dx = -\frac{1}{x}\cos x\Big|_{\pi/4}^{+\infty} - \int_{\pi/4}^{+\infty} \frac{\cos x}{x^2} dx.$$

Since the last integral converges ($\left|\frac{\cos x}{x^2}\right| \le \frac{1}{x^2}$ and $\int_{\pi/4}^{+\infty} \frac{1}{x^2} dx$ converges) and $\lim\limits_{x\to+\infty} \frac{\cos x}{x} = 0$ ($\left|\frac{\cos x}{x}\right| \le \frac{1}{x} \underset{x\to+\infty}{\to} 0$), it follows that $\int_{\pi/4}^{+\infty} \frac{\sin x}{x} dx$ converges. Now let us evaluate the improper integral for $|f(x)| = \left|\frac{\sin x}{x}\right|$:

$$\int_{\pi/4}^{+\infty} \left|\frac{\sin x}{x}\right| dx \ge \int_{\pi/4}^{3\pi/4} \frac{1/2}{x} dx + \int_{5\pi/4}^{7\pi/4} \frac{1/2}{x} dx + \dots$$

$$\ge \int_{\pi/4}^{3\pi/4} \frac{1/4}{x} dx + \int_{3\pi/4}^{5\pi/4} \frac{1/4}{x} dx + \int_{5\pi/4}^{7\pi/4} \frac{1/4}{x} dx + \int_{7\pi/4}^{9\pi/4} \frac{1/4}{x} dx + \dots = \int_{\pi/4}^{+\infty} \frac{1/4}{x} dx.$$

Since the last improper integral diverges, it follows that $\int_{\pi/4}^{+\infty} \left|\frac{\sin x}{x}\right| dx$ is divergent.

Remark. The converse is correct.

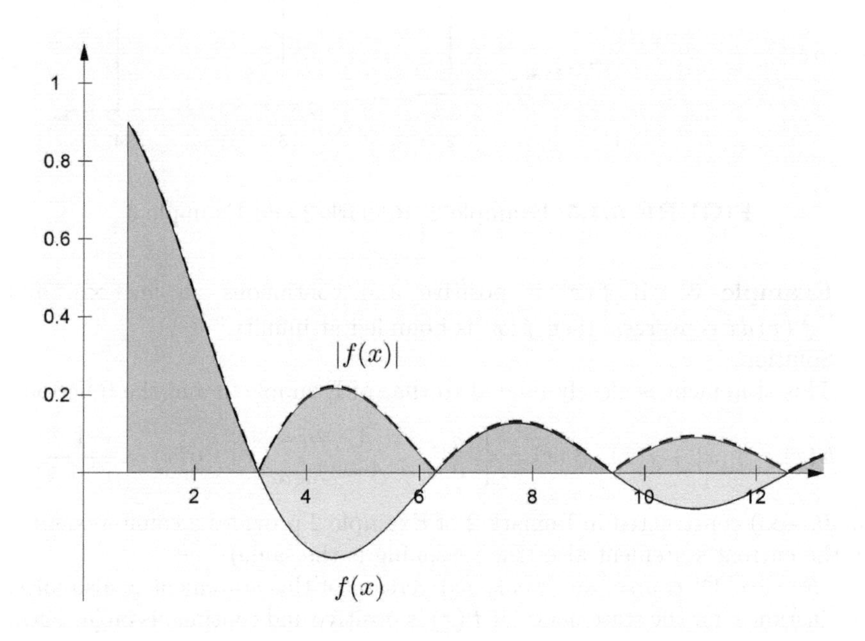

FIGURE 5.4.6: Example 4 and Example 5

Example 5. "If $\int_a^{+\infty} f(x)\, dx$ converges and $g(x)$ is bounded on $[a, +\infty)$, then $\int_a^{+\infty} f(x)\, g(x)\, dx$ also converges."

Solution.

We can reduce this case to that of Example 4 by choosing $g(x)$ in such a way that it will mimic the action of the absolute value. Let $f(x) = \frac{\sin x}{x}$ on $\left[\frac{\pi}{4}, +\infty\right)$. It was already shown in Example 4 that $\int_{\pi/4}^{+\infty} \frac{\sin x}{x} dx$ converges.

Choose now

$$g(x) = \begin{cases} 1, \; x \in [2k\pi, \; \pi + 2k\pi), \; k \in \mathbb{N} \\ -1, \; x \in [\pi + 2k\pi, \; 2\pi + 2k\pi), \; k \in \mathbb{N} \end{cases}.$$

Then $f(x)\, g(x) = \left|\frac{\sin x}{x}\right|$ and it was shown above that $\int_{\pi/4}^{+\infty} \left|\frac{\sin x}{x}\right| dx$ diverges.

Remark. If additionally $\int_a^{+\infty} |f(x)|\, dx$ converges, then the statement will be true.

Example 6. "If both $\int_a^{+\infty} f(x)\, dx$ and $\int_a^{+\infty} g(x)\, dx$ converge, then $\int_a^{+\infty} f(x)\, g(x)\, dx$ also converges."

Solution.

The counterexample is: $f(x) = \frac{\cos x}{\sqrt{x}}$ and $g(x) = \frac{\cos x}{\sqrt[3]{x}}$ on $[1, +\infty)$. Following the reasoning of Example 4, it can be shown that $\int_1^{+\infty} \frac{\cos x}{\sqrt{x}} dx$ and $\int_1^{+\infty} \frac{\cos x}{\sqrt[3]{x}} dx$ are convergent. At the same time, $\int_1^{+\infty} \frac{\cos x}{\sqrt{x}} \cdot \frac{\cos x}{\sqrt[3]{x}} dx$ is divergent due to the following considerations: $\frac{\cos^2 x}{x^{5/6}} = \frac{1}{2x^{5/6}} + \frac{\cos 2x}{2x^{5/6}}$, and the integral $\int_1^{+\infty} \frac{1}{2x^{5/6}} dx$ diverges, while $\int_1^{+\infty} \frac{\cos 2x}{2x^{5/6}} dx$ converges.

Remark. The following statement is also false: "if $\int_a^{+\infty} f(x)\, dx$ converges, then $\int_a^{+\infty} f^2(x)\, dx$ also converges". The corresponding counterexample can be given with the function $f(x) = \frac{\cos x}{\sqrt{x}}$ considered on $[1, +\infty)$.

Example 7. "If $\int_a^{+\infty} f(x)\, dx$ diverges and $g(x)$ is unbounded on $[a, +\infty)$, then $\int_a^{+\infty} f(x)\, g(x)\, dx$ also diverges."

Solution.

Let us consider $f(x) = \frac{1}{x}$ and $g(x) = \sqrt{x} \sin x$ on $[1, +\infty)$. $\int_1^{+\infty} \frac{1}{x} dx$ diverges and $g(x) = \sqrt{x} \sin x$ is unbounded on $[1, +\infty)$, since for $x_n = \frac{\pi}{2} + 2n\pi$, $n \in \mathbb{N}$ it follows that $\lim_{x_n \to +\infty} g(x_n) = \lim_{n \to +\infty} \sqrt{\frac{\pi}{2} + 2n\pi} = +\infty$. So the conditions of the statement are satisfied. Now, let us evaluate the improper integral of the product:

$$\int_1^{+\infty} f(x)\, g(x)\, dx = \int_1^{+\infty} \frac{\sin x}{\sqrt{x}} dx = -\frac{\cos x}{\sqrt{x}}\bigg|_1^{+\infty} - \frac{1}{2}\int_1^{+\infty} \frac{\cos x}{x^{3/2}} dx.$$

Since $\lim_{x \to +\infty} \frac{\cos x}{\sqrt{x}} = 0$ ($\left|\frac{\cos x}{\sqrt{x}}\right| \le \frac{1}{\sqrt{x}}$) and $\int_1^{+\infty} \frac{\cos x}{x^{3/2}} dx$ converges ($\left|\frac{\cos x}{x^{3/2}}\right| \le \frac{1}{x^{3/2}}$ and $\int_1^{+\infty} \frac{1}{x^{3/2}} dx$ converges), we can conclude that $\int_1^{+\infty} \frac{1}{x} \sqrt{x} \sin x dx$ converges.

Example 8. "If $\int_a^{+\infty} f(x) + g(x)\, dx$ converges, then $\int_a^{+\infty} f(x)\, dx$ and $\int_a^{+\infty} g(x)\, dx$ also converge."

Solution.

Both $\int_1^{+\infty} \frac{1}{x} dx$ and $\int_1^{+\infty} \frac{-1}{x} dx$ diverge, while $\int_1^{+\infty} \frac{1}{x} + \frac{-1}{x} dx = \int_1^{+\infty} 0 dx = 0$ converges.

Remark. Similar examples can be given for other arithmetic operations.

For instance, for the product the same functions can be used: $\int_1^{+\infty} \frac{1}{x} \cdot \frac{-1}{x} dx = -\int_1^{+\infty} \frac{1}{x^2} dx$ is a convergent integral.

Remark. We have focused here on improper integrals of the first kind, that is the integrals considered on infinite intervals on the x-axis (the integrand domain is unbounded). Another kind (the second) of improper integrals is related to infinite intervals on the y-axis (the integrand image is unbounded), while the domain is bounded. Since two kinds of improper integrals are intimately connected (there is a simple change of variable that transforms the second kind of improper integrals to the first kind and vice-versa), we do not repeat here similar counterexamples, which can be constructed for improper integrals of the second kind as well. To take the advantage of the counterexamples constructed for integrals of the first kind, just recall the formulas that connect two kinds of integrals: if $f(x)$ is continuous on $[a, +\infty)$ (without a loss of generality we can assume that $a > 0$), then by changing the variable $t = \frac{1}{x}$, an improper integral of the first kind $\int_a^{+\infty} f(x)\, dx$ can be reduced to the improper integral of the second kind $\int_0^b f\left(\frac{1}{t}\right) \frac{1}{t^2} dt$, $b = \frac{1}{a}$, where the lower limit is a singular point (that is the integrand is unbounded in any right-hand neighborhood of 0, but it is bounded on any interval $[c, b]$, $0 < c < b$). Moreover, both integrals converge (or diverge) simultaneously and are equal: $\int_a^{+\infty} f(x)\, dx = \int_0^b f\left(\frac{1}{t}\right) \frac{1}{t^2} dt$. Using these formulas, some of the presented examples can be easily reformulated for improper integrals of the second kind. For instance, Example 1 can be reformulated as follows: "if $\int_0^b g(t)\, dt$ converges, then $\lim\limits_{t \to 0_+} g(t) \cdot t^2 = 0$", with the corresponding counterexample $g(t) = \frac{1}{t^2} \cos \frac{1}{t^2}$ defined on $(0, 1]$. At the same time, some examples assume simpler form when solved directly for improper integrals of the second kind. For instance, Example 6 can be reformulated in the form: "if both $\int_0^b f(t)\, dt$ and $\int_0^b g(t)\, dt$ converge, then $\int_0^b f(t)\, g(t)\, dt$ also converges". However, the counterexample with the functions $f(t) = \frac{1}{t^{3/2}} \cos \frac{1}{t}$ and $g(t) = \frac{1}{t^{5/3}} \cos \frac{1}{t}$, obtained by transformation of the functions in Example 6, is somewhat involved. A much simpler counterexample for this statement can be provided with the functions $f(t) = \frac{1}{\sqrt{t}}$ and $g(t) = \frac{1}{\sqrt[3]{t^2}}$ on $(0, 1]$.

5.5 Applications

Example 1. "If a function is continuous on $[a, b]$ and continuously differentiable on (a, b), then its graph is a rectifiable curve."

Solution.

The function $f(x) = \begin{cases} x \sin \frac{1}{x}, \ x \in (0, 1] \\ 0, \ x = 0 \end{cases}$ is differentiable on $(0, 1)$

according to the arithmetic and chain rules, and the derivative $f'(x) = \sin\frac{1}{x} - \frac{1}{x}\cos\frac{1}{x}$ is continuous on $(0,1)$, according to the arithmetic and composition rules. The function is also continuous at the endpoints of $[0,1]$: at 1 by the arithmetic and composition rules and at 0 by the definition - $\lim\limits_{x\to 0_+} x\sin\frac{1}{x} = 0 = f(0)$. Hence, $f(x)$ satisfies the conditions of the statement on $[0,1]$. However, the graph of $f(x)$ is not a rectifiable curve. Indeed, the graph represents infinitely many oscillations with decreasing heights toward to 0, accumulating near the y-axis. The locations of the crests of these oscillations are given by the formula $x_n = \frac{2}{\pi+4n\pi}$, $n \in \mathbb{N}$ and their heights are $h_n = x_n$, $n \in \mathbb{N}$. It is obvious that the length of one oscillation L_n (the length of the part of the curve between two crests) is greater than the corresponding wave height h_n. So the sum of the lengths of the first N oscillations can be evaluated as follows:

$$\sum_{n=1}^{N} L_n > \sum_{n=1}^{N} h_n = \sum_{n=1}^{N} \frac{2}{\pi(1+4n)} > \frac{2}{5\pi}\sum_{n=1}^{N}\frac{1}{n}.$$

Choosing $N = 2^k$, we obtain the following evaluation for the last sum:

$$S_N = \sum_{n=1}^{N}\frac{1}{n} = 1 + \frac{1}{2} + \ldots + \frac{1}{2^k}$$

$$= 1 + \frac{1}{2} + \left(\frac{1}{3}+\frac{1}{4}\right) + \left(\frac{1}{5}+\frac{1}{6}+\frac{1}{7}+\frac{1}{8}\right) + \ldots$$

$$+ \left(\frac{1}{2^{k-1}+1} + \frac{1}{2^{k-1}+2} + \ldots + \frac{1}{2^k}\right)$$

$$> 1 + \frac{1}{2} + 2\cdot\frac{1}{4} + 4\cdot\frac{1}{8} + \ldots + 2^{k-1}\cdot\frac{1}{2^k} = 1 + k\frac{1}{2} \underset{k\to\infty}{\to} +\infty.$$

Therefore, the chosen subsequence of the sums S_N, $N = 2^k$, is divergent, that implies the divergence of the sequence S_N. Since $\sum_{n=1}^{N} L_n > \frac{2}{5\pi}S_N$, it follows that $\sum_{n=1}^{N} L_n \underset{N\to\infty}{\to} +\infty$, that is the length of the graph is infinite and the curve is not rectifiable. (The sums S_N are the partial sums of the harmonic series; see more about numerical series in section 6.3).

Remark 1. Another interesting counterexample is the function $g(x) = \begin{cases} x^2\sin\frac{1}{x^2}, \ x \in (0,1] \\ 0, \ x = 0 \end{cases}$, which is even smoother than $f(x)$ above, because it is differentiable on $[0,1]$ and continuously differentiable on $(0,1)$. However the derivative $g'(x) = \begin{cases} 2x\sin\frac{1}{x^2} - \frac{2}{x}\cos\frac{1}{x^2}, \ x \in (0,1] \\ 0, \ x = 0 \end{cases}$ is unbounded near 0, that opens the possibility for the curve to be non-rectifiable, and really the curve has an infinite length, that can be shown by applying the same arguments as for $f(x)$.

Remark 2. This statement is a wrongly "weakened" version of one of the theorems on rectifiable curves: if $f(x)$ is continuously differentiable on $[a,b]$, then its graph is a rectifiable curve. Of course, this is a sufficient condition, but we can see that in some cases the violation of the differentiability condition only at one of the endpoints can result in a non-rectifiable curve.

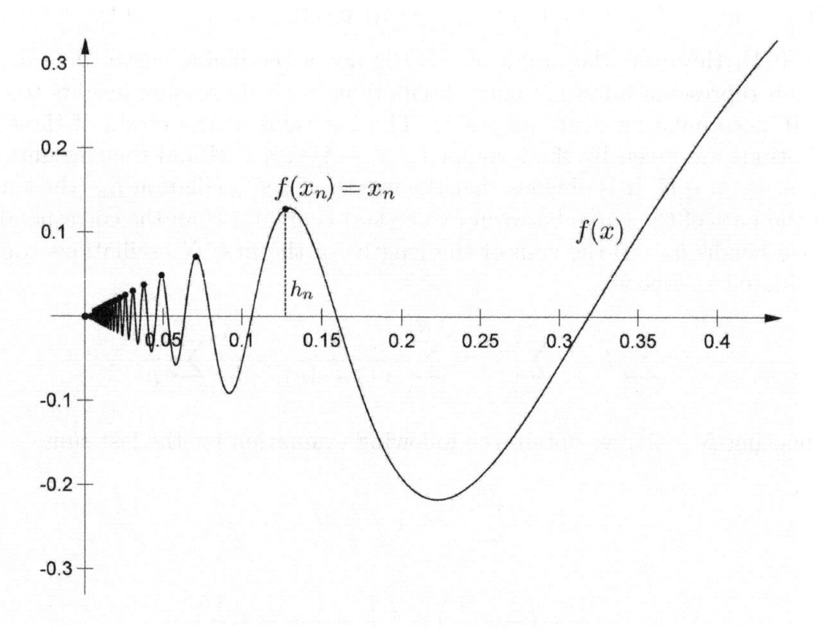

FIGURE 5.5.1: Example 1

Example 2. "If the graph of a continuous on $[a,b]$ function has a finite length, then this function is continuously differentiable on $[a,b]$."

Solution.

First of all, a very simple counterexample is $f(x) = |x|$ on $[-1,1]$, with the function non-differentiable at 0, but the length of the graph is well-defined and equal to $2\sqrt{2}$.

A much more interesting example occurs if we require that a function be non-differentiable at infinitely many points in $[a,b]$. To construct such example, consider a "saw" function located in the interval $[0,1]$ with "teeth" of variable height and width. More precisely, the n-th "tooth" is defined by the formula:

$$f_n(x) = \begin{cases} \frac{h_n}{x_{n-1/2}-x_n}(x-x_n), & x \in (x_n, x_{n-1/2}) \\ \frac{h_n}{x_{n-1/2}-x_{n-1}}(x-x_{n-1}), & x \in (x_{n-1/2}, x_{n-1}) \end{cases},$$

where h_n is the height, $[x_n, x_{n-1}]$ is the base ($x_{n-1} - x_n$ being the width), and $x_{n-1/2}$ is the centerpoint (note that $x_{n-1} > x_n$). The entire function is defined as follows: consider an infinite partition of the interval $[0,1]$

in the form $1 = x_0 > x_1 > \ldots > x_n > x_{n+1} > \ldots > 0$, with the only limit point 0 ($\lim_{n\to\infty} x_n = 0$); define the saw function in the form $f(x) = \begin{cases} f_n,\ x \in [x_n, x_{n-1}]\ ,\ n \in \mathbb{N} \\ 0\ ,\ x = 0 \end{cases}$. By construction, $f(x)$ is continuously differentiable on every interval $(x_n, x_{n-1/2})$ and $(x_{n-1/2}, x_{n-1})$, $\forall n \in \mathbb{N}$, but it is not differentiable at the points $x_{n-1/2}$ and x_n, $\forall n \in \mathbb{N}$. Moreover, if a sequence of the heights h_n is chosen in such a way that $h_n \underset{n\to\infty}{\to} 0$, then $f(x)$ is continuous on $[0,1]$. To specify the saw function we choose $x_{n-1} = \frac{1}{n}$, $x_{n-1/2} = \frac{x_{n-1}+x_n}{2}$, and $h_n = \frac{1}{2(n+1)^2}$, $\forall n \in \mathbb{N}$. Let us show that the graph of $f(x)$ is a rectifiable curve. Noting that the half-width of each tooth is $\Delta_n = x_{n-1/2} - x_n = x_{n-1} - x_{n-1/2} = \frac{1}{2n(n+1)}$, we obtain for the length of each tooth:

$$L_n = \sqrt{\left(x_{n-1} - x_{n-1/2}\right)^2 + h_n^2} + \sqrt{\left(x_{n-1/2} - x_n\right)^2 + h_n^2}$$

$$= 2\sqrt{\Delta_n^2 + h_n^2} = \sqrt{\frac{1}{n^2(n+1)^2} + \frac{1}{(n+1)^4}} \le \frac{\sqrt{2}}{n(n+1)}.$$

The sum of the lengths of the first N teeth is readily evaluated in the form:

$$\sum_{n=1}^{N} L_n \le \sum_{n=1}^{N} \frac{\sqrt{2}}{n(n+1)} = \sqrt{2}\left(1 - \frac{1}{N+1}\right) \le \sqrt{2}.$$

Since the inequality holds for $\forall N \in \mathbb{N}$, it implies that the length of the graph is limited by $\sqrt{2}$, and we conclude that the curve is rectifiable.

Remark. We see that $f(x)$ is not even piecewise smooth function (that is a function which is continuously differentiable on all the closed subintervals, which form a finite partition of $[a,b]$).

Example 3. "If a plane figure is quadrable (that is the area of a plane figure exists and is finite), then its boundary is a rectifiable curve."

Solution.

A simple example is related to improper integrals, involving a figure unbounded in the horizontal (improper integral of the first kind) or vertical (improper integral of the second kind) direction. Let us consider $f(x) = e^{-x}$ on $[0, +\infty)$. The graph of $f(x)$ is an unbounded curve, which has an infinite length:

$$L = \int_0^{+\infty} \sqrt{1 + e^{-2x}}dx = \int_1^{\sqrt{2}} \frac{t^2}{t^2 - 1}dt$$

$$= \left(t + \frac{1}{2}\ln\left|\frac{t-1}{t+1}\right|\right)\Bigg|_1^{\sqrt{2}} = \sqrt{2} - 1 + \frac{1}{2}\ln\frac{\sqrt{2}-1}{\sqrt{2}+1} - \frac{1}{2}\lim_{t\to 1_+}\ln\left|\frac{t-1}{t+1}\right| = +\infty$$

(to calculate this integral we have used the change of the variable: $t =$

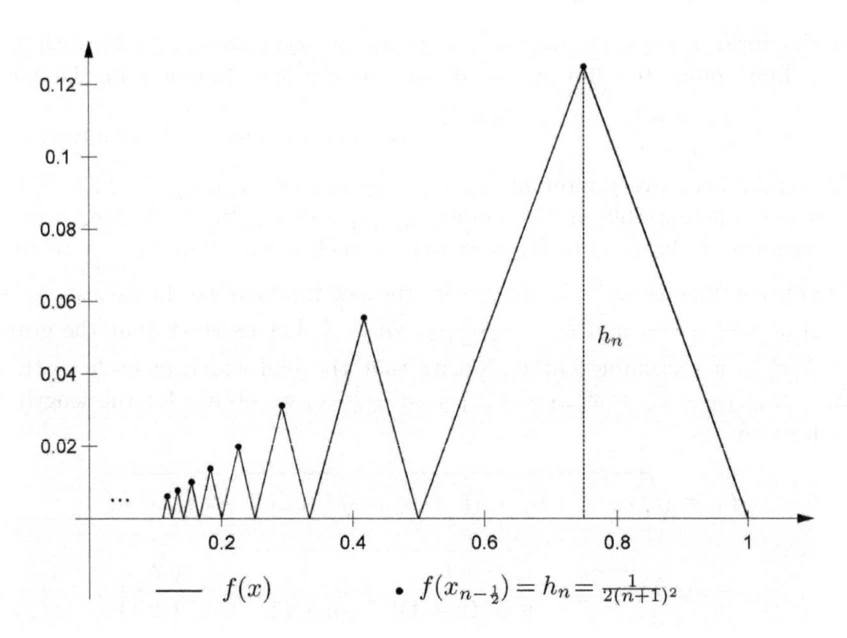

FIGURE 5.5.2: Example 2

$\sqrt{1+e^{-2x}}$, $x = -\frac{1}{2}\ln\left(t^2-1\right)$, $dx = -\frac{tdt}{t^2-1}$). However, the area of the figure below the graph of $f(x)$ (above the x-axis and at the left of the y-axis) is finite:

$$A = \int_0^{+\infty} e^{-x}dx = \left(-e^{-x}\right)\Big|_0^{+\infty} = 1.$$

More interesting, albeit a bit more involved example, is related to a bounded figure, defined by a bounded function on a finite interval $[a, b]$. Let us consider the saw function of the previous Example 2 with the following specification of the heights and widths of teeth: $x_{n-1} = \frac{1}{n}$, $x_{n-1/2} = \frac{x_{n-1}+x_n}{2}$, and $h_n = \frac{1}{2n}$, $\forall n \in \mathbb{N}$. Like in Example 2, $f(x)$ is continuous on $[0, 1]$ and is continuously differentiable on every interval $\left(x_n, x_{n-1/2}\right)$ and $\left(x_{n-1/2}, x_{n-1}\right)$, $\forall n \in \mathbb{N}$, but it is not differentiable at the points $x_{n-1/2}$ and x_n, $\forall n \in \mathbb{N}$. Let us show that the graph of $f(x)$ is a non-rectifiable curve. Indeed, for the length of each tooth we have: $L_n \geq 2h_n = \frac{1}{n}$, and consequently for the sum of the lengths of the first N teeth we obtain: $\sum_{n=1}^{N} L_n \geq \sum_{n=1}^{N} \frac{1}{n}$. Using the evaluation made in Example 1 for the sequence $S_N = \sum_{n=1}^{N} \frac{1}{n}$ we conclude that this sequence diverges $\lim\limits_{N\to\infty} S_N = +\infty$. Hence, the constructed saw curve is non-rectifiable. At the same time, continuity and non-negativity of the function $f(x)$ guarantee that the area between the graph of $f(x)$ and the x-axis (with the lateral sides $x = 0$ and $x = 1$) exists and can be expressed in the form of the Riemann integral $A = \int_0^1 f(x)\,dx$.

Remark 1. For those familiar with infinite series or willing to consult the required results in the theory of series, the last integral can be easily calculated: $A = \int_0^1 f(x)\,dx = \sum_{n=1}^{\infty} A_n$, where $A_n = \frac{1}{2} 2\Delta_n \cdot h_n = \frac{1}{4}\left(\frac{1}{n^2} - \frac{1}{n(n+1)}\right)$ is the area of the n-th tooth. Since $\sum_{n=1}^{\infty} \frac{1}{n^2} = \frac{\pi^2}{6}$ and $\sum_{n=1}^{\infty} \frac{1}{n(n+1)} = 1$, we get $A = \sum_{n=1}^{\infty} A_n = \frac{1}{4}\left(\frac{\pi^2}{6} - 1\right)$. (Infinite series will be considered in section 6.3.)

Remark 2. Another interesting example is a slight modification of the function in Example 1: $f(x) = \begin{cases} x \sin^2 \frac{1}{x}, \ x \in (0,1] \\ 0, \ x = 0 \end{cases}$. Just like in Example 1, it can be readily shown that this function is continuous on $[0,1]$ and differentiable on $(0,1]$. Also the graph of $f(x)$ has the crests located at $x_n = \frac{2}{\pi + 2n\pi}$, $n \in \mathbb{N}$ and heights $h_n = x_n$, $n \in \mathbb{N}$. So the same evaluation for the sum of the lengths of the first N oscillations can be used:

$$\sum_{n=1}^{N} L_n > \sum_{n=1}^{N} h_n = \sum_{n=1}^{N} \frac{2}{\pi(1+2n)} > \frac{2}{3\pi} \sum_{n=1}^{N} \frac{1}{n}.$$

Recalling that $S_N = \sum_{n=1}^{N} \frac{1}{n} \underset{n \to \infty}{\to} +\infty$, we conclude that the graph of this function is a non-rectifiable curve. On the other hand, the area bounded by the graph of $f(x)$ and the x-axis (with the lateral sides $x = 0$ and $x = 1$) exists and is equal to $A = \int_0^1 f(x)\,dx$, because the function $f(x)$ is continuous and non-negative on $[0,1]$. It is easy to see that this area is less then $1/2$, since the figure is contained within the triangle with the vertices $(0,0)$, $(1,0)$ and $(1,1)$.

Remark 3. According to the topics of this section we confine ourselves to the cases when a plane figure is defined by a real function of one variable $f(x)$. However, there is a variety of other examples related to the definition of plane curves in another form (parametric, geometric, etc.). For example, the Koch snowflake, T-Square, Dragon curve and many other fractals have these properties, some of them (for example, the Sierpinski gasket) even have a zero area along with infinite boundary length.

Example 4. "If $f(x)$ is continuous on $[a,b]$, then the area between the graph of $f(x)$ and the x-axis (and the lateral walls $x = a$ and $x = b$) can be found by the formula $A = \int_a^b f(x)\,dx$."

Solution.

A simple counterexample is $f(x) = x$ on $[-1,1]$. In this case the area of the figure is the sum of the areas of two equal triangles: $A = 2 \cdot \frac{1}{2} = 1$, but the Riemann formula gives $\int_{-1}^{1} x dx = 0$.

Remark. The condition of non-negativity of $f(x)$ is missing in the statement conditions, or without non-negativity the correct formula is $A = \int_a^b |f(x)|\,dx$.

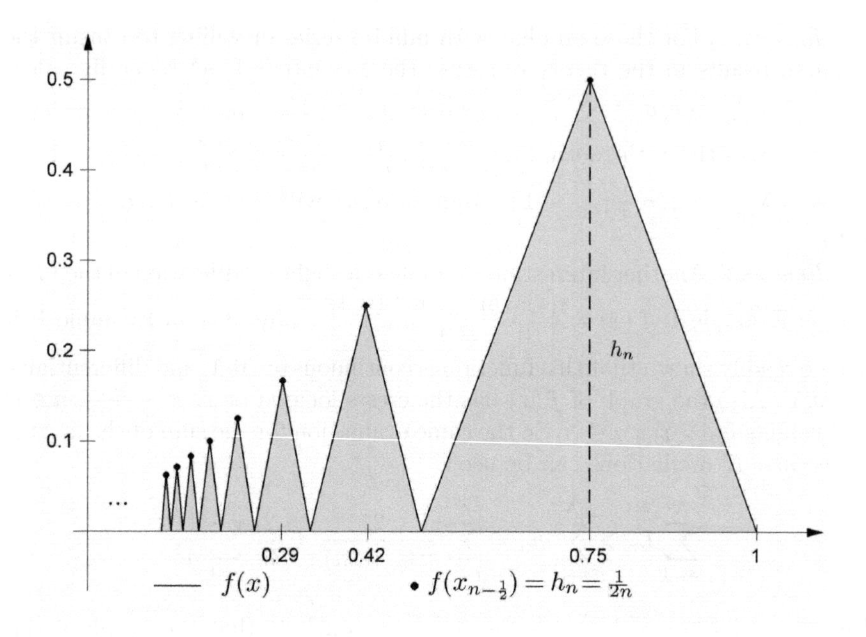

FIGURE 5.5.3: Example 3

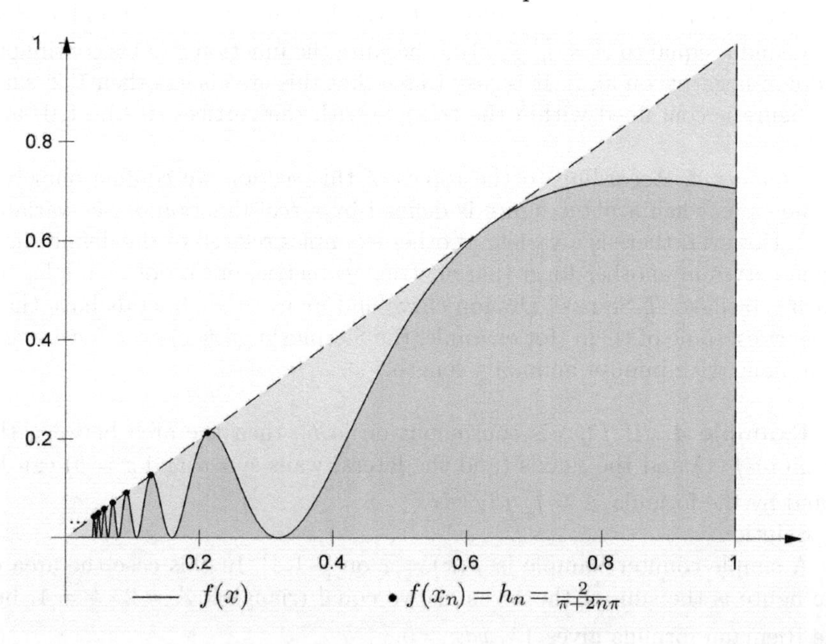

FIGURE 5.5.4: Example 3, Remark 2

Example 5. "If a plane figure has a finite area, then the corresponding solid of revolution has a finite surface area."

Solution.

Let us consider the plane figure between the graph of the function $f(x) = -\ln x$ considered on $(0, 1]$, the segment $[0, 1]$ of the x-axis and the y-axis. This unbounded figure has a finite area:

$$A = \int_0^1 -\ln x dx = -x\ln x|_0^1 + \int_0^1 dx$$

$$= 1 + \lim_{x\to 0_+} x\ln x = 1 + \lim_{x\to 0_+} \frac{\ln x}{1/x} = 1 + \lim_{x\to 0_+} \frac{1/x}{-1/x^2} = 1$$

(the integration by parts was applied to the first integral and L'Hospital's rule was applied to calculate an indeterminate form). At the same time, the solid of revolution, obtained by rotating this plane figure about the x-axis, has an infinite surface area:

$$S = 2\pi \int_0^1 (-\ln x)\sqrt{1+\frac{1}{x^2}}dx > -2\pi \int_0^1 \frac{\ln x}{x}dx$$

$$= -\pi \ln^2 x\Big|_0^1 = \pi \lim_{x\to 0_+} \ln^2 x = +\infty.$$

Remark 1. It is simple to show that the volume of this solid of revolution is finite:

$$V = \pi\int_0^1 \ln^2 x dx = \pi x\ln^2 x\Big|_0^1 - 2\pi\int_0^1 \ln x dx = \pi x\ln^2 x\Big|_0^1 - 2\pi x\ln x|_0^1 + 2\pi\int_0^1 dx$$

$$= 2\pi - \pi \lim_{x\to 0_+} \frac{\ln^2 x - 2\ln x}{1/x} = 2\pi - \pi \lim_{x\to 0_+} \frac{2\ln x\cdot 1/x - 2\cdot 1/x}{-1/x^2}$$

$$= 2\pi - 2\pi \lim_{x\to 0_+} \frac{\ln x - 1}{-1/x} = 2\pi - 2\pi \lim_{x\to 0_+} \frac{1/x}{1/x^2} = 2\pi - 2\pi \lim_{x\to 0_+} x = 2\pi$$

(the integration by parts was applied twice to find the integral, and then L'Hospital's rule was used twice to calculate an indeterminate form).

Remark 2. It is easy to see that the length of the corresponding curve is infinite:

$$L = \int_0^1 \sqrt{1+\frac{1}{x^2}}dx = \int_0^1 \frac{x^2+1}{x\sqrt{x^2+1}}dx = \int_0^1 \frac{x}{\sqrt{x^2+1}}dx + \int_0^1 \frac{dx}{x\sqrt{x^2+1}}$$

$$= \sqrt{x^2+1}\Big|_0^1 + \int_1^{+\infty} \frac{dt}{\sqrt{t^2+1}} = \sqrt{2} - 1 + \ln\left(t+\sqrt{t^2+1}\right)\Big|_1^{+\infty}$$

$$= \sqrt{2} - 1 - \ln\left(1+\sqrt{2}\right) + \lim_{t\to+\infty} \ln\left(t+\sqrt{t^2+1}\right) = +\infty$$

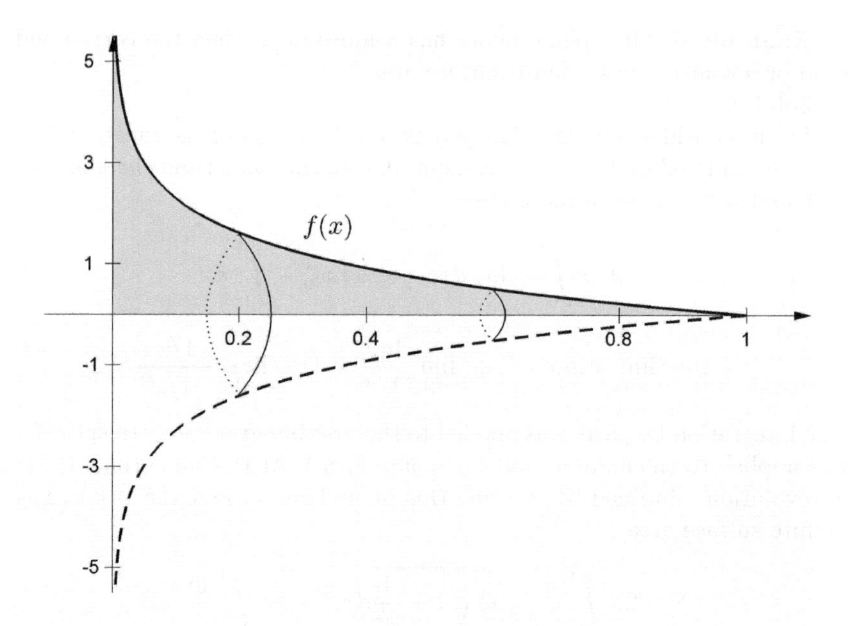

FIGURE 5.5.5: Example 5

(the change of variable $x = 1/t$ was applied to one of the integrals).

Example 6. "If a solid of revolution has a finite volume, then the plane figure, which gave rise to this solid, is quadrable."

Solution.

Let us consider $f(x) = \frac{1}{x}$ on $[1, +\infty)$. For any $b > 1$, the curve on the subinterval $[1, b]$ has the length $L(b) = \int_1^b \sqrt{1 + \frac{1}{x^4}} dx$, and the area of the region between the graph and the x-axis is $A(b) = \int_1^b \frac{1}{x} dx = \ln b$. For the same subinterval $[1, b]$, the surface area and the volume of the solid of revolution (obtained by rotating this region around the x-axis) are, respectively:

$$S(b) = 2\pi \int_1^b \frac{1}{x} \sqrt{1 + \frac{1}{x^4}} dx > 2\pi \ln b$$

and

$$V(b) = \pi \int_1^b \frac{1}{x^2} dx = \pi \left(1 - \frac{1}{b}\right).$$

Letting $b \to +\infty$, we obtain the infinite plane region (with an infinite length of two boundaries) with an infinite area $A = \lim_{b \to +\infty} A(b) = +\infty$, which generates the solid of revolution with an infinite surface area $S = \lim_{b \to +\infty} S(b) = +\infty$, but with the finite volume $V = \lim_{b\to+\infty} V(b) = \pi$. (This solid of revolution is called Gabriel's horn or Torricelli's trumpet).

Remark 1. The integral of the surface area can be calculated directly:

$$S = 2\pi \int_1^{+\infty} \frac{1}{x}\sqrt{1+\frac{1}{x^4}}dx = \frac{\pi}{2}\int_0^1 \frac{1}{t}\sqrt{1+t}dt$$

$$= \pi \int_1^{\sqrt{2}} \frac{u^2}{u^2-1}du = \pi\left(u + \frac{1}{2}\ln\left|\frac{u-1}{u+1}\right|\right)\Bigg|_1^{\sqrt{2}}$$

$$= \pi\left(\sqrt{2}-1\right) + \frac{\pi}{2}\ln\left|\frac{\sqrt{2}-1}{\sqrt{2}+1}\right| - \frac{\pi}{2}\lim_{u\to1_+}\ln\left|\frac{u-1}{u+1}\right| = +\infty.$$

Here, in the first integral we have applied the change of the variable $t = \frac{1}{x^4}$, $dx = -\frac{1}{4}t^{-5/4}dt$ and in the second integral one more change of the variable $u = \sqrt{1+t}$, $t = u^2-1$, $dt = 2udu$.

Remark 2. Similar results can be obtained for the family of functions $f(x) = \frac{1}{x^\alpha}$, $\alpha \in \left(\frac{1}{2}, 1\right]$ on $[1, +\infty)$.

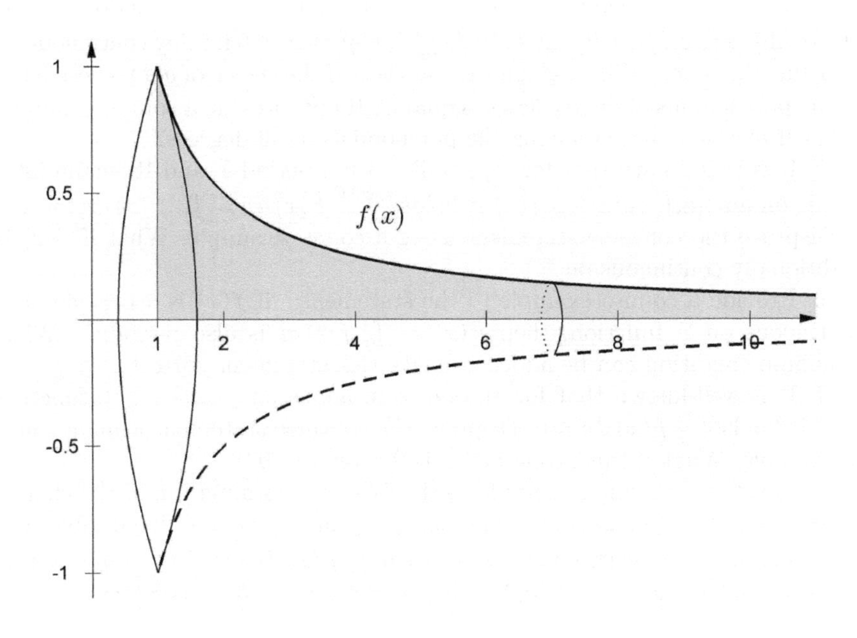

FIGURE 5.5.6: Example 6

Exercises

1. Suppose that the Riemann integral of $f(x)$ on $[a,b]$ is defined as follows: $\int_a^b f(x)\,dx = \lim\limits_{n\to+\infty} S(f)$, where $S(f)$ is the Riemann sum corresponding to a partition with n subintervals. Is this definition correct? If not, give a counterexample. For what kind of partitions are the tendencies $n \to +\infty$ and $\Delta \to 0$ equivalent?

2. Suppose that the condition of independence of the limit of the Riemann sums from the choice of points $c_i \in [x_{i-1}, x_i]$ is not required. Show by example that this will violate the definition of the Riemann integral.

3. Give a counterexample to the statement: "if $f(x)$ is Riemann integrable and non-zero on $[a,b]$, then $\frac{1}{f(x)}$ is also integrable on $[a,b]$". What condition should be added to make this statement correct? Generalize the statement, counterexample and additional correctness condition on the case of the ratio of two Riemann integrable functions.

4. The known theorem states that a continuous on $[a,b]$ function $f(x)$ is identically zero on $[a,b]$ if, and only if, $\int_a^b f(x)g(x)dx = 0$ for any continuous on $[a,b]$ function $g(x)$. Will this result remain true if the choice of $g(x)$ is restricted to the polynomials of degree less or equal n? If not, provide a counterexample. What if $g(x)$ are chosen among the polynomials of all degrees?

5. It is well-known that for a periodic with a period T and Riemann integrable on any $[a,b]$ function $f(x)$ it holds $\int_c^{c+T} f(x)\,dx = \int_0^T f(x)\,dx$ for any c. Disprove the converse statement using a counterexample. What if $f(x)$ is additionally continuous on $\mathbb{R}$?

6. Provide a counterexample to the statement: "if $f(x)$ is a periodic and continuous on $\mathbb{R}$ function, then $F(x) = \int_0^x f(t)\,dt$ is also periodic". What minimum condition can be added to make this statement correct?

7. It is well-known that for an odd, continuous on $[-c,c], c>0$ function $f(x)$ it holds $\int_{-c}^{c} f(x)\,dx = 0$. Disprove the converse statement using a counterexample. What if this property holds for any $c>0$?

8. Construct a counterexample to the following formulation of the change of variable: "if $f(x)$ is continuous on $[a,b]$, and $\varphi(t)$ is differentiable and increasing on $[\alpha,\beta]$ with image $[a,b]$, then $\int_a^b f(x)\,dx = \int_\alpha^\beta f(\varphi(t))\,\varphi'(t)\,dt$. (Hint: consider $f(x) \equiv 1$ on $[0,1]$ and $\varphi(t)$ defined on $[0,1]$ as follows

$$\varphi(t) = \begin{cases} 0,\ t=0 \\ \frac{1}{3^n},\ t \in \left(\frac{1}{2^n}, \frac{1}{2^{n-1}} - \frac{1}{n3^n}\right],\ n \in \mathbb{N} \\ \frac{1}{3^n} + \frac{2}{3^n}\cos^2\frac{\pi n 3^n (t-1/2^{n-1})}{2},\ t \in \left(\frac{1}{2^{n-1}} - \frac{1}{n3^n}, \frac{1}{2^{n-1}}\right],\ n \in \mathbb{N} \end{cases}.$$

It can be shown that $\varphi(t)$ is continuous, differentiable and increasing on $[0,1]$, however $f(\varphi(t))\varphi'(t) = \varphi'(t)$ is not integrable since it is not bounded on $[0,1]$.)

9. It is well-known that for a continuous on $[a,b]$ function $f(x)$ the integral

with the variable upper limit $F(x) = \int_a^x f(t)dt$ is a differentiable function on $[a, b]$ and $F'(x) = f(x)$. Show that if the condition of continuity is relaxed to integrability on $[a, b]$, then the statement is not true anymore. On the other hand, if $f(x)$ is not continuous at $c \in [a, b]$, it still may happen that $F(x)$ is differentiable at c. Formulate both results as counterexamples. (Hint: use the Heaviside function on $[-1, 1]$ and show that $F(x)$ is not differentiable at 0; use $f(x) = \begin{cases} 0, \ x = 0 \\ \cos\frac{1}{x}, \ x \neq 0 \end{cases}$ on $[-1, 1]$ and show that $F(x)$ is differentiable at 0 .)

10. Use $\operatorname{sgn} x$ on a corresponding interval as another counterexample to the statement in Example 15, section 5.3.

11. The generalized Mean Value theorem states that if $f(x)$ is continuous and $g(x)$ is integrable and sign-preserving on $[a, b]$, then there exists $c \in [a, b]$ such that $\int_a^b f(x)g(x)dx = f(c)\int_a^b g(x)dx$. Show that the condition of continuity of $f(x)$ cannot be relaxed and the condition of sign-preserving cannot be omitted even for continuous $g(x)$. Formulate these results in the form of counterexamples. (Hint: for continuous but not sign-preserving function use $g(x) = \sin x$, $f(x) = x$ on $[-\pi, \pi]$.)

12. It is well-known that for a Riemann integrable on $[a, b]$ function $f(x)$ there is a point $c \in [a, b]$ such that $\int_a^c f(x)dx = \int_c^b f(x)dx$. Use an example to show that it is not always possible to choose $c \in (a, b)$. (Hint: consider $f(x) = \sin x$ on $[0, 2\pi]$.)

13. If $f(x)$ is a convex function on $[a, b]$ then $f\left(\frac{a+b}{2}\right)(b-a) \leq \int_a^b f(x)dx \leq \frac{f(a)+f(b)}{2}(b-a)$. Show that the converse is not true.

14. Show that the statement "if $\int_a^{+\infty} f(x)\,dx$ diverges, then $\lim\limits_{x\to+\infty} f(x) = A \neq 0$ " is not correct.

15. Suppose $f(x)$ is continuous and monotone on $[a, +\infty)$. Show that the condition $\lim\limits_{x\to+\infty} xf(x) = 0$ does not imply the convergence of $\int_a^{+\infty} f(x)\,dx$. (Notice that the converse is true.)

16. Suppose $f(x)$ is continuous and monotone on $(0, b]$ and 0 is a singular point of $f(x)$. Show that the condition $\lim\limits_{x\to 0_+} xf(x) = 0$ does not imply the convergence of $\int_0^b f(x)\,dx$. (Notice that the converse is true.) Compare with the previous exercise.

17. Provide a counterexample to the following statement: "if $f(x) \leq g(x)$ on $[a, +\infty)$ and $\int_a^{+\infty} g(x)dx$ converges, then $\int_a^{+\infty} f(x)dx$ also converges". What if the condition $f(x) \leq g(x)$ is substituted by $|f(x)| \leq |g(x)|$ on $[a, +\infty)$? Compare with the General Comparison theorem. (Hint: for the first case choose $f(x) = -x$ and $g(x) = \frac{1}{x^2}$ on $[1, +\infty)$; and for the second try $f(x) = \frac{|\sin x|}{2x}$ and $g(x) = \frac{\sin x}{x}$ on $[1, +\infty)$.)

18. Provide a counterexample to the following statement: "if $f(x)$ and $g(x)$ are integrable on any $[c, b] \subset (a, b]$, $f(x) \leq g(x)$ on $(a, b]$ and $\int_a^b g(x)dx$ converges, then $\int_a^b f(x)dx$ also converges". What if the condition $f(x) \leq g(x)$

is substituted by $|f(x)| \le |g(x)|$ on $(a,b]$? Give a correct formulation of the General Comparison theorem for the improper integrals of the second type. (Hint: for simplicity suppose that $f(x)$ and $g(x)$ are continuous on $(a,b]$ and follow the suggestions in the previous exercise.)

19. Show that the following statement is false: "let $f(x)$ be integrable on any $[c,b]$, $c \in (0,b)$ and 0 be a singular point of $f(x)$; if $\int_0^b f(x)dx$ converges, then $|f(x)| < \frac{1}{x}$" (Hint: try $f(x) = \frac{\sin(1/x)}{x^{3/2}}$ on $(0,1]$.)

20. Show that the following statement is false: "let $f(x)$ be integrable on any $[c,b]$, $c \in (0,b)$ and 0 be a singular point of $f(x)$; if $\int_0^b f(x)dx$ diverges, then $|f(x)| > \frac{1}{x}$" (Hint: consider $f(x) = \frac{1}{x\ln x}$ on $(0,\frac{1}{e}$.)

21. Suppose $f(x)$ is continuous on $(a,b]$, the improper integral $\int_a^b f(x)\,dx$ converges and $g(x)$ is continuous on $[a,b]$. Is it sufficient to guarantee the convergence of $\int_a^b f(x)g(x)dx$? (Hint: consider $f(x) = \frac{1}{\sqrt{x^3}} \sin\frac{1}{x}$ and $g(x) = \begin{cases} 0, \, x=0 \\ \sqrt{x}\sin\frac{1}{x}, \, x \neq 0 \end{cases}$ on $[0,1]$.)

22. Does the convergence of the improper integral $\int_a^b f(x)dx$ imply the convergence of $\int_a^b |f(x)|dx$? If not, formulate in the form suitable for a counterexample and provide one.

23. Show by example that the convergence of the improper integrals $\int_a^b f(x)dx$ and $\int_a^b g(x)dx$ does not guarantee the convergence of $\int_a^b f(x)g(x)dx$.

24. Show by example that the divergence of the improper integrals $\int_a^b f(x)dx$ and $\int_a^b g(x)dx$ does not imply the divergence of $\int_a^b f(x)g(x)dx$. (Hint: choose $x \in (0,1]$ and consider the functions

$$f(x) = \begin{cases} f_n(x), \, x \in [x_n - \delta_n, x_n + \delta_n] \\ 0, \, \textit{otherwise} \end{cases}$$

and

$$g(x) = \begin{cases} g_n(x), \, x \in [x_{n+1/2} - \delta_n, x_{n+1/2} + \delta_n] \\ 0, \, \textit{otherwise} \end{cases},$$

where $x_n = \frac{1}{n}$, $\delta_n = \frac{1}{8n^2}$, $x_{n+1/2} = \frac{x_n + x_{n+1}}{2}$,

$$f_n(x) = 2n\pi\cos(\frac{\pi}{2\delta_n}(x - x_n)) \cdot H(x - x_n + \delta_n)H(x_n + \delta_n - x),$$

$$g_n(x) = 2n\pi\cos(\frac{\pi}{2\delta_n}(x - x_{n+1/2})) \cdot H(x - x_{n+1/2} + \delta_n)H(x_{n+1/2} + \delta_n - x),$$

$\forall n \in \mathbb{N}$, and $H(x)$ is the Heaviside function.)

25. By definition, the principal value of the improper integral $\int_{-\infty}^{+\infty} f(x)dx$ is $\lim\limits_{a\to+\infty} \int_{-a}^{a} f(x)dx$. It is evident, that the convergence of $\int_{-\infty}^{+\infty} f(x)dx$ imply the convergence of its principal value. Show by example that the converse is not true. Construct a similar definition for the principal value of the improper integral of the second type and give a corresponding counterexample.

26. Elaborate technical details of the proof for the counterexample in Example 1, section 5.5.

27. Show that the graph of the function $f(x) = \begin{cases} x\cos\frac{1}{x}, & x \in (0,1] \\ 0, & x = 0 \end{cases}$ on $[0,1]$ has an infinite length, but the area of the corresponding plane figure between the graph and x-axis is finite. Thus, this is one more counterexample to the statement in Example 3, section 5.5.

28. Show that the plane region between the graph of $f(x) = \frac{1}{\sqrt{x}}$ and x-axis on the interval $(0,1]$ has the finite area, but the corresponding solid formed by revolving this region around x-axis has an infinite volume. Formulate this result as a counterexample. Show that this solid of revolution has an infinite surface area. Therefore, it also provides a counterexample to the statement in Example 5, section 5.5 (although in Example 5 the volume is finite).

29. Show that the graph of $f(x) = \frac{x}{x^2+1}$ considered on $[0,+\infty)$ generates the plane figure (between the graph and x-axis) with infinite area and corresponding solid of revolution (rotating around x-axis) with infinite surface area, but finite volume. Thus, this is one more counterexample to the statement in Example 6, section 5.5.

Chapter 6

Sequences and series

6.1 Elements of theory

Numerical sequences

Remark. We will consider here only infinite real sequences.

Sequence definition. A *sequence* is a function whose domain is $\mathbb{N}$. The standard notation is: $a_1, a_2, \ldots, a_n, \ldots$ or $\{a_n\}_{n=1}^{\infty}$, or a_n, $n \in \mathbb{N}$ or simply a_n when this notation is clear within a specific context.

Remark. The definition is usually extended to domains which include additional integers or exclude some naturals. The point in these extensions is to maintain the domain with the properties of ordering of $\mathbb{N}$: all the elements of a sequence should be indexed (with an integer index), the first element with an initial index must exist, and each following element has the index equal to the index of the preceding element plus one. (Evidently the number of non-natural indices is finite or empty).

Convergent sequence. A sequence a_n is *convergent* if there exists a finite limit $\lim\limits_{n\to\infty} a_n = A$. Otherwise a sequence is *divergent*.

Remark 1. A standard concept of the limit at infinity, slightly adopted to sequences, is used here: for every $\varepsilon > 0$ there exists a natural number (index) N such that for all natural numbers (indices) $n > N$ it follows that $|a_n - A| < \varepsilon$.

Remark 2. For the sequences the only limit point in domain is positive infinity, so the notation $n \to +\infty$ is frequently abbreviated to $n \to \infty$.

Subsequences. The *subsequence* of a sequence a_n is a new sequence composed of some elements of the original sequence, keeping their order.

Monotonicity and boundedness. In the same way as for general functions, the properties of *monotonicity (increasing/decreasing) and boundedness/unboundedness* are defined for sequences.

The usual properties (arithmetic and comparison) of the limits are valid for the convergent sequences (see the list of properties in subsection 2.1.2).

Additional (specific) properties of convergent sequences:
1) If $\lim_{x\to+\infty} f(x) = A$ and $a_n = f(n)$, then $\lim_{n\to\infty} a_n = A$.
2) If a sequence a_n is increasing (decreasing) and bounded above (below), then it is convergent.
3) A convergent sequence a_n is bounded.
4) Any bounded sequence has a convergent subsequence.

Cauchy's criterion. A sequence a_n converges if, and only if, for any $\varepsilon > 0$ there exists N such that for all $n > N$ and all $p \in \mathbb{N}$ it holds $|a_{n+p} - a_n| < \varepsilon$.

Numerical series: convergence and elementary properties

Remark. We will consider here only infinite real series.

Series definition. The sum of all the elements of a sequence is called *series.* The standard notation is: $\sum_{n=1}^{+\infty} a_n$ or $\sum_{n=i}^{+\infty} a_n$, where i is the initial index, or simply $\sum a_n$.

Partial sums. The sum of the first k terms of a series is called the k-th *partial sum.* Notation: $S_k = \sum_{n=1}^{k} a_n$ or $S_k = \sum_{n=i}^{k+i-1} a_n$.

Remark. Even when a series starts from the index i, the k-th partial sum can be defined as $S_k = \sum_{n=i}^{k} a_n$. The choice of one of two options for nominating the partial sums has no influence on the main properties of series, including convergence/divergence.

Convergence. A series is *convergent* if the sequence of its partial sums is convergent, that is $\lim_{k\to\infty} S_k = S$, where S is called the *sum* of the series. The used notation for a convergent series: $\sum a_n = S$. Otherwise, the series is *divergent.*

Divergence test. If $\sum a_n$ is convergent, then $\lim_{n\to\infty} a_n = 0$.

Arithmetic properties. If $\sum a_n$ and $\sum b_n$ are convergent, then so are the series $\sum ca_n (c$ is a constant) and $\sum (a_n + b_n)$ and, furthermore,

1) $\sum ca_n = c\sum a_n$;
2) $\sum (a_n + b_n) = \sum a_n + \sum b_n$.

Theorem. Removal or addition of a finite number of terms does not affect the convergence or divergence of a series.

Cauchy criterion. A series $\sum a_n$ is convergent if, and only if, for $\forall \varepsilon > 0$ there is a natural number N such that for $\forall n > N$ and $\forall p \in \mathbb{N}$ it follows that $\left|\sum_{k=n+1}^{n+p} a_k\right| < \varepsilon$.

Numerical series: convergence tests

Convergence of geometric series. A geometric series $\sum q^n$ $(q = const)$ is convergent if $|q| < 1$ and divergent if $|q| \geq 1$.

Convergence of p-series. A p-series $\sum \frac{1}{n^p}$ $(p = const)$ is convergent if $p > 1$ and divergent if $p \leq 1$.

The Integral test. Suppose $f(x)$ is positive and decreasing function on $[1, +\infty)$ and $a_n = f(n)$, $n \in \mathbb{N}$. Then the improper integral $\int_1^{+\infty} f(x)\,dx$ is convergent if, and only if, the series $\sum a_n$ is convergent.

The Comparison test. Suppose $\sum a_n$ and $\sum b_n$ are *positive series* (that is, series with positive terms). Then
1) if $\sum b_n$ is convergent and $a_n \leq b_n$, $n \in \mathbb{N}$, then $\sum a_n$ is also convergent;
2) if $\sum b_n$ is divergent and $a_n \geq b_n$, $n \in \mathbb{N}$, then $\sum a_n$ is also divergent.

The Limit Comparison test. Suppose $\sum a_n$ and $\sum b_n$ are positive series. If $\lim\limits_{n\to\infty} \frac{a_n}{b_n} = c$, where c is a positive constant, then either both series converge or both diverge.

The Leibniz (alternating series) test. If an *alternating series* $\sum (-1)^{n-1} b_n$, $b_n > 0$, $n \in \mathbb{N}$ satisfies the two conditions
1) $b_{n+1} \leq b_n$, $n \in \mathbb{N}$
2) $\lim\limits_{n\to\infty} b_n = 0$,
then the series is convergent.

Absolute convergence. A series $\sum a_n$ is called *absolutely convergent* if the series of the absolute values $\sum |a_n|$ is convergent.

Theorem. If a series is absolutely convergent, then it is convergent.

Conditional convergence. A series is *conditionally convergent* if it is convergent, but not absolutely convergent.

The Cauchy (root) test. Let $\sum a_n$ be a positive series and $\lim\limits_{n\to\infty} a_n^{\frac{1}{n}} = C$. Then:
1) if $C < 1$, then $\sum a_n$converges;
2) if $C > 1$, then $\sum a_n$diverges;
3) if $C = 1$ or the limit does not exist, then the test is inconclusive.

The D'Alembert (ratio) test. Let $\sum a_n$ be a positive series and $\lim\limits_{n\to\infty} \frac{a_n}{a_{n+1}} = D$. Then:
1) if $D > 1$, then $\sum a_n$ is convergent;
2) if $D < 1$, then $\sum a_n$ is divergent;
3) if $D = 1$ or the limit does not exist, then the test is inconclusive.

The Raabe test. Consider a series $\sum a_n$, $a_n > 0$ and suppose that $\lim\limits_{n\to\infty} n \cdot \left(\frac{a_n}{a_{n+1}} - 1\right) = R$. Then:
1) if $R > 1$, then $\sum a_n$ is convergent;
2) if $R < 1$, then $\sum a_n$ is divergent;
3) if $R = 1$ or the limit does not exist, then the test is inconclusive.

The Bertrand test. Consider a positive series $\sum a_n$ and assume that $\lim\limits_{n\to\infty} \ln n \cdot \left(n \cdot \left(\frac{a_n}{a_{n+1}} - 1\right) - 1\right) = B$. Then:
1) if $B > 1$, then $\sum a_n$ converges;
2) if $B < 1$, then $\sum a_n$ diverges;
3) if $B = 1$ or the limit does not exist, then the test is inconclusive.

Remark 1. All the tests for positive series can also be used for checking the absolute convergence of arbitrary series.

Remark 2. In all convergence results, the properties required to be satisfied for all indices, can be actually satisfied for all indices starting from some index N.

Power series

Power series. The series $\sum_{n=0}^{\infty} c_n (x-a)^n$, where c_n, $n \in \mathbb{N} \cup \{0\}$ and a are real numbers, is called a *power series* centered at a (or simply a power series). The numbers c_n are called the *coefficients* and a the *centerpoint* of the series.

Convergence theorem. For a given power series $\sum_{n=0}^{\infty} c_n (x-a)^n$ there are only three options:
1) the series converges only when $x = a$,

2) the series converges for all x,
3) there is a positive number R such that the series converges for $|x-a|<R$ and diverges for $|x-a|>R$.

The number R is called the *radius of convergence*. By convention, the radius of convergence is $R=0$ in the first case and $R=\infty$ in the second case.

Differentiation and integration. If the power series $\sum_{n=0}^{\infty} c_n (x-a)^n$ has radius of convergence $R>0$, then the function $f(x)$ defined by $f(x)=\sum_{n=0}^{\infty} c_n (x-a)^n$ is differentiable on $(a-R,\, a+R)$ and

$$f'(x)=\sum_{n=1}^{\infty} nc_n (x-a)^{n-1},$$

$$\int f(x)\,dx=\sum_{n=0}^{\infty} c_n \frac{(x-a)^{n+1}}{n+1}+C.$$

The last two formulas show that the power series can be differentiated and integrated term by term within the interval of convergence. The radius of convergence of the last two series is R. The immediate consequence of this result is that $f(x)$ represented by a convergent power series is infinitely differentiable inside the interval of convergence.

Theorem. If two power series $\sum_{n=0}^{+\infty} c_n (x-a)^n$ and $\sum_{n=0}^{+\infty} b_n (x-a)^n$ are convergent on the same interval and the set of the points of this interval, on which they take the same values, has a finite limit point contained within the interval of convergence, then $a_n=b_n$, $\forall n$.

Theorem. A function can have only one expansion in a power series centered at the chosen point.

Theorem. If the radii of convergence of $\sum_{n=0}^{+\infty} a_n x^n$ and $\sum_{n=0}^{+\infty} b_n x^n$ are R_a and R_b, respectively, then the radius of convergence of $\sum_{n=0}^{+\infty} (a_n+b_n) x^n$ is

1) $R_c=\min\{R_a, R_b\}$ if $R_a\neq R_b$;
2) $R_c\geq \min\{R_a, R_b\}$ if $R_a=R_b$.

Taylor coefficients. If $f(x)=\sum_{n=0}^{\infty} c_n (x-a)^n$ is convergent for $|x-a|<R$, $R>0$, then $c_n=\frac{f^{(n)}(a)}{n!}$, $\forall n\in\mathbb{N}\cup\{0\}$.

Taylor series. The power series representation $f(x)=\sum_{n=0}^{\infty}\frac{f^{(n)}(a)}{n!}(x-a)^n$ is called the Taylor series of $f(x)$ (centered at a).

6.2 Numerical sequences

Remark. A numerical sequence is a function with a special domain. Therefore, many properties of sequences related to monotonicity, boundedness and limits (considering limits at infinity) are just repetitions of the corresponding properties of general functions. Hence, many counterexamples presented for functions in sections 1.3, 1.6, 2.2 and 2.3 can be easily adopted for the case of sequences. For this reason, we will not present in this section all such reformulations, but just give a few of them and also provide some additional examples specific for the case of sequences.

Example 1. "If a sequence is bounded, then it is convergent."
Solution.

Consider the sequence $a_n = (-1)^n$, $n \in \mathbb{N}$. This is a bounded sequence ($|(-1)^n| \le 1$, $\forall n \in \mathbb{N}$), but it has no limit, because all even elements of the sequence are equal to 1 and odd elements to -1. So whatever number a would be chosen as a candidate for the limit value, there are elements of the sequence with arbitrary large indices that have distance to a greater than $\varepsilon = 1/2$, that is the definition of the limit is not satisfied.

Another proof of the non-existence of limit can be made by using the relation between a general limit of a sequence and limits of its subsequences (partial limits): since the limits of two subsequences are different $\lim\limits_{n\to\infty} a_{2n} = \lim\limits_{n\to\infty} 1 = 1$ and $\lim\limits_{n\to\infty} a_{2n+1} = \lim\limits_{n\to\infty} (-1) = -1$ (the first containing the even terms and the second - the odd terms), the limit of the original sequence does not exist.

Remark. The converse statement is true: if a sequence has a finite limit, then it is bounded.

Example 2. "If a sequence a_n is unbounded, then $\lim\limits_{n\to\infty} a_n = (\pm)\,\infty$."
(see Example 9, section 2.3 for comparison)
Solution.

The sequence $a_n = (1 + (-1)^n)\, n$ is unbounded, because $\lim\limits_{n\to\infty} a_{2n} = \lim\limits_{n\to\infty} 4n = +\infty$, but it has no limit (either finite or infinite), because another partial limit is zero: $\lim\limits_{n\to\infty} a_{2n+1} = \lim\limits_{n\to\infty} 0 = 0$.

Remark. The converse is true.

Example 3. "If both sequences a_n and b_n are divergent, then the sequence $a_n + b_n$ is also divergent."
(see Example 1, section 2.3 for comparison)
Solution.

Let us choose $a_n = (-1)^n$ and $b_n = (-1)^{n+1}$. Then both sequences diverge (see Example 1 for details), but the sequence $c_n = a_n + b_n = 0$ is convergent to zero.

Remark 1. Analogous constructions can be made for other arithmetic operations. In particular, the product of the same divergent sequences is the convergent sequence $d_n = (-1)^n \cdot (-1)^{n+1} = -1$.

Remark 2. Under the conditions of the statement, the opposite conclusion is also false. For instance, if $a_n = b_n = (-1)^n$, then the sequence $c_n = a_n + b_n = 2 \cdot (-1)^n$ is also divergent.

Example 4. "If a sequence a_n diverges and b_n converges, then the sequence $a_n \cdot b_n$ is divergent."

(see Examples 5 and 6, section 2.3 for comparison)

Solution.

For the divergent $a_n = (-1)^n$ and convergent $b_n = \frac{1}{n}$ sequences, their product $c_n = \frac{(-1)^n}{n}$ is a convergent sequence, since the even and odd partial limits are equal:

$$\lim_{n\to\infty} c_{2n} = \lim_{n\to\infty} \frac{1}{2n} = 0 = \lim_{n\to\infty} \frac{-1}{2n+1} = \lim_{n\to\infty} c_{2n+1}.$$

Remark 1. The simplest example is when $b_n = 0$.

Remark 2. The statement will be true if to add that $\lim\limits_{n\to\infty} b_n \neq 0$.

Remark 3. Actually, it is easy to prove that for any choice of convergent and divergent sequences their sum is always divergent. The same is true for the difference, but not for the product and ratio as it is shown in the provided counterexample.

Example 5. "If the ranges of two convergent sequences a_n and b_n coincide, then $\lim\limits_{n\to\infty} a_n = \lim\limits_{n\to\infty} b_n$."

Solution.

The sequences $a_n = \begin{cases} 0, \ n=1 \\ 1, \ n>1 \end{cases}$ and $b_n = \begin{cases} 1, \ n=1 \\ 0, \ n>1 \end{cases}$ are convergent and have the same ranges, but $\lim\limits_{n\to\infty} a_n = 1 \neq 0 = \lim\limits_{n\to\infty} b_n$.

Example 6. "If a sequence a_n is bounded, positive and $(a_{n+1} - a_n) \underset{n\to\infty}{\to} 0$, then a_n is convergent."

Solution.

The sequence $a_n = 2 + \sin\sqrt{n}\frac{\pi}{2}$ is bounded (located between 1 and 3) and positive, and

$$a_{n+1} - a_n = \sin\sqrt{n+1}\frac{\pi}{2} - \sin\sqrt{n}\frac{\pi}{2}$$
$$= 2\sin\frac{\pi}{4}\left(\sqrt{n+1} - \sqrt{n}\right)\cos\frac{\pi}{4}\left(\sqrt{n+1} + \sqrt{n}\right)$$
$$= 2\sin\frac{\pi}{4}\frac{1}{\sqrt{n+1}+\sqrt{n}}\cos\frac{\pi}{4}\left(\sqrt{n+1} + \sqrt{n}\right) \underset{n\to\infty}{\to} 0$$

(the first term $\sin\frac{\pi}{4}\frac{1}{\sqrt{n+1}+\sqrt{n}}$ approaches 0, when n approaches infinity, and the second term $\cos\frac{\pi}{4}\left(\sqrt{n+1}+\sqrt{n}\right)$ is bounded). At the same time, this sequence is divergent, because for $n=(2k)^2$, $k\in\mathbb{N}$ one has $a_n = 2+\sin 2k\frac{\pi}{2} = 2$, while for $n=(4m+1)^2$, $m\in\mathbb{N}$ one gets $a_n = 2+\sin(4m+1)\frac{\pi}{2} = 3$.

Remark. The converse is true: if a_n is convergent, then a_n is bounded and $(a_{n+1}-a_n)\underset{n\to\infty}{\to} 0$.

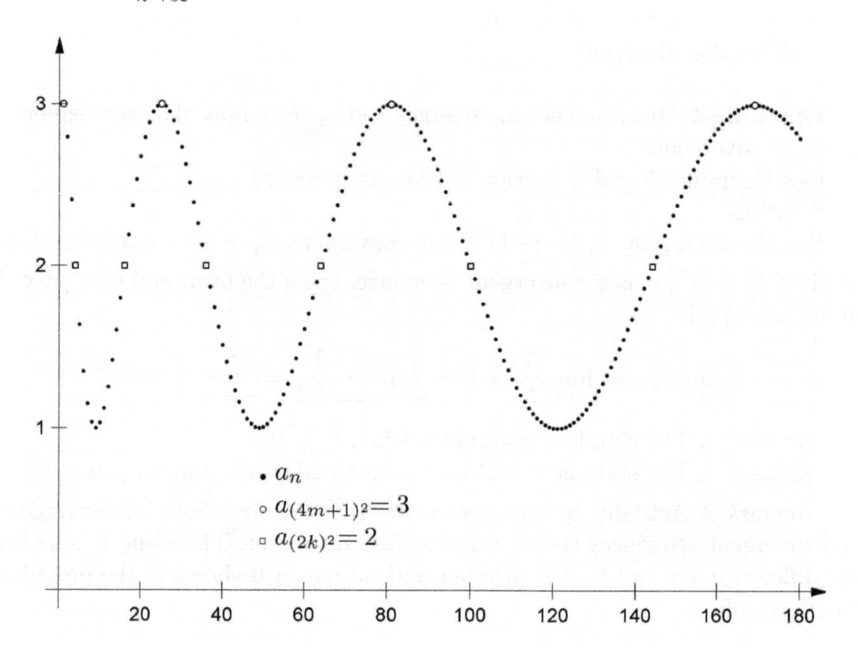

FIGURE 6.2.1: Example 6 and Example 7

Example 7. "If a sequence a_n is bounded, positive and $\frac{a_{n+1}}{a_n}\underset{n\to\infty}{\to} 1$, then a_n is convergent."

Solution.

The sequence $a_n = 2+\sin\sqrt{n}\frac{\pi}{2}$ of the previous Example 6 is bounded and positive, and satisfies the property

$$\left|\frac{a_{n+1}}{a_n}-1\right| = \left|\frac{\sin\sqrt{n+1}\frac{\pi}{2}-\sin\sqrt{n}\frac{\pi}{2}}{2+\sin\sqrt{n}\frac{\pi}{2}}\right|$$

$$= \left|\frac{2\sin\frac{\pi}{4}\frac{1}{\sqrt{n+1}+\sqrt{n}}\cos\frac{\pi}{4}\left(\sqrt{n+1}+\sqrt{n}\right)}{2+\sin\sqrt{n}\frac{\pi}{2}}\right| \le 2\left|\sin\frac{\pi}{4}\frac{1}{\sqrt{n+1}+\sqrt{n}}\right|\underset{n\to\infty}{\to} 0.$$

However, as it was shown in Example 6, this sequence is divergent.

Remark. The converse is also false. Indeed, the sequence $a_n =$

$\begin{cases} 1/n\,,\ n = 2k \\ 2/n,\ n = 2k+1 \end{cases}$, $k \in \mathbb{N}$ is convergent to zero and positive. However, the limit of $\frac{a_{n+1}}{a_n}$ does not exist, because $\lim\limits_{k\to\infty} \frac{a_{2k+1}}{a_{2k}} = \lim\limits_{k\to\infty} \frac{2\cdot 2k}{2k+1} = 2$ and $\lim\limits_{k\to\infty} \frac{a_{2k}}{a_{2k-1}} = \lim\limits_{k\to\infty} \frac{2k-1}{2k\cdot 2} = \frac{1}{2}$.

Example 8. "If $\lim\limits_{n\to\infty} a_n b_n = 0$, then $\lim\limits_{n\to\infty} a_n = 0$ or $\lim\limits_{n\to\infty} b_n = 0$ (or both)."

Solution.

The sequences $a_n = 1 + (-1)^n$ and $b_n = 1 - (-1)^n$ have no limit (see a similar Example 1 for details), but the sequence $a_n b_n$ approaches zero: $\lim\limits_{n\to\infty} a_n b_n = \lim\limits_{n\to\infty} \left(1 - (-1)^{2n}\right) = 0$.

Example 9. "If a sequence a_n has the property that $\lim\limits_{n\to\infty} (a_{n+p} - a_n) = 0\,,\ \forall p \in \mathbb{N}$ (p is fixed), then a_n converges."

Solution.

Consider $a_n = 1 + \frac{1}{2} + \ldots + \frac{1}{n}$. Let us evaluate the following subsequence

$$a_{2^k} = 1 + \frac{1}{2} + \ldots + \frac{1}{2^k}$$

$$= 1 + \frac{1}{2} + \left(\frac{1}{3} + \frac{1}{4}\right) + \left(\frac{1}{5} + \frac{1}{6} + \frac{1}{7} + \frac{1}{8}\right) + \ldots$$

$$+ \left(\frac{1}{2^{k-1}+1} + \frac{1}{2^{k-1}+2} + \ldots + \frac{1}{2^k}\right)$$

$$> 1 + \frac{1}{2} + 2\cdot\frac{1}{4} + 4\cdot\frac{1}{8} + \ldots + 2^{k-1}\cdot\frac{1}{2^k} = 1 + k\frac{1}{2} \underset{k\to\infty}{\to} +\infty.$$

Therefore, the chosen subsequence is divergent, that implies the divergence of the original sequence. At the same time, the condition of the statement is satisfied:

$$\lim_{n\to\infty} (a_{n+p} - a_n) = \lim_{n\to\infty} \left(\frac{1}{n+1} + \frac{1}{n+2} + \ldots + \frac{1}{n+p}\right) = 0$$

for any fixed natural number p.

Remark 1. Another interesting example, involving a bounded sequence, is $a_n = \cos\pi\sqrt{n}$. For any fixed $p \in \mathbb{N}$ one has

$$\lim_{n\to\infty} (a_{n+p} - a_n) = \lim_{n\to\infty} \left(-2\sin\pi\frac{\sqrt{n+p} - \sqrt{n}}{2}\,\sin\pi\frac{\sqrt{n+p} + \sqrt{n}}{2}\right) = 0,$$

since the last multiplier is bounded and

$$\lim_{n\to\infty} \sin\pi\frac{\sqrt{n+p} - \sqrt{n}}{2} = \lim_{n\to\infty} \sin\pi\frac{(\sqrt{n+p} - \sqrt{n})(\sqrt{n+p} + \sqrt{n})}{2(\sqrt{n+p} + \sqrt{n})}$$

$$= \lim_{n\to\infty} \sin \frac{p\pi}{2\left(\sqrt{n+p}+\sqrt{n}\right)} = 0.$$

However, the sequence is divergent because for $n = 4m^2$, $m \in \mathbb{N}$ one has $\lim_{m\to\infty} a_{4m^2} = \lim_{m\to\infty} \cos 2m\pi = 1$, while for $n = (2m+1)^2$, $m \in \mathbb{N}$ it follows that $\lim_{m\to\infty} a_{(2m+1)^2} = \lim_{m\to\infty} \cos(2m+1)\pi = -1$.

Remark 2. This is a wrong reformulation of Cauchy's criterion: a sequence a_n converges if, and only if, for any $\varepsilon > 0$ there exists N such that for all $n > N$ and all $p \in \mathbb{N}$ it holds $|a_{n+p} - a_n| < \varepsilon$.

Example 10. "If $\lim_{n\to\infty} a_n = 0$ and $\lim_{n\to\infty} b_n = +\infty$, then $\lim_{n\to\infty} a_n b_n = 0$." (see Examples 2 and 5, section 2.3 for comparison)

Solution.

Let $a_n = \frac{1}{n}$ and $b_n = n$. Then $\lim_{n\to\infty} a_n = 0$ and $\lim_{n\to\infty} b_n = +\infty$, but $\lim_{n\to\infty} a_n b_n = \lim_{n\to\infty} \frac{1}{n} n = 1$.

Remark 1. This is a false extension of the product rule for limits.

Remark 2. This Example is similar to Example 2, section 2.3, and an indeterminate form $0 \cdot \infty$ treated here is one of seven indeterminate forms mentioned in Remark 3 to that Example. Let us see that the situation is really undefined, that is the limit of the product strongly depends on a specification of the sequences a_n and b_n:

1) if $a_n = \frac{c}{n}$ ($c = const$) and $b_n = n$, then $\lim_{n\to\infty} a_n = 0$, $\lim_{n\to\infty} b_n = +\infty$ and $\lim_{n\to\infty} a_n b_n = \lim_{n\to\infty} c = c$.

2) if $a_n = \frac{1}{n}$ and $b_n = n^2$, then $\lim_{n\to\infty} a_n = 0$, $\lim_{n\to\infty} b_n = +\infty$ and $\lim_{n\to\infty} a_n b_n = \lim_{n\to\infty} n = +\infty$.

3) if $a_n = -\frac{1}{n}$ and $b_n = n^2$, then $\lim_{n\to\infty} a_n = 0$, $\lim_{n\to\infty} b_n = +\infty$ and $\lim_{n\to\infty} a_n b_n = \lim_{n\to\infty} (-n) = -\infty$.

4) if $a_n = \frac{(-1)^n}{n}$ and $b_n = n$, then $\lim_{n\to\infty} a_n = 0$, $\lim_{n\to\infty} b_n = +\infty$ and $\lim_{n\to\infty} a_n b_n = \lim_{n\to\infty} (-1)^n$ does not exist.

So, just like in Example 2, section 2.3, an indeterminate form $0 \cdot \infty$ can result in an arbitrary constant (including zero), infinity or even non-existence of the limit.

6.3 Numerical series: convergence and elementary properties

Example 1. "If $\lim\limits_{n\to\infty} a_n = 0$, then $\sum a_n$ converges."

Solution.

The harmonic series $\sum \frac{1}{n}$ diverges, but $\lim\limits_{n\to\infty} \frac{1}{n} = 0$. The divergence can be shown by evaluating the partial sums $H_N = \sum_{n=1}^{N} a_n$ with the indices $N = 2^k$:

$$H_{2^k} = 1 + \frac{1}{2} + \ldots + \frac{1}{2^k}$$

$$= 1 + \frac{1}{2} + \left(\frac{1}{3} + \frac{1}{4}\right) + \left(\frac{1}{5} + \frac{1}{6} + \frac{1}{7} + \frac{1}{8}\right) + \ldots$$

$$+ \left(\frac{1}{2^{k-1}+1} + \frac{1}{2^{k-1}+2} + \ldots + \frac{1}{2^k}\right)$$

$$> 1 + \frac{1}{2} + 2\cdot\frac{1}{4} + 4\cdot\frac{1}{8} + \ldots + 2^{k-1}\cdot\frac{1}{2^k} = 1 + k\frac{1}{2} \underset{k\to\infty}{\to} +\infty.$$

Remark 1. The converse is true and represents the necessary condition of convergence for numerical series. Formulated in an equivalent form: if $\lim\limits_{n\to\infty} a_n \neq 0$, then $\sum a_n$ diverges, the converse is called the Divergence test.

Remark 2. The following strengthened version is also false: "if $\lim\limits_{n\to\infty} a_n = 0$ and the sequence a_n is decreasing, then $\sum a_n$ is convergent". The same harmonic series gives a counterexample.

Example 2. "If $\lim\limits_{n\to\infty} a_n = 0$ and the sequence of the partial sums is bounded, then $\sum a_n$ is convergent."

Solution.

Let us consider the following sequence:

$$a_1 = 1\,,\ a_2 = a_3 = -\frac{1}{2}\,,\ a_4 = a_5 = a_6 = \frac{1}{3}\,,$$

$$\ldots,\ a_{\frac{n(n-1)}{2}+1} = \ldots = a_{\frac{n(n+1)}{2}} = (-1)^{n-1}\frac{1}{n}\,,\ldots.$$

We have $\lim\limits_{n\to\infty} a_n = 0$ and the sequence of the partial sums S_k is bounded by 0 and 1, but the series $\sum a_n$ diverges, because the partial sums have no limit: for the partial sums with indices $k = 1, 6, \ldots, \frac{(2n-1)2n}{2}, \ldots, \forall n \in \mathbb{N}$ one gets $S_k = 1 \underset{k\to\infty}{\to} 1$, while for the partial sums with indices $k = 3, 10, \ldots, \frac{2n(2n+1)}{2}, \ldots, \forall n \in \mathbb{N}$ one obtains $S_k = 0 \underset{k\to\infty}{\to} 0$.

Remark 1. The converse is true.

Remark 2. For positive series, the condition of boundedness of partial sums is necessary and sufficient for series convergence.

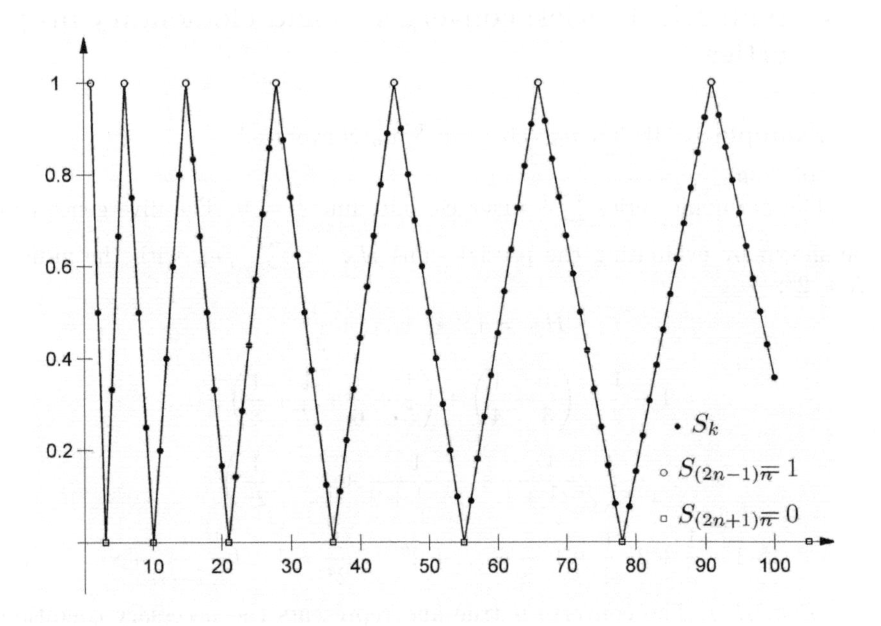

FIGURE 6.3.1: Example 2

Example 3. "If $\lim\limits_{n\to\infty} na_n = 0$, then $\sum a_n$ converges."

Solution.

Consider $a_n = \frac{1}{n\ln n}$, $n > 2$. Evidently, $\lim\limits_{n\to\infty} na_n = \lim\limits_{n\to\infty} \frac{1}{\ln n} = 0$, but the series $\sum \frac{1}{n\ln n}$ is divergent according to the integral test: $f(x) = \frac{1}{x\ln x}$ is positive and decreasing on $[3, +\infty)$ and the improper integral

$$\int_3^{+\infty} f(x)\,dx = \int_3^{+\infty} \frac{1}{x\ln x} dx = \ln(\ln x)\big|_3^{+\infty} = +\infty$$

is divergent.

Example 4. "If $\sum a_n$ converges , then $\lim\limits_{n\to\infty} na_n = 0$."

Solution.

The alternating series $\sum (-1)^n \frac{1}{\sqrt{n}}$ is convergent (both conditions of the Leibniz test are satisfied), but $\lim\limits_{n\to\infty} na_n = \lim\limits_{n\to\infty} (-1)^n \sqrt{n} = \infty$ (the limit does not exist if one considers the specific infinity - positive or negative).

Remark 1. The statement is still false if the condition of the positivity of a_n is added. In this case the counterexample is provided by the sequence

$a_n = \begin{cases} 1/n, \ n = k^2, \ k \in \mathbb{N} \\ 1/n^2, \ otherwise \end{cases}$. Indeed, $\lim\limits_{n\to\infty} na_n$ does not exist, since for $n = k^2$, $k \in \mathbb{N}$, we have $\lim\limits_{n\to\infty} na_n = \lim\limits_{k\to\infty} k^2 \frac{1}{k^2} = 1$, while for $n = k^2 + 1$, $k \in \mathbb{N}$, we get $\lim\limits_{n\to\infty} na_n = \lim\limits_{k\to\infty} \left(k^2+1\right) \frac{1}{(k^2+1)^2} = 0$. At the same time, the following evaluation for partial sums holds:

$$S_{k^2} = 1 + \frac{1}{2^2} + \frac{1}{3^2} + \frac{1}{4} + \ldots + \frac{1}{\left(\left(k-1\right)^2+1\right)^2} + \ldots + \frac{1}{\left(k^2-1\right)^2} + \frac{1}{k^2}$$

$$< \sum_{n=1}^{k^2} \frac{1}{n^2} + \sum_{n=1}^{k} \frac{1}{n^2} < 2\sum_{n=1}^{\infty} \frac{1}{n^2} = 2\frac{\pi^2}{6}.$$

Since the series is positive, it follows that $S_k < \frac{\pi^2}{3}$, $\forall k \in \mathbb{N}$, that implies convergence of the series.

Remark 2. The statement becomes true if a_n is a non-negative decreasing sequence.

Example 5. "If $\sum a_n$ and $\sum b_n$ are convergent, then $\sum a_n b_n$ is also convergent."

Solution.

If $a_n = (-1)^n \frac{1}{\sqrt{n}}$ and $b_n = (-1)^n \frac{1}{\sqrt[3]{n}}$, then the alternating series $\sum (-1)^n \frac{1}{\sqrt{n}}$ and $\sum (-1)^n \frac{1}{\sqrt[3]{n}}$ are convergent since both conditions of the Leibniz test are satisfied for these two series:

$$\lim_{n\to\infty} |a_n| = \lim_{n\to\infty} \frac{1}{\sqrt{n}} = 0, |a_{n+1}| = \frac{1}{\sqrt{n+1}} < \frac{1}{\sqrt{n}} = |a_n|, \forall n \in \mathbb{N},$$

and

$$\lim_{n\to\infty} |b_n| = \lim_{n\to\infty} \frac{1}{\sqrt[3]{n}} = 0, |b_{n+1}| = \frac{1}{\sqrt[3]{n+1}} < \frac{1}{\sqrt[3]{n}} = |b_n|, \forall n \in \mathbb{N}.$$

However the series $\sum a_n b_n = \sum \frac{1}{n^{5/6}}$ is divergent (it is a p-series with $p = \frac{5}{6}$).

Remark 1. As a particular case of this example, one can also formulate the following false statement: "if a series $\sum a_n$ converges then $\sum a_n^2$ also converges". The corresponding counterexample is in line with the above: if $a_n = (-1)^n \frac{1}{\sqrt{n}}$, then the alternating series $\sum (-1)^n \frac{1}{\sqrt{n}}$ is convergent, but the series of the squares $\sum \frac{1}{n}$ is harmonic, that is divergent.

Remark 2. For positive series this statement is true.

Example 6. "If $\sum a_n$ and $\sum b_n$ are divergent, then $\sum a_n b_n$ is also divergent."

Solution.

If $a_n = \frac{1}{n}$ and $b_n = \frac{1}{\sqrt{n}}$, then $\sum a_n$ and $\sum b_n$ are divergent (p-series with $p \le 1$), but $\sum \frac{1}{n^{3/2}}$ is convergent (a p-series with $p > 1$).

Remark. As a particular case of this example, one can also formulate the following false statement: "if a series $\sum a_n$ diverges then $\sum a_n^2$ also diverges". The corresponding counterexample is similar to the above: the harmonic series $\sum \frac{1}{n}$ is divergent , but the series of the squares $\sum \frac{1}{n^2}$ is convergent (this is a p-series with $p > 1$).

Example 7. "If $\sum a_n$ and $\sum b_n$ are divergent, then $\sum (a_n + b_n)$ is also divergent."

Solution.

If $a_n = -b_n = \frac{1}{n}$, then $\sum a_n$ and $\sum b_n$ diverge, but $\sum (a_n + b_n) = \sum 0$ converges.

Remark 1. This is a wrong "negative" version of the following arithmetic rule: if $\sum a_n$ and $\sum b_n$ are convergent, then $\sum (a_n + b_n)$ is also convergent and $\sum (a_n + b_n) = \sum a_n + \sum b_n$.

Remark 2. If additionally both series are positive, then the statement is true.

Example 8. "If $\sum a_n$ converges and $\sum b_n$ diverges, then $\sum a_n b_n$ diverges."

Solution.

If $a_n = \frac{1}{n^2}$ and $b_n = \frac{1}{\sqrt{n}}$, then $\sum \frac{1}{n^2}$ converges (a p-series with $p > 1$), and $\sum \frac{1}{\sqrt{n}}$ diverges (a p-series with $p \leq 1$), but $\sum \frac{1}{n^{5/2}}$ converges (a p-series with $p > 1$).

Remark 1. The statement with the opposite conclusion is also false: if $a_n = (-1)^n \frac{1}{n}$ and $b_n = (-1)^n$, then $\sum (-1)^n \frac{1}{n}$ converges (by Leibniz test), and $\sum (-1)^n$ diverges (the general term does not approach zero), but $\sum \frac{1}{n}$ diverges (the harmonic series).

Remark 2. The corresponding statement for the sum and difference is true: if $\sum a_n$ converges and $\sum b_n$ diverges, then $\sum (a_n \pm b_n)$ diverges.

Example 9. "If $\sum a_n$ diverges and b_n is unbounded, then $\sum a_n b_n$ diverges."

Solution.

Let $a_n = \frac{1}{n}$ and $b_n = (-1)^n \sqrt{n}$. The harmonic series $\sum \frac{1}{n}$ is divergent and the sequence b_n is unbounded, because $\lim\limits_{n\to\infty} |b_n| = \lim\limits_{n\to\infty} \sqrt{n} = +\infty$. However the series $\sum a_n b_n = \sum \frac{(-1)^n}{\sqrt{n}}$ is a convergent alternating series (according to the Leibniz test).

Example 10. "Let us consider two series $\sum_{n=1}^{\infty} a_n$ and $\sum_{k=1}^{\infty} b_k$, where $b_k = a_{2k-1} + a_{2k}$, $\forall k \in \mathbb{N}$. If $\sum_{k=1}^{\infty} b_k$ converges, then so does $\sum_{n=1}^{\infty} a_n$."

Solution.

Consider the divergent series $\sum_{n=1}^{\infty} a_n = \sum_{n=1}^{\infty} (-1)^{n+1}$. The elements $b_k = 1 - 1 = 0$, $\forall k \in \mathbb{N}$. So the series $\sum_{k=1}^{\infty} b_k = \sum_{k=1}^{\infty} 0 = 0$ converges.

Remark 1. This counterexample shows that the associative property does not work for series in the same way as for finite sums. However, if series $\sum_{n=1}^{\infty} a_n$ is positive, then the statement is true.

Remark 2. The converse is true.

Example 11. "If $\sum a_n$ and $\sum b_n$ diverge, and $a_n \leq c_n \leq b_n$, $\forall n \in \mathbb{N}$, then $\sum c_n$ also diverges."

Solution.

If $a_n = -\frac{1}{n}$ and $b_n = 1$, then the series $\sum a_n$ and $\sum b_n$ diverge. Choosing $c_n = \frac{1}{n^2}$ the last condition of the statement is also satisfied: $-\frac{1}{n} \leq \frac{1}{n^2} \leq 1$, but the series $\sum \frac{1}{n^2}$ is convergent.

Remark 1. Under the given conditions, the statement with the opposite conclusion is also false. For instant, keeping the same $a_n = -\frac{1}{n}$ and $b_n = 1$, but choosing $c_n = \frac{1}{\sqrt{n}}$, all the conditions holds, but $\sum c_n$ diverges.

Remark 2. This statement is a wrong "negative" reformulation of the following correct property: if $\sum a_n$ and $\sum b_n$ converge, and $a_n \leq c_n \leq b_n$, $\forall n \in \mathbb{N}$, then $\sum c_n$ also converges.

6.4 Numerical series: convergence tests

Example 1. "If $\sum a_n$ converges and $\sum b_n$ diverges, then $a_n < b_n$, $\forall n$."

Solution.

If $a_n = \frac{1}{n^2}$ and $b_n = -\frac{1}{n}$, then $\sum \frac{1}{n^2}$ is a convergent p-series (with $p > 1$) and $\sum \frac{-1}{n}$ is the divergent negative harmonic series, but $\frac{1}{n^2} > -\frac{1}{n}$, $\forall n$.

Remark. The statement is true for positive series (that is, series $\sum a_n$ with positive terms $a_n > 0$, $\forall n$), and in this case it represents a part of the Comparison test.

Example 2. "If $\sum a_n$ converges and $\sum b_n$ diverges, then $|a_n| < |b_n|$, $\forall n$."

Solution.

If $a_n = (-1)^n \frac{1}{\sqrt{n}}$ and $b_n = \frac{1}{n}$, then the series $\sum (-1)^n \frac{1}{\sqrt{n}}$ converges (both conditions of the Leibniz test are satisfied: $\lim_{n\to\infty} |a_n| = \lim_{n\to\infty} \frac{1}{\sqrt{n}} = 0$ and $|a_{n+1}| = \frac{1}{\sqrt{n+1}} < \frac{1}{\sqrt{n}} = |a_n|$, $\forall n \in \mathbb{N}$) and $\sum \frac{1}{n}$ diverges (the harmonic series), but $\frac{1}{\sqrt{n}} \geq \frac{1}{n}$, $\forall n$.

Remark. The statement is true for positive series, and in this case it represents a part of the Comparison test.

Example 3. "Let $\sum (-1)^{n+1} a_n$, $a_n > 0$ be an alternating series. If $\lim_{n\to\infty} a_n = 0$, then the series converges."

Solution.

Consider the alternating series with $a_n = \begin{cases} 2/k, & n = 2k-1 \\ 1/k, & n = 2k \end{cases}, k \in \mathbb{N}$. Since $\lim_{k\to\infty} a_{2k-1} = \lim_{k\to\infty} \frac{2}{k} = 0$ and $\lim_{k\to\infty} a_{2k} = \lim_{k\to\infty} \frac{1}{k} = 0$, the condition $\lim_{n\to\infty} a_n = 0$ is satisfied. However, the series is divergent, because the even partial sums are equal to those of the harmonic series:

$$S_{2k} = \sum_{n=1}^{2k} (-1)^{n+1} a_n = \left(\frac{2}{1} - \frac{1}{1}\right) + \left(\frac{2}{2} - \frac{1}{2}\right) + \ldots + \left(\frac{2}{k} - \frac{1}{k}\right)$$

$$= 1 + \frac{1}{2} + \ldots + \frac{1}{k} = \sum_{n=1}^{k} \frac{1}{n} = H_k \underset{k\to\infty}{\to} +\infty.$$

Remark. This is a weakened false version of the convergence test for alternating series (the Leibniz test) where the condition $a_{n+1} < a_n$, $\forall n \in \mathbb{N}$ is missing.

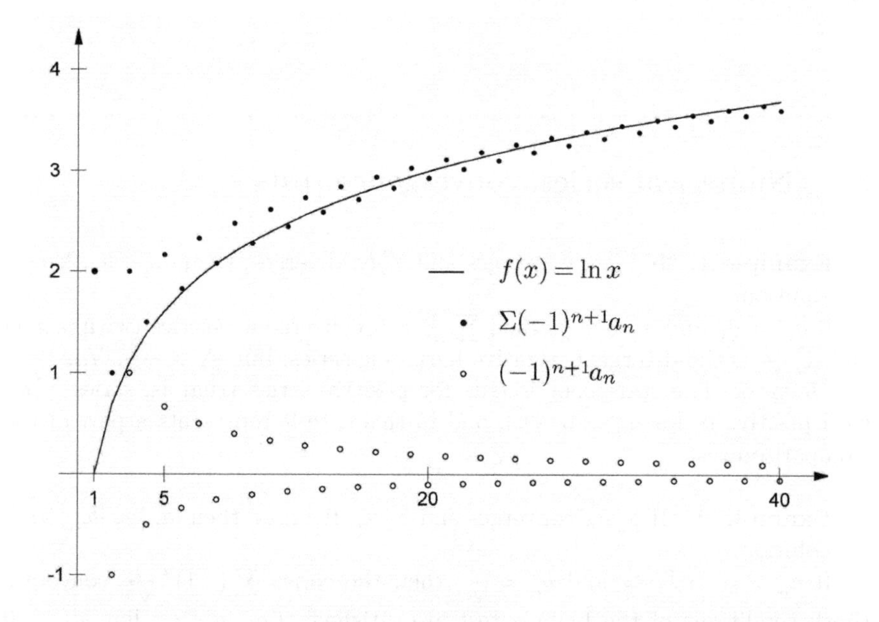

FIGURE 6.4.1: Example 3

Example 4. "Let $\sum (-1)^{n+1} a_n$, $a_n > 0$ be an alternating series. If $a_{n+1} < a_n$, $\forall n \in \mathbb{N}$, then the series converges."

Solution.

Consider the alternating series with $a_n = 1 + \frac{1}{n}$, $n \in \mathbb{N}$. Although $1 + \frac{1}{n+1} < 1 + \frac{1}{n}$, $\forall n \in \mathbb{N}$, but the series is divergent, because the necessary condition of convergence is not satisfied: $\lim_{n\to\infty} a_n = \lim_{n\to\infty} \left(1 + \frac{1}{n}\right) = 1 \neq 0$.

Remark. This is a weakened false version of the convergence test for alternating series (the Leibniz test) where the condition $\lim_{n\to\infty} a_n = 0$ is missing.

Example 5. "If $\sum (-1)^n a_n$ is convergent and $0 \le b_n \le a_n$ for all $n \in \mathbb{N}$, then $\sum (-1)^n b_n$ is also convergent."

Solution.

The series $\sum_{n=1}^{\infty} \frac{(-1)^n}{\sqrt{n}}$ converges according to the Leibniz test: $a_n = \frac{1}{\sqrt{n}} \underset{n\to\infty}{\to} 0$ and $a_{n+1} = \frac{1}{\sqrt{n+1}} \le a_n = \frac{1}{\sqrt{n}}$ for each $n \in \mathbb{N}$. Also, $0 \le b_n = \frac{1}{2\sqrt{n}+(-1)^n} \le \frac{1}{\sqrt{n}} = a_n$, for all $n \in \mathbb{N}$. Nevertheless, the series $\sum_{n=1}^{\infty} \frac{(-1)^n}{2\sqrt{n}+(-1)^n}$ diverges due to the following arguments: first rewrite this series in the form

$$\sum_{n=1}^{\infty} \frac{(-1)^n}{2\sqrt{n}+(-1)^n} = \sum_{n=1}^{\infty} \frac{1}{2}\left(\frac{(-1)^n}{\sqrt{n}} - \frac{1}{\sqrt{n}\,(2\sqrt{n}+(-1))^n}\right),$$

and next note that $\sum_{n=1}^{\infty} \frac{(-1)^n}{\sqrt{n}}$ converges, while $\sum_{n=1}^{\infty} \frac{1}{2n+(-1)^n\sqrt{n}}$ diverges according to the comparison test for positive series, because $\frac{1}{2n+(-1)^n\sqrt{n}} \ge \frac{1}{3n}$ for all $n \in \mathbb{N}$.

Remark. This is a wrong extension of the comparison test (valid for positive series) onto the case of alternating series.

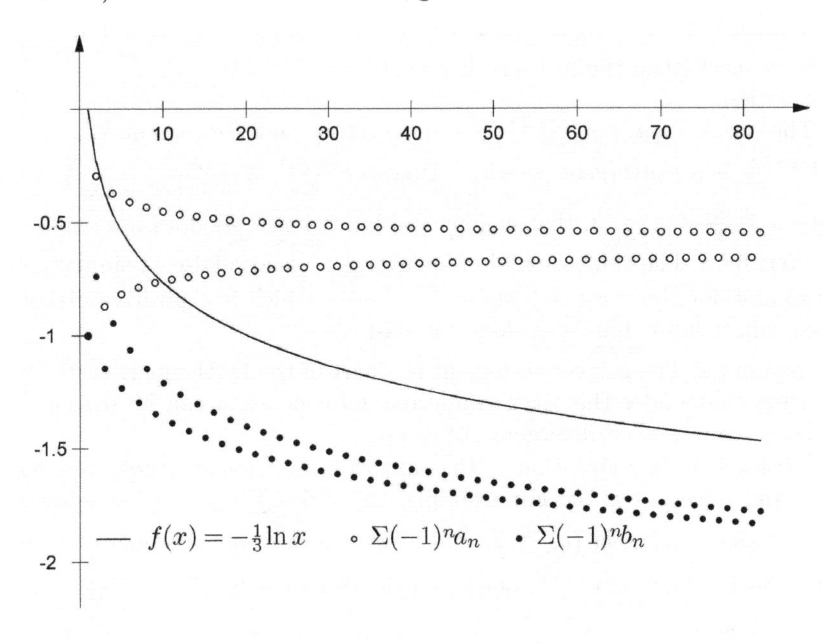

FIGURE 6.4.2: Example 5

Example 6. "If $\lim_{n\to\infty}(a_{n+1}+a_{n+2}+\ldots+a_{n+p})=0$ for $p=1, 2, 3, \ldots$, then the series $\sum a_n$ converges."

Solution.

The harmonic series $\sum\frac{1}{n}$ is divergent, but $\lim_{n\to\infty}\left(\frac{1}{n+1}+\frac{1}{n+2}+\ldots+\frac{1}{n+p}\right)=0$ for $p=1, 2, 3, \ldots$

Remark. This is a wrong reformulation of Cauchy's criterion for convergence of series.

Example 7. "If $\sum a_n$ converges and $\lim_{n\to\infty}\frac{b_n}{a_n}=1$, then $\sum b_n$ also converges."

Solution.

Let us consider the two alternating series $\sum a_n=\sum\frac{(-1)^n}{\sqrt{n}}$ and $\sum b_n=\sum\left(\frac{(-1)^n}{\sqrt{n}}+\frac{1}{n}\right)$. The former is convergent according to the Leibniz test, and the latter is divergent, since its general term is the sum of the terms of the convergent and divergent series. At the same time, $\lim_{n\to\infty}\frac{b_n}{a_n}=\lim_{n\to\infty}\left(1+\frac{(-1)^n}{\sqrt{n}}\right)=1$.

Remark. The statement is true for positive series, and in this case it represents a particular form of the Limit Comparison test.

Example 8. "If $\sum a_n$ is a positive series (that is $a_n>0, \forall n$) and $\lim_{n\to\infty}\frac{a_{n+1}}{a_n}$ does not exist, then the series is divergent."

Solution.

The series $\sum a_n=\sum\frac{2+(-1)^n}{n^2}$ is positive and convergent ($a_n\le\frac{3}{n^2}$ and $\sum\frac{3}{n^2}$ is a convergent p-series). However $\frac{a_{2n+1}}{a_{2n}}=\frac{(2n)^2}{3(2n+1)^2}\underset{n\to\infty}{\to}\frac{1}{3}$, while $\frac{a_{2n}}{a_{2n-1}}=\frac{3(2n-1)^2}{(2n)^2}\underset{n\to\infty}{\to}3$, that is a general limit $\lim_{n\to\infty}\frac{a_{n+1}}{a_n}$ does not exist.

Remark 1. The conclusion about convergence is also false. A similar counterexample for this case is $\sum a_n=\sum\frac{2+(-1)^n}{n}$ which is a positive divergent series, whose limit $\lim_{n\to\infty}\frac{a_{n+1}}{a_n}$ does not exist.

Remark 2. The correct statement is a part of the D'Alembert (ratio) test asserting that under the given conditions no conclusion can be drawn with regard to convergence/divergence of series.

Remark 3. The situation is the same for the Cauchy (root) test when $\lim_{n\to\infty}\sqrt[n]{a_n}$ does not exist. For example, the series $\sum a_n=\sum\frac{1}{(3+(-1)^n)^n}$ is positive and convergent ($a_n\le\frac{1}{2^n}$ and $\sum\frac{1}{2^n}$ is a convergent geometric series), but a general limit $\lim_{n\to\infty}\sqrt[n]{a_n}$ does not exist, because $\sqrt[2n]{a_{2n}}=\sqrt[2n]{\frac{1}{4^{2n}}}\underset{n\to\infty}{\to}\frac{1}{4}$, while $\sqrt[2n+1]{a_{2n+1}}=\sqrt[2n+1]{\frac{1}{2^{2n+1}}}\underset{n\to\infty}{\to}\frac{1}{2}$. On the other hand, the series $\sum a_n=\sum(3+(-1)^n)^n$ is positive and divergent (the general term $a_n=(3+(-1)^n)^n$ does not approach zero), but again a general limit $\lim_{n\to\infty}\sqrt[n]{a_n}$ does not exist, since $\sqrt[2n]{a_{2n}}=\sqrt[2n]{4^{2n}}\underset{n\to\infty}{\to}4$, while $\sqrt[2n+1]{a_{2n+1}}=\sqrt[2n+1]{2^{2n+1}}\underset{n\to\infty}{\to}2$.

Example 9. "Let $\sum a_n$ be a positive series. If the D'Alembert (ratio) test is not applicable, then the Cauchy (root) test does not work as well."

Solution.

Let us consider the series $\sum_{n=1}^{+\infty} 2^{(-1)^n - n}$. The ratio test does not work, because the limit in the test $\lim\limits_{n\to\infty} \frac{a_{n+1}}{a_n}$ does not exist: two partial sequences provide different results - $\frac{a_{2n}}{a_{2n-1}} = \frac{2^{1-2n}}{2^{-1-2n+1}} = 2 > 1$ and $\frac{a_{2n+1}}{a_{2n}} = \frac{2^{-1-2n-1}}{2^{1-2n}} = \frac{1}{8} < 1$. At the same time, the root test is applicable: $\lim\limits_{n\to\infty} \sqrt[n]{a_n} = \lim\limits_{n\to\infty} 2^{((-1)^n - n)/n} = \frac{1}{2} < 1$, that is the series converges.

Another example can be given for the divergent series $\sum_{n=1}^{+\infty} 2^{n-(-1)^n}$. Again the D'Alembert test is not applicable, because $\lim\limits_{n\to\infty} \frac{a_{n+1}}{a_n}$ does not exist: $\frac{a_{2n}}{a_{2n-1}} = \frac{2^{2n-1}}{2^{2n-1+1}} = \frac{1}{2} < 1$ and $\frac{a_{2n+1}}{a_{2n}} = \frac{2^{2n+1+1}}{2^{2n-1}} = 8 > 1$. However, the Cauchy test shows divergence of the series: $\lim\limits_{n\to\infty} \sqrt[n]{a_n} = \lim\limits_{n\to\infty} 2^{(n-(-1)^n)/n} = 2 > 1$.

Remark. The converse is true: if the Cauchy test is not conclusive, then neither is the D'Alembert test. This is because the existence of $\lim\limits_{n\to\infty} \frac{a_{n+1}}{a_n} = a$ implies that $\lim\limits_{n\to\infty} \sqrt[n]{a_n} = a$, but the converse does not hold.

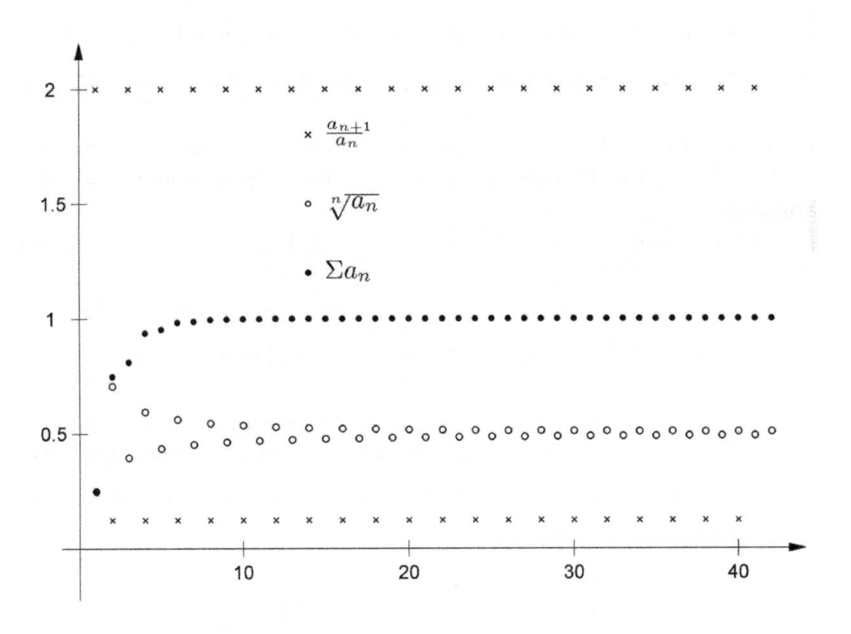

FIGURE 6.4.3: Example 9, the first counterexample

Example 10. "Let $\sum a_n$ be a positive series. If the ratio and root tests are not conclusive, then neither is any other test."

Solution.

Let us consider the convergent p-series ($p=2$) $\sum_{n=1}^{+\infty}\frac{1}{n^2}$. The limits of the ratio and root tests are equal to 1: $\lim\limits_{n\to\infty}\frac{a_{n+1}}{a_n}=\lim\limits_{n\to\infty}\frac{n^2}{(n+1)^2}=1$ and $\lim\limits_{n\to\infty}\sqrt[n]{a_n}=\lim\limits_{n\to\infty}\sqrt[n]{\frac{1}{n^2}}=1$. So both tests are inconclusive. However, convergence can be established by applying the integral test with the positive decreasing function $f(x)=\frac{1}{x^2}$, $x\in[1,+\infty)$, whose improper integral is convergent:

$$\int_1^{+\infty} f(x)\,dx=\int_1^{+\infty}\frac{1}{x^2}dx=-\left.\frac{1}{x}\right|_1^{+\infty}=1.$$

Or one can use the Raabe test, which also reveals convergence:

$$\lim_{n\to\infty} n\left(\frac{a_n}{a_{n+1}}-1\right)=\lim_{n\to\infty} n\left(\frac{(n+1)^2}{n^2}-1\right)=\lim_{n\to\infty} n\left(\frac{2}{n}+\frac{1}{n^2}\right)=2>1.$$

Remark. Similar example can be provided for divergent series: divergence of the p-series ($p=1/2$) $\sum_{n=1}^{+\infty}\frac{1}{\sqrt{n}}$ can be shown by applying the integral test with $f(x)=\frac{1}{\sqrt{x}}$, $x\in[1,+\infty)$ or the Raabe test

$$\lim_{n\to\infty} n\left(\frac{a_n}{a_{n+1}}-1\right)=\lim_{n\to\infty} n\left(\frac{\sqrt{n+1}}{\sqrt{n}}-1\right)=\lim_{n\to\infty}\frac{\sqrt{n}}{\sqrt{n+1}+\sqrt{n}}=\frac{1}{2}<1,$$

but both the ratio and root tests do not work: $\lim\limits_{n\to\infty}\frac{a_{n+1}}{a_n}=\lim\limits_{n\to\infty}\sqrt[n]{a_n}=1$.

Example 11. "Let $\sum a_n$ be a positive series. If the Cauchy, D'Alembert, Raabe and Integral tests are not conclusive, then neither is any other test."

Solution.

Let us consider $a_n=e^{-\left(1+\frac{1}{2}+\ldots+\frac{1}{n-1}\right)}$ and the corresponding series $\sum_{n=2}^{+\infty}e^{-\left(1+\frac{1}{2}+\ldots+\frac{1}{n-1}\right)}$. The limit of the D'Alembert test gives

$$\lim_{n\to\infty}\frac{a_{n+1}}{a_n}=\lim_{n\to\infty}\frac{e^{-\left(1+\frac{1}{2}+\ldots+\frac{1}{n}\right)}}{e^{-\left(1+\frac{1}{2}+\ldots+\frac{1}{n-1}\right)}}=\lim_{n\to\infty}e^{-\frac{1}{n}}=1,$$

that is the D'Alembert test is inconclusive. Since this limit exists, so does the limit of the Cauchy test and the former and latter coincide $\lim\limits_{n\to\infty}\sqrt[n]{a_n}=\lim\limits_{n\to\infty}\frac{a_{n+1}}{a_n}=1$, which means that the Cauchy test is also inconclusive. Moreover, the Raabe test also fails:

$$\lim_{n\to+\infty} n\cdot\left(\frac{a_n}{a_{n+1}}-1\right)=\lim_{n\to+\infty} n\left(e^{\frac{1}{n}}-1\right)=\lim_{x\to 0+}\frac{e^x-1}{x}=\lim_{x\to 0+}e^x=1.$$

Finally, the Integral test is not applicable for this series (one cannot generate an integrable function $f(x)$ such that $f(n)=a_n$). Nevertheless, the Bertrand test shows that the series diverges:

$$\lim_{n\to+\infty}\ln n\cdot\left(n\cdot\left(\frac{a_n}{a_{n+1}}-1\right)-1\right)=\lim_{n\to+\infty}\ln n\left(ne^{\frac{1}{n}}-(n+1)\right)$$

$$= \lim_{x\to 0+} (-\ln x)\left(\frac{1}{x}e^x - \frac{1}{x} - 1\right) = -\lim_{x\to 0+} \frac{e^x - 1 - x}{x(\ln x)^{-1}}$$

$$= -\lim_{x\to 0+} \frac{e^x - 1}{(\ln x)^{-1}} \cdot \lim_{x\to 0+} \frac{1}{1-(\ln x)^{-1}} = -\lim_{x\to 0+} \frac{e^x - 1}{(\ln x)^{-1}}$$

$$= \lim_{x\to 0+} e^x \cdot \lim_{x\to 0+} \frac{(\ln x)^2}{1/x} = \lim_{x\to 0+} \frac{2(\ln x)}{-x^{-1}} = \lim_{x\to 0+} \frac{2.x}{1} = 0 < 1$$

(here we change the discrete index n by the continuous variable x in order to apply L'Hospital's rule).

Remark. There is the well-known result in the theory of series stating that there is no definitive test for verification of convergence or divergence of all series.

Example 12. "Let $f(x)$ be continuous and non-negative on $[1, +\infty)$. If the improper integral $\int_1^{+\infty} f(x)\,dx$ converges, then the series $\sum_{n=1}^{\infty} f(n)$ also converges."

Solution

Consider the continuous function $f(x) = \begin{cases} f_n(x), \ x \in X_n, \ n \in \mathbb{N} \\ 0, \ otherwise \end{cases}$ with $f_n(x)$ defined on $X_n = \left[n - \frac{1}{2n^2}, \, n + \frac{1}{2n^2}\right]$ in the form

$$f_n(x) = \begin{cases} 2n^2 x + \left(1 - 2n^3\right), \ x \in \left[n - \frac{1}{2n^2}, \, n\right] \\ -2n^2 x + \left(1 + 2n^3\right), \ x \in \left[n, \, n + \frac{1}{2n^2}\right] \end{cases}.$$

The improper integral converges:

$$\int_1^{+\infty} f(x)\,dx = \frac{1}{2}\cdot 1 \cdot \frac{1}{2} + \sum_{n=2}^{+\infty} 2 \cdot \frac{1}{2} \cdot 1 \cdot \frac{1}{2n^2}$$

$$= \frac{1}{4} + \sum_{n=2}^{+\infty} \frac{1}{2n^2} = \frac{1}{4} + \frac{1}{2}\left(\frac{\pi^2}{6} - 1\right) = \frac{\pi^2}{12} - \frac{1}{4}.$$

However, the corresponding series is divergent: $\sum_{n=1}^{\infty} f(n) = \sum_{n=1}^{\infty} 1$.

Remark 1. The function in the statement can even be made positive: for the positive continuous function $g(x) = f(x) + \frac{1}{x^2}$ the improper integral

$$\int_1^{+\infty} g(x)\,dx = \int_1^{+\infty} f(x)\,dx + \int_1^{+\infty} \frac{1}{x^2}dx$$

converges as the sum of two convergent integrals, but the series diverges $\sum_{n=1}^{\infty} g(n) = \sum_{n=1}^{\infty} \left(1 + \frac{1}{n^2}\right)$, since the general term does not approach 0.

Remark 2. Analogous statement about divergence: "if $f(x)$ is continuous and non-negative on $[1, +\infty)$, and the improper integral $\int_1^{+\infty} f(x)\,dx$ diverges, then the series $\sum_{n=1}^{\infty} f(n)$ also diverges", is also false. The following counterexample can be considered. Let $f(x) = \cos(\pi + 2\pi x) + 1$. Then the series is convergent $\sum_{n=1}^{\infty} f(n) = \sum_{n=1}^{\infty} 0 = 0$, but the improper integral diverges:

$$\int_{1}^{+\infty} f(x)\,dx = \int_{1}^{+\infty} 1 + \cos(\pi + 2\pi x)\,dx$$

$$= \lim_{x\to+\infty} \left(x + \frac{1}{2\pi}\sin(\pi + 2\pi x) - 1 \right) = +\infty.$$

Again the positivity of $f(x)$ does not save the situation - just consider the same modification as in Remark 1: $g(x) = f(x) + \frac{1}{x^2} = \cos(\pi + 2\pi x) + 1 + \frac{1}{x^2}$.

Remark 3. It is possible to construct counterexamples to the main statement and to the statement in Remark 2 with functions unbounded on $[1, +\infty)$.

Remark 4. The main statement and the statement in Remark 2 are weakened false versions of the Integral test missing the condition of monotonicity of the function $f(x)$ on $[1, +\infty)$.

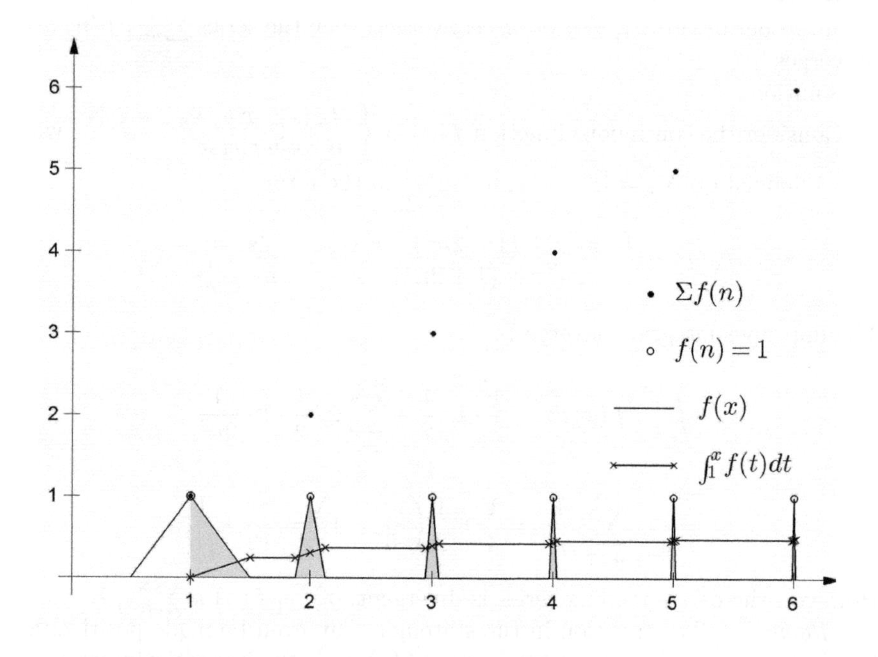

FIGURE 6.4.4: Example 12

Example 13. "Let $f(x)$ be continuous and non-negative on $[1, +\infty)$. If the improper integral $\int_1^{+\infty} f(x)\,dx$ converges and the series $\sum_{n=1}^{\infty} f(n)$ also converges, then $f(x) \underset{x\to+\infty}{\to} 0$."

Solution.

Consider the function constructed in Example 12 and move it the distance of a half of unit to the right along the x-axis. The resulting function is deter-

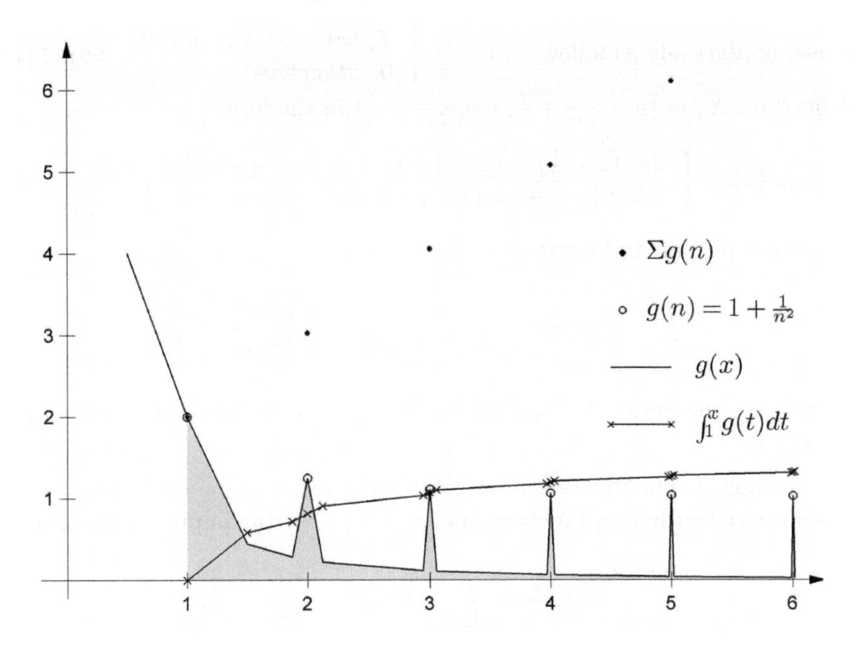

FIGURE 6.4.5: Example 12, Remark 1

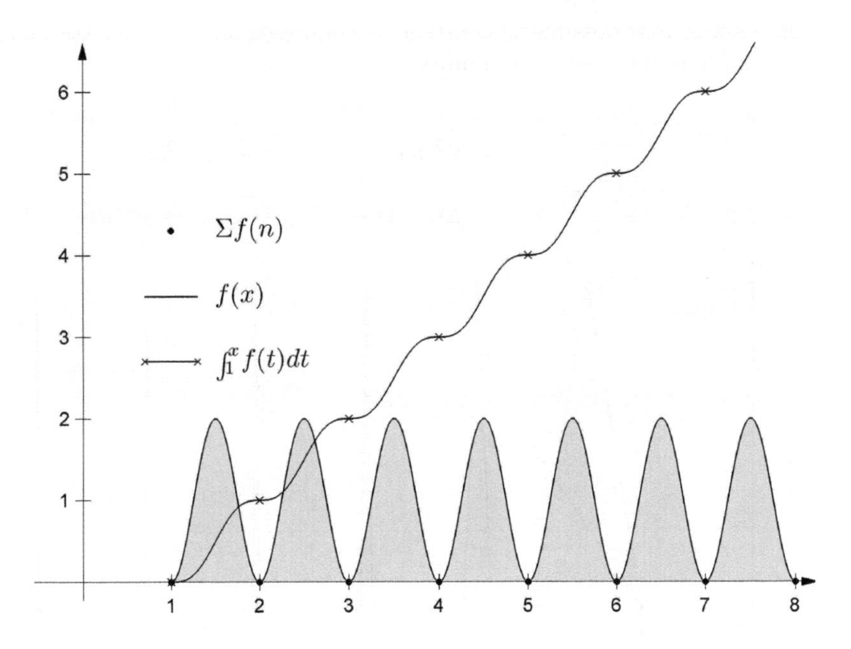

FIGURE 6.4.6: Example 12, Remark 2

mined analytically as follows: $\tilde{f}(x) = \begin{cases} \tilde{f}_n(x), \ x \in \tilde{X}_n, \ n \in \mathbb{N} \\ 0, \ otherwise \end{cases}$ with $\tilde{f}_n(x)$ defined on $\tilde{X}_n = \left[n - \frac{1}{2n^2} + \frac{1}{2}, n + \frac{1}{2n^2} + \frac{1}{2}\right]$ in the form

$$\tilde{f}_n(x) = \begin{cases} 2n^2\left(x - \frac{1}{2}\right) + \left(1 - 2n^3\right), \ x \in \left[n - \frac{1}{2n^2} + \frac{1}{2}, n + \frac{1}{2}\right] \\ -2n^2\left(x - \frac{1}{2}\right) + \left(1 + 2n^3\right), \ x \in \left[n + \frac{1}{2}, n + \frac{1}{2n^2} + \frac{1}{2}\right] \end{cases}.$$

The improper integral converges

$$\int_1^{+\infty} \tilde{f}(x)\,dx = \frac{1}{2} + \sum_{n=2}^{+\infty} \frac{1}{2n^2} = \frac{1}{2}\sum_{n=1}^{+\infty} \frac{1}{n^2} = \frac{\pi^2}{12}$$

as well as the series $\sum_{n=1}^{\infty} \tilde{f}(n) = \sum_{n=1}^{\infty} 0 = 0$. However, $\tilde{f}\left(n + \frac{1}{2}\right) = 1 \underset{n\to+\infty}{\not\to} 0$.

Remark 1. The function in the statement can even be made positive: for the positive continuous function $\tilde{g}(x) = \tilde{f}(x) + \frac{1}{x^2}$ the improper integral

$$\int_1^{+\infty} \tilde{g}(x)\,dx = \int_1^{+\infty} \tilde{f}(x)\,dx + \int_1^{+\infty} \frac{1}{x^2}dx$$

converges as the sum of two convergent integrals, and the p-series ($p = 2$) $\sum_{n=1}^{\infty} \tilde{g}(n) = \sum_{n=1}^{\infty} \frac{1}{n^2}$ is also convergent.

Remark 2. It is possible to construct a counterexample to this statement with a function unbounded on infinity.

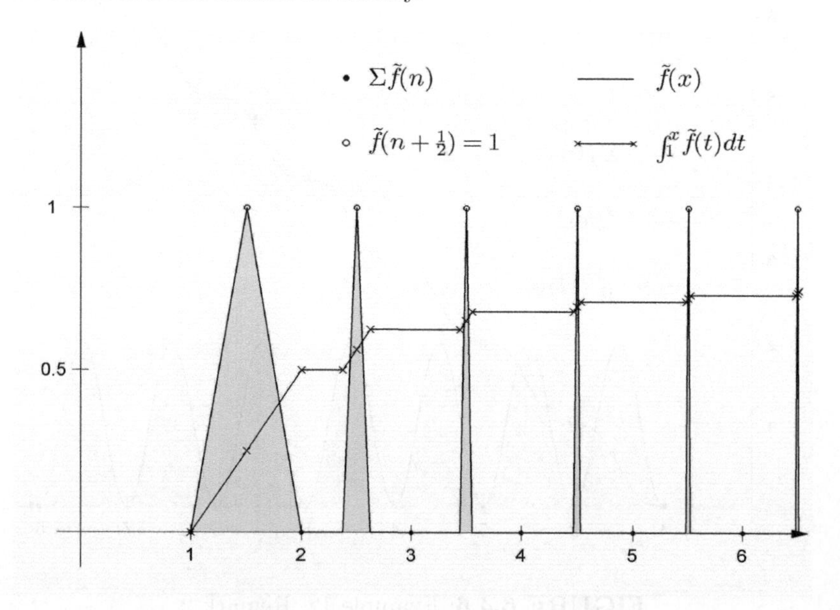

FIGURE 6.4.7: Example 13

6.5 Power series

Example 1. "A power series $\sum_{n=0}^{+\infty} c_n x^n$ always has an interval of convergence."

Solution.

The power series $\sum_{n=1}^{+\infty} n^n x^n$ converges only at the origin.

Example 2. "If series $\sum_{n=0}^{+\infty} a_n x^n$ converges at $x = c$, then it also converges at $x = -c$."

Solution.

The series $\sum_{n=0}^{+\infty} \frac{x^n}{n}$ converges at $x = -1$, but diverges at $x = 1$.

Example 3. "If a function $f(x)$ is infinitely differentiable on $\mathbb{R}$, then the power series $\sum_{n=0}^{+\infty} \frac{f^{(n)}(0)}{n!} x^n$ converges on $\mathbb{R}$."

Solution.

The function $f(x) = \frac{1}{1+x^2}$ is infinitely differentiable on $\mathbb{R}$, but its Taylor series $\sum_{n=0}^{+\infty} (-1)^n x^{2n}$ converges only on $(-1, 1)$.

Example 4. "If a function $f(x)$ is infinitely differentiable on $\mathbb{R}$, and the series $\sum_{n=0}^{+\infty} \frac{f^{(n)}(0)}{n!} x^n$ converges on $\mathbb{R}$, then $\sum_{n=0}^{+\infty} \frac{f^{(n)}(0)}{n!} x^n = f(x)$, $\forall x \in \mathbb{R}$."

Solution.

The function $f(x) = \begin{cases} e^{-1/x^2}, & x \neq 0 \\ 0, & x = 0 \end{cases}$ is infinitely differentiable on $\mathbb{R}$ and $f^{(n)}(0) = 0$, $\forall n \in \mathbb{N}$ (see Example 6, section 4.2 for details). The Taylor series $\sum_{n=0}^{+\infty} \frac{f^{(n)}(0)}{n!} x^n = \sum_{n=0}^{+\infty} 0 \cdot x^n = 0$ converges on $\mathbb{R}$, but it coincides with $f(x)$ only at the origin.

Remark. The same counterexample works also for the following false statement: "if a function $f(x)$ is infinitely differentiable on $\mathbb{R}$, then there exists at least one interval where $\sum_{n=0}^{+\infty} \frac{f^{(n)}(0)}{n!} x^n = f(x)$ ".

Example 5. "If two power series $\sum_{n=0}^{+\infty} a_n x^n$ and $\sum_{n=0}^{+\infty} b_n x^n$ take the same values at infinitely many points, then $a_n = b_n$, $\forall n$."

Solution.

The series $\sin x = \sum_{n=0}^{+\infty} (-1)^n \frac{x^{2n+1}}{(2n+1)!}$ and $\sin 2x = \sum_{n=0}^{+\infty} (-1)^n \frac{(2x)^{2n+1}}{(2n+1)!}$ converge on $\mathbb{R}$ and take the same zero values at all points $x_k = k\pi$, $\forall k \in \mathbb{Z}$. However, their coefficients are different and the series do not coincide.

Remark. The correct statement is the following: if two power series $\sum_{n=0}^{+\infty} a_n x^n$ and $\sum_{n=0}^{+\infty} b_n x^n$ are convergent on the same interval $(-R, R)$, and the set of the points of this interval, on which they take the same values, has a limit point in $(-R, R)$, then $a_n = b_n$, $\forall n$.

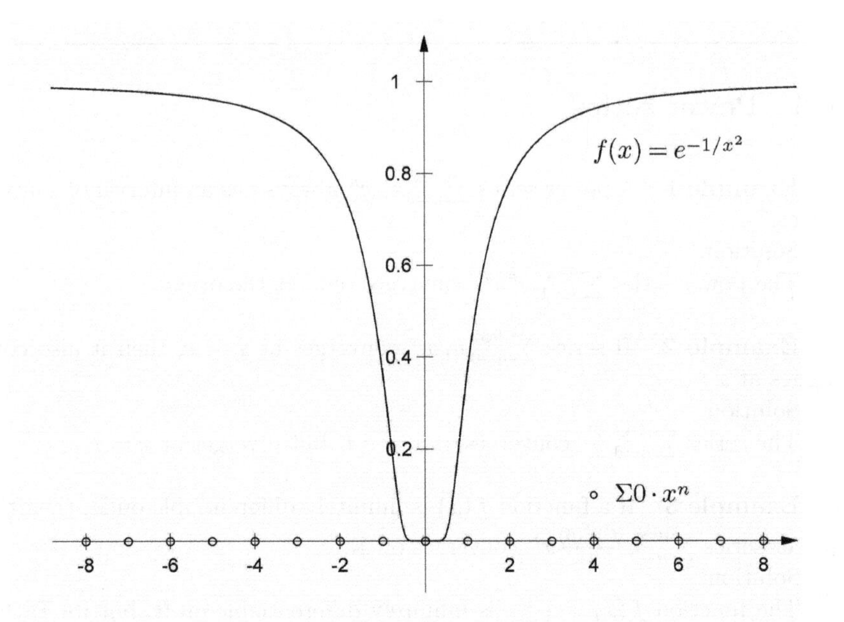

FIGURE 6.5.1: Example 4

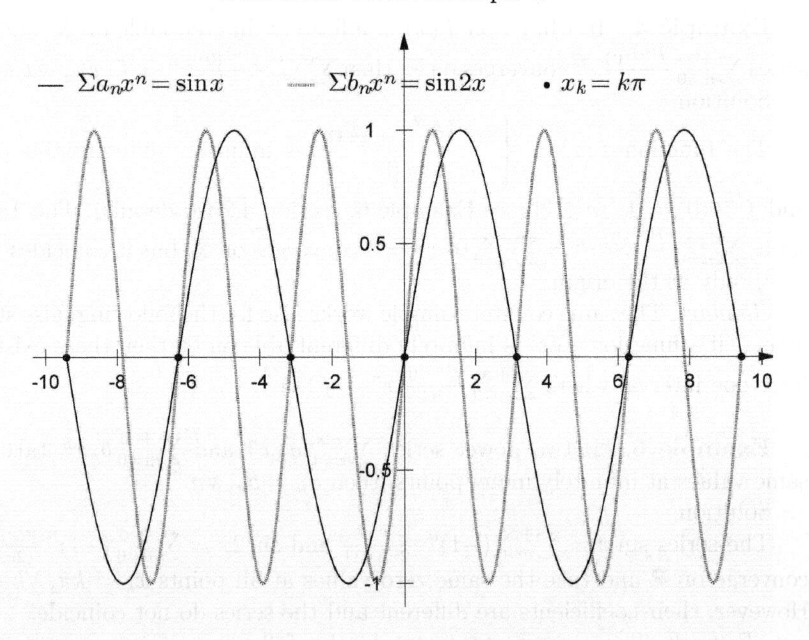

FIGURE 6.5.2: Example 5

Example 6. "If two power series, convergent on the same interval, have different coefficients, then they assume different values at least at one point."

Solution.

Two power series $\sum_{n=0}^{+\infty} \frac{x^n}{n!}$ and $\sum_{n=0}^{+\infty} e\frac{(x-1)^n}{n!}$ converge on $\mathbb{R}$ and have different coefficients, but they take the same values at every $x \in \mathbb{R}$ (they are the Taylor series expansions for e^x centered at 0 and 1, respectively).

Remark. The correct statement asserts that a function can have only one representation as a power series centered at the same point.

Example 7. "If two power series $\sum_{n=0}^{+\infty} a_n x^n$ and $\sum_{n=0}^{+\infty} b_n x^n$ have the same radius of convergence, then the series $\sum_{n=0}^{+\infty} (a_n + b_n) x^n$ also has this radius of convergence."

Solution.

The series $\sum_{n=0}^{+\infty} (-1)^n x^n$ and $\sum_{n=0}^{+\infty} (-1)^{n+1} x^n$ have the radius of convergence equal to 1: $R_a = \lim\limits_{n\to\infty} \left|\frac{a_n}{a_{n+1}}\right| = 1$ and $R_b = \lim\limits_{n\to\infty} \left|\frac{b_n}{b_{n+1}}\right| = 1$. However, the series $\sum_{n=0}^{+\infty} (a_n + b_n) x^n = \sum_{n=0}^{+\infty} 0 \cdot x^n$ is convergent on $\mathbb{R}$, that is $R_c = \infty$.

Remark 1. Of course, the resulting series can be different from zero. For example, choosing the second series in the form $\sum_{n=0}^{+\infty} (-1)^{n+1} \left(1 - \frac{1}{3^n}\right) x^n$ we keep the same radius of convergence $R_b = \lim\limits_{n\to\infty} \left|\frac{b_n}{b_{n+1}}\right| = \lim\limits_{n\to\infty} \frac{1-1/3^n}{1-1/3^{n+1}} = 1$, but the third series $\sum_{n=0}^{+\infty} (a_n + b_n) x^n = \sum_{n=0}^{+\infty} \frac{(-1)^n}{3^n} \cdot x^n$ has a different radius of convergence $R_c = \lim\limits_{n\to\infty} \left|\frac{c_n}{c_{n+1}}\right| = 3$.

Remark 2. The corresponding correct result is as follows: if power series $\sum_{n=0}^{+\infty} a_n x^n$ and $\sum_{n=0}^{+\infty} b_n x^n$ have the radii of convergence R_a and R_b, respectively, then the radius of convergence of the series $\sum_{n=0}^{+\infty} (a_n + b_n) x^n$ satisfies the inequality $R_c \geq \min\{R_a, R_b\}$. If additionally $R_a \neq R_b$, then $R_c = \min\{R_a, R_b\}$.

Exercises

1. Show that the following condition: "there exists $\varepsilon > 0$ such that $\forall n \in \mathbb{N}$ it holds $|a_n - A| < \varepsilon$" does not define A as a limit of a_n. What property of a_n is defined by this condition?

2. Provide a counterexample to the statement: "any subsequence of an unbounded sequence is also unbounded". (Recall that any subsequence of a bounded sequence is bounded.)

3. Give an example of a divergent sequence a_n such that $|a_n|$ and a_n^2 converges. Formulate as a counterexample.

4. Show that $\lim\limits_{n\to\infty} a_n = +\infty$ does not imply $\lim\limits_{n\to\infty}(a_{n+1} - a_n) = +\infty$ neiher $\lim\limits_{n\to\infty} \frac{a_{n+1}}{a_n)} = +\infty$. State in the form of counterexamples.

5. Give a counterexample to the statement "if both sequences a_n and b_n are divergent, then the sequence $\frac{a_n}{b_n}$ is also divergent". Compare with Example 3 in section 6.2.

6. Give a counterexample to the statement "if sequences a_n is convergent and b_n is divergent, then the sequence $\frac{a_n}{b_n}$ is divergent".

7. Illustrate the indeterminate form $\infty - \infty$ with different examples of sequences, that is, show that when $\lim\limits_{n\to\infty} a_n = +\infty$ and $\lim\limits_{n\to\infty} b_n = +\infty$, then any statement about existence/non-existence or (in the case of the existence) about a specific value of $\lim\limits_{n\to\infty}(a_n - b_n)$ is not true.

8. Consider the sequence of arithmetic means $y_n = \frac{x_1+\ldots+x_n}{n}$ of the first n terms of a sequence x_n. Show that the statement "if y_n converges then x_n converges also" is false. (Notice, that the converse is true: if x_n converges then y_n converges to the same limit.)

9. Consider the sequence of geometric means $z_n = \sqrt[n]{x_1 \cdot \ldots \cdot x_n}$ of the first n terms of a positive sequence x_n. Provide a counterexample to the statement "if z_n converges then x_n converges also". (Notice, that the converse is true: if x_n converges then z_n converges to the same limit.)

10. Let x_n be a positive sequence. What can be said about the relation between convergence of arithmetic and geometric means? (Hint: consider $x_n = \begin{cases} \frac{1}{(n+1)^2}, & n - \text{impar} \\ n^2, & n - \text{par} \end{cases}$.)

What about simultaneous convergence of arithmetic and geometric means? Does it guarantee the convergence of the original sequence? (Hint: consider $x_n = 2 + (-1)^n$.)

11. Show that the following statement is false: "if $\sum a_n$ converges and $b_n \le a_n, \forall n$, then $\sum b_n$ also converges". Does the statement become true if the condition $b_n \le a_n$ is substituted by $|b_n| \le |a_n|, \forall n$? Compare with Examples 1 and 2 in section 6.4.

12. Give an example of a divergent series $\sum a_n$ such that $\sum \sqrt{a_n a_{n+1}}$ is convergent.

13. Give an example of divergent positive series $\sum a_n$ and $\sum b_n$ such that $\sum \sqrt{a_n b_n}$ is convergent. Compare with the previous exercise. (Hint: consider $a_n = \begin{cases} \frac{1}{n}, & n - \text{impar} \\ \frac{1}{n^3}, & n - \text{par} \end{cases}$ and $b_n = \begin{cases} \frac{1}{n}, & n - \text{par} \\ \frac{1}{n^3}, & n - \text{impar} \end{cases}$.)

14. Show that the series $\sum \left(\frac{1}{\sqrt{n}} - \frac{(-1)^n}{n} \right)$ is another counterexample to the statement in Example 3, section 6.4.

15. Provide a counterexample to the following statement: "if both positive series $\sum a_n$ and $\sum b_n$ are divergent, then the series $\sum \min(a_n, b_n)$ also diverges".

16. Show that the divergence of the series $\sum_{n=1}^{+\infty} \frac{(2n)!!}{(2n+1)!!}$ cannot be established by the root, ratio and Integral tests, but it can be proved by the Raabe test. Compare this result with Examples 10 and 11 in section 6.4.

17. Show that the convergence of the series $\sum_{n=2}^{+\infty} \frac{1}{(n-\sqrt{n}) \log^2 n}$ cannot be established by the root, ratio and Raabe tests, but it can be proved by the Integral or Bertrand tests. Compare this result with Examples 10 and 11 in section 6.4.

18. Show that the root, ratio, Raabe and Bertrand tests fail for the series $\sum_{n=1}^{+\infty} 2^{(-1)^n - n}$. However, the simple application of the Comparison test gives the result. The last three exercises illustrates the conclusion in Remark to Example 11, section 6.4.

19. Let $\sum a_n$ be an arbitrary series with partial sums S_n. Consider the sequence of arithmetic means of the partial sums $C_n = \frac{S_1 + \ldots + S_n}{n}$. If the last sequence converges then the series $\sum a_n$ is called convergent by Cesàro. It is well-known that the convergence of the original series implies the convergence by Cesàro. Show that the converse is not true.

20. Give a counterexample to the statement "if $\sum a_n$ is positive and the sequence na_n is unbounded, then the series diverges". (Hint: try $a_n = \begin{cases} \frac{1}{\sqrt{n}}, & n = k^4 \\ \frac{1}{n^2}, & n \neq k^4 \end{cases}$, $k \in \mathbb{N}$) (Notice, that the following statement is true: if $\sum a_n$ is positive and $\lim\limits_{n\to\infty} na_n = A$, $0 < A \leq +\infty$, then the series diverges.)

21. Show that the statement "if $\sum a_n$ is positive and $\lim\limits_{n\to\infty} \sqrt{n} a_n = 0$, then the series is convergent" is false.

22. Show that the statement "if $\sum a_n$ is positive and $\lim\limits_{n\to\infty} n \log n a_n = 0$, then the series is convergent" is false. (Hint: consider $a_n = \frac{1}{n \log n \log \log n}$.)

23. Provide counterexamples mentioned in Remark 3 to Example 12, section 6.4.

24. Provide a counterexample mentioned in Remark 2 to Example 13, section 6.4.

25. Suppose $f(x)$ can be developed in the series $\sum_{n=0}^{+\infty} a_n x^n$ on $(-1, 1)$.

Show by example that the following statement is false: "if $\lim\limits_{x\to 1_-} f(x) = A$, then the numerical series $\sum_{n=0}^{+\infty} a_n$ converges to A". (Hint: consider $f(x) = \frac{1}{1+x}$.)

Part II

Functions of two real variables

Chapter 7

Limits and continuity

General Remark. Part II (chapters 7-9) contains the topics of the functions of two variables, which are quite representative for the most topics of the multivariable functions. The counterexamples presented in this part of the work are separated into two groups. The first one includes those examples that have intimate connections with counterexamples of one-variable functions considered in Part I. They show how the ideas applied in the one-dimensional case can be generalized/extended to many variables. Accordingly, the examples of this group are placed in the sections titled "one-dimensional links" in each of chapters 7-9. The examples of the second group have a weak or no connection with one-dimensional case, highlighting a specificity of concepts and results for multivariable functions. Some of them illustrate the situations that are feasible for two-variable functions but cannot happen in the case of one-variable functions. In each chapter, all the examples of the second group are collected in sections titled "multidimensional essentials".

7.1 Elements of theory

Limits. Concepts

Limit (general limit). Let $f(x,y)$ be defined on X and (a,b) be a limit point of X. We say that the *limit* of $f(x,y)$, as (x,y) approaches (a,b), exists and equals A if for every $\varepsilon > 0$ there exists $\delta > 0$ such that for all $(x,y) \in X$ such that $0 < \sqrt{(x-a)^2 + (y-b)^2} < \delta$ it follows that $|f(x,y) - A| < \varepsilon$. The usual notations are $\lim\limits_{(x,y)\to(a,b)} f(x,y) = A$ and $f(x,y) \underset{(x,y)\to(a,b)}{\to} A$.

Remark. In calculus, a non-essential simplification that $f(x,y)$ is defined in some deleted neighborhood of (a,b) is frequently used.

Partial limit. Let $f(x,y)$ be defined on X, S be a subset of X and (a,b) be a limit point of S. The *partial limit* of $f(x,y)$ on X at (a,b) is the limit of $f(x,y)$, as (x,y) approaches (a,b), on a subset S.

Infinite limit. Let $f(x,y)$ be defined on X and (a,b) be a limit point of X. We say that the limit of $f(x,y)$, as (x,y) approaches (a,b), is $+\infty$ $(-\infty)$ if for every $E > 0$ there exists $\delta > 0$ such that for all $(x,y) \in X$ such that $0 < \sqrt{(x-a)^2 + (y-b)^2} < \delta$ it follows that $f(x,y) > E$ $(f(x,y) < -E)$. The usual notations are $\lim\limits_{(x,y)\to(a,b)} f(x,y) = +\infty$ $(\lim\limits_{(x,y)\to(a,b)} f(x,y) = -\infty)$ and $f(x,y) \underset{(x,y)\to(a,b)}{\to} +\infty$ $(f(x,y) \underset{(x,y)\to(a,b)}{\to} -\infty)$.

Sometimes it is also considered a "general" infinite limit in the following sense:

The limit of $f(x,y)$, as (x,y) approaches (a,b), is ∞ if for every $E > 0$ there exists $\delta > 0$ such that for all $(x,y) \in X$ such that $0 < \sqrt{(x-a)^2 + (y-b)^2} < \delta$ it follows that $|f(x,y)| > E$. The standard notations are $\lim\limits_{(x,y)\to(a,b)} f(x,y) = \infty$ and $f(x,y) \underset{(x,y)\to(a,b)}{\to} \infty$.

If any of the above three infinite limits is admitted at the same time the notation used is $\lim\limits_{(x,y)\to(a,b)} f(x,y) = (\pm)\,\infty$.

Limit at infinity. Let $f(x,y)$ be defined on an unbounded set X. We say that the limit of$f(x,y)$, as (x,y) approaches ∞, exists and equals A , if for every $\varepsilon > 0$ there exists $D > 0$ such that for all $(x,y) \in X$ such that $\sqrt{x^2 + y^2} > D$ it follows that $|f(x) - A| < \varepsilon$. The standard notations are $\lim\limits_{(x,y)\to\infty} f(x,y) = A$ and $f(x,y) \underset{(x,y)\to\infty}{\to} A$.

Iterated limits. Let $f(x,y)$ be defined on X, and (a,b) be a limit point of X. Let $f(x,y)$ possesses the limits $\lim\limits_{x\to a} f(x,y) = g(y)$ for any fixed y close to b, $y \neq b$, and such that $(x,y) \in X$. In this way, a new function $g(y)$ is determined. If the following limit exists $\lim\limits_{y\to b} g(y) = A$, then it is called the *iterated limit* of the original function $f(x,y)$ and denoted $\lim\limits_{y\to b}\lim\limits_{x\to a} f(x,y) = A$. Similarly can be defined the second iterated limit $\lim\limits_{x\to a}\lim\limits_{y\to b} f(x,y) = B$.

Path limits. The important case of partial limits for the functions of two variables is the *path limits*, when the partial limit is considered along a chosen curve, which allows a reduction of the analysis of multivariate limits to the investigation of one-dimensional limits. Let (a,b) be a limit point of the domain X of the function $f(x,y)$, and $(x,y) = (\varphi(t), \psi(t))$, $t \in [\alpha,\beta]$ be a parametric definition of the curve that belongs to X and passes through (a,b), that is, $(\varphi(t_0), \psi(t_0)) = (a,b)$ for some t_0 in $[\alpha,\beta]$. Then the limit $\lim\limits_{t\to t_0} g(t) \equiv \lim\limits_{t\to t_0} f(\varphi(t), \psi(t))$ is called the path limit as (x,y) approaches

(a,b) along the chosen curve. Usually, it is sufficient to consider continuous one-to-one curves passing through (a,b), and we will consider path limits under this restriction.

Limits. Elementary properties

Unicity. If the limit exists, then it is unique.

Remark 1. In the properties below it is supposed that $f(x,y)$ and $g(x,y)$ are defined on the same domain.

Remark 2. The point (a,b) can be finite or infinite.

Comparative properties

If $\lim\limits_{(x,y)\to(a,b)} f(x,y) = A$, $\lim\limits_{(x,y)\to(a,b)} g(x,y) = B$ then

1) if $f(x,y) \le g(x,y)$ for all $(x,y) \in X$ in a deleted neighborhood of (a,b), then $A \le B$.

2) if $A < B$, then $f(x,y) < g(x,y)$ for all $(x,y) \in X$ in a deleted neighborhood of (a,b).

Arithmetic (algebraic) properties

If $\lim\limits_{(x,y)\to(a,b)} f(x,y) = A$, $\lim\limits_{(x,y)\to(a,b)} g(x,y) = B$, then

1) $\lim\limits_{(x,y)\to(a,b)} (f(x,y) + g(x,y)) = A + B$

2) $\lim\limits_{(x,y)\to(a,b)} (f(x,y) - g(x,y)) = A - B$

3) $\lim\limits_{(x,y)\to(a,b)} (f(x,y) \cdot g(x,y)) = A \cdot B$

4) $\lim\limits_{(x,y)\to(a,b)} \frac{f(x,y)}{g(x,y)} = \frac{A}{B}$ (under the additional condition $B \ne 0$)

If $\lim\limits_{(x,y)\to(a,b)} f(x,y) = A$, then

5) $\lim\limits_{(x,y)\to(a,b)} |f(x,y)| = |A|$

6) $\lim\limits_{(x,y)\to(a,b)} (f(x,y))^{\alpha} = A^{\alpha}$ (under the condition $\alpha \in \mathbb{N}$; or under the condition $\alpha \in \mathbb{Z}$ and $A \ne 0$; or under the condition $A > 0$)

7) $\lim\limits_{(x,y)\to(a,b)} \alpha^{f(x,y)} = \alpha^{A}$ (under the assumption $\alpha > 0$)

The squeeze theorem. If $\lim\limits_{(x,y)\to(a,b)} f(x,y) = \lim\limits_{(x,y)\to(a,b)} g(x,y) = A$ and the inequality $f(x,y) \le h(x,y) \le g(x,y)$ holds for all $x \in X$ in a deleted neighborhood of (a,b), then $\lim\limits_{(x,y)\to(a,b)} h(x,y) = A$.

Continuity. Concepts

Continuity at a point. A function $f(x,y)$ defined on X is *continuous* at a point $(a,b) \in X$ if for every $\varepsilon > 0$ there exists $\delta > 0$ such that whenever $(x,y) \in X$ and $\sqrt{(x-a)^2 + (y-b)^2} < \delta$ it follows that $|f(x,y) - f(a,b)| < \varepsilon$.

Remark 1. Frequently it is helpful to consider the definition of the continuity written in the limit form $\lim\limits_{(x,y)\to(a,b)} f(x,y) = f(a,b)$ or $f(x,y) \underset{(x,y)\to(a,b)}{\to} f(a,b)$. Actually the original definition of continuity and the definition by limit differ only in one "pathological" situation of an isolated point (a,b) of X, because the limit definition requires (a,b) to be a limit point of X, while the general definition of continuity does not contain such a requirement. It means that at any isolated point (a,b) of X, a function $f(x,y)$ is continuous according to the general definition, but not continuous according to the definition by limit. Since the behavior of a function at an isolated point is hardly of any interest, we will usually consider the definition of continuity by limit as a complete definition.

Remark 2. In calculus, a simplification that $f(x,y)$ is defined in some neighborhood of (a,b) is frequently used. Evidently, in this case the general and limit definitions of continuity coincide.

Continuity with respect to a separate variable. A function $f(x,y)$ defined on X is *continuous in x* at a point $(a,b) \in X$ if the function of one variable $f(x,b)$ is continuous at a. Continuity in y is defined similarly.

Continuity on a set. A function $f(x,y)$ defined on X is *continuous on a set* $S \subset X$, if $f(x,y)$ is continuous at every point of S.

Remark. It is worth to notice that in the case when a boundary point (a,b) belongs to set S, continuity on S implies continuity at (a,b) for the original function considered only on S.

Discontinuity point. Let (a,b) be a limit point of the domain of $f(x,y)$. If $f(x,y)$ is not continuous at (a,b), then (a,b) is a *point of discontinuity* of $f(x,y)$ (or equivalently, $f(x,y)$ has a discontinuity at (a,b)).

Remark 1. Sometimes it is required that (a,b) should be a point of the domain of $f(x,y)$. We will not impose this restriction.

Remark 2. Classification of discontinuities is usually not considered for functions of several variables.

Continuity. Local properties

Comparative properties

If $f(x,y)$ and $g(x,y)$ are continuous at (a,b) then
1) if $f(x) \leq g(x)$ for all $(x,y) \in X$ in a deleted neighborhood of (a,b), then $f(a,b) \leq g(a,b)$.
2) if $f(a,b) < g(a,b)$, then $f(x,y) < g(x,y)$ for all $(x,y) \in X$ in a neighborhood of (a,b).

Arithmetic (algebraic) properties

If $f(x,y)$ and $g(x,y)$ are continuous at (a,b), then
1) $f(x,y) + g(x,y)$ is continuous at (a,b)
2) $f(x,y) - g(x,y)$ is continuous at (a,b)
3) $f(x,y) \cdot g(x,y)$ is continuous at (a,b)
4) $\frac{f(x,y)}{g(x,y)}$ is continuous at (a,b) (under the additional condition $g(a,b) \neq 0$)

If $f(x,y)$ is continuous at (a,b), and $\alpha \in \mathbb{R}$, then
5) $|f(x,y)|$ is continuous at (a,b)
6) $(f(x,y))^{\alpha}$ is continuous at (a,b) (under the condition $\alpha \in \mathbb{N}$; or under the condition $f(a,b) > 0$)
7) $\alpha^{f(x,y)}$ is continuous at (a,b) (under the assumption $\alpha > 0$)

Composite function theorem. If $f(x,y)$ and $g(x,y)$ are continuous at (a,b), $h(p,q)$ is continuous at $(c,d) = (f(a,b), g(a,b))$, and the composite function $h(f(x,y), g(x,y))$ is defined in a neighborhood of (a,b), then $h(f(x,y), g(x,y))$ is continuous at (a,b).

Continuity. Global properties

Open Set Characterization. $f(x,y)$ is continuous on a domain X if, and only if, the inverse image $f^{-1}(S)$ of any open set $S \subset f(X)$ is open.

Closed Set Characterization. $f(x,y)$ is continuous on a domain X if, and only if, the inverse image $f^{-1}(S)$ of any closed set $S \subset f(X)$ is closed.

The Compact Set Theorem. If $f(x,y)$ is continuous on a compact set S, then its image $f(S)$ is also a compact set.

The First Weierstrass Theorem. If $f(x,y)$ is continuous on a compact set, then $f(x,y)$ is bounded on this set.

The Second Weierstrass Theorem. If $f(x,y)$ is continuous on a compact set, then $f(x,y)$ attains its global maximum and minimum values on this set.

Intermediate Value property. A function $f(x,y)$ satisfies the intermediate value property on a set $S \subset X$, if $f(x,y)$ takes on every value in between $f(x_1,y_1)$ and $f(x_2,y_2)$ for any pair (x_1,y_1), $(x_2,y_2) \in S$ (that is, if $f(x,y)$ takes on two values somewhere in S, it also takes on every value in between).

The Intermediate Value Theorem. If $f(x,y)$ is continuous on a connected set S, then $f(x,y)$ satisfies the intermediate value property on S. In particular, if $f(x,y)$ is continuous on a connected set S, then its image $f(S)$ is also a connected set.

Uniform Continuity. $f(x,y)$ is *uniformly continuous* on a set S if for every $\varepsilon > 0$ there exists $\delta > 0$ such that whenever (x_1,y_1), $(x_2,y_2) \in S$ and $\sqrt{(x_1-x_2)^2+(y_1-y_2)^2} < \delta$ it follows that $|f(x_1,y_1) - f(x_2,y_2)| < \varepsilon$.

Nonuniform Continuity. $f(x,y)$ fails to be uniformly continuous on a set S if there exists $\varepsilon_0 > 0$ such that for any $\delta > 0$ one can found two points (x_δ,y_δ) and $(\tilde{x}_\delta,\tilde{y}_\delta)$ in S such that $\sqrt{(x_\delta-\tilde{x}_\delta)^2+(y_\delta-\tilde{y}_\delta)^2} < \delta$, but $|f(x_\delta,y_\delta) - f(\tilde{x}_\delta,\tilde{y}_\delta)| \geq \varepsilon_0$.

Criterion for Nonuniform Continuity. $f(x,y)$ fails to be uniformly continuous on a set S if there exists $\varepsilon_0 > 0$ and two sequences (x_n,y_n) and $(\tilde{x}_n,\tilde{y}_n)$ in S such that $\sqrt{(x_n-\tilde{x}_n)^2+(y_n-\tilde{y}_n)^2} \underset{n\to\infty}{\to} 0$, but $|f(x_n,y_n) - f(\tilde{x}_n,\tilde{y}_n)| \geq \varepsilon_0$.

The Cantor Theorem. If $f(x,y)$ is continuous on a compact set S, then $f(x,y)$ is uniformly continuous on S.

7.2 One-dimensional links

General Remark. In this section, we gather the examples that have direct connections with those considered in chapters 2 and 3 for functions of one variable. It means that both the formulation of examples and their solutions can be found as a straightforward generalization of the corresponding one-dimensional cases. The aim is to show interconnection between the functions of one and two variables, and illustrate different ways of how the one-dimensional constructions and results can be extended to multidimensional case. Each example in this section has the references to the corresponding one-dimensional counterpart.

7.2.1 Concepts and local properties

Example 1. "If $\lim\limits_{(x,y)\to(a,b)} f(x,y)$ does not exist, then $\lim\limits_{(x,y)\to(a,b)} |f(x,y)|$ does not exist as well."

(compare with Example 2, section 2.2)

Solution.

Consider the two-dimensional modified Dirichlet's function $f(x,y) = \tilde{D}(x,y) = \begin{cases} 1, & \sqrt{x^2+y^2} \in \mathbb{Q} \\ -1, & \sqrt{x^2+y^2} \in \mathbb{I} \end{cases}$. The restriction of this function to x-axis gives the one-dimensional modified Dirichlet's function, which does not have a limit at any point in $\mathbb{R}$ (see Example 2 in section 2.2). By the same arguments as in one-dimensional case, the function $\tilde{D}(x,y)$ does not have a limit at any point in $\mathbb{R}^2$. On the other hand, $\left|\tilde{D}(x,y)\right| = 1$ and this function has the limit 1 at any point in $\mathbb{R}^2$.

Remark 1. The Remarks to Example 2 in section 2.2 adjusted to the case of $\mathbb{R}^2$ are applicable here also.

Remark 2. There are many other two-dimensional modifications of Dirichlet's function suitable for this counterexample. For instance, the functions $\bar{D}(x,y) = \begin{cases} 1, & x \in \mathbb{Q} \\ -1, & x \in \mathbb{I} \end{cases}$ and $\hat{D}(x,y) = \begin{cases} 1, & |x|+|y| \in \mathbb{Q} \\ -1, & |x|+|y| \in \mathbb{I} \end{cases}$ can be used.

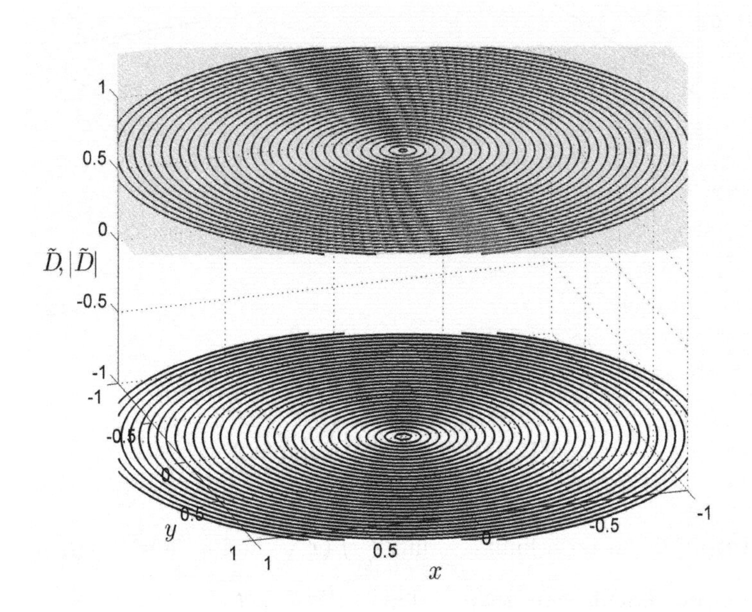

FIGURE 7.2.1: Example 1

Example 2. "If there exists a sequence (x_n, y_n) such that $\lim_{(x_n,y_n)\to(a,b)} f(x_n, y_n) = A$, then $\lim_{(x,y)\to(a,b)} f(x,y) = A$."
(compare with the second counterexample in Example 4, section 2.2)
Solution.

Let us consider the function $f(x,y) = \sin\frac{1}{|x|+|y|}$ defined on $\mathbb{R}^2 \setminus \{(0,0)\}$. On the one hand, using the points with the coordinates $x_k = y_k = \frac{1}{2k\pi}$, $k \in \mathbb{N}$ one has the sequence $(x_k, y_k) \underset{k\to+\infty}{\to} (0,0)$ such that $\lim_{(x_k,y_k)\to(0,0)} f(x_k, y_k) = \lim_{k\to+\infty} \sin(k\pi) = 0$. On the other hand, the points with the coordinates $x_n = y_n = \frac{1}{\pi+4n\pi}$, $n \in \mathbb{N}$ form the sequence approaching the origin such that $\lim_{(x_n,y_n)\to(0,0)} f(x_n, y_n) = \lim_{n\to+\infty} \sin\left(\frac{\pi}{2} + 2n\pi\right) = 1$. Therefore, a general limit does not exist.

Remark. All Remarks to Example 4 of section 2.2 are applicable here with the corresponding adjustments to the case of $\mathbb{R}^2$.

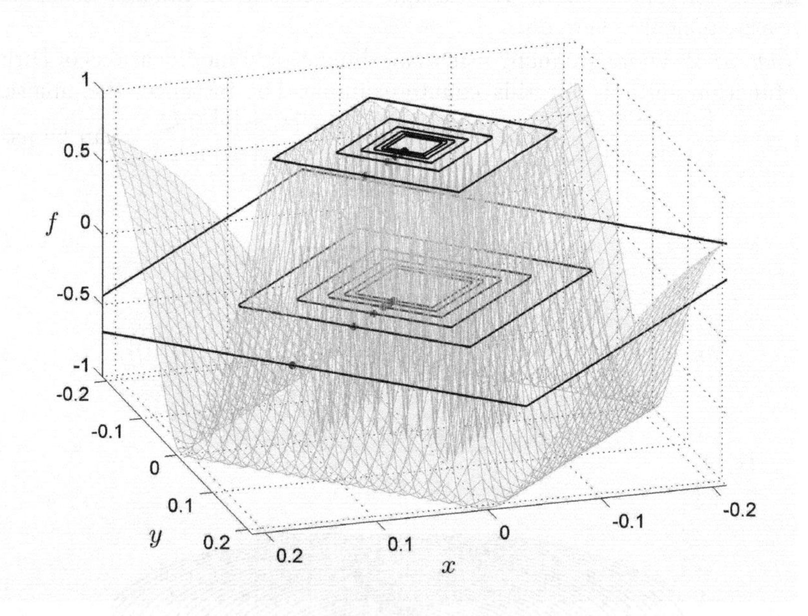

FIGURE 7.2.2: Example 2

Example 3. "If both limits $\lim_{(x,y)\to(a,b)} f(x,y)$ and $\lim_{(x,y)\to(a,b)} g(x,y)$ exist and the former equals zero, then $\lim_{(x,y)\to(a,b)} \frac{f(x,y)}{g(x,y)} = 0$."
(compare with Example 2, section 2.3)
Solution.

Let $f(x) = g(x) = \sqrt{x^2 + y^2}$ and $(a, b) = (0, 0)$. Then $\lim_{(x,y)\to(0,0)} f(x, y) = \lim_{(x,y)\to(0,0)} g(x, y) = 0$, but $\lim_{(x,y)\to(0,0)} \frac{f(x,y)}{g(x,y)} = 1$.

Remark. All Remarks to Example 2 of section 2.3 are applicable here with corresponding adjustments to the case of $\mathbb{R}^2$. In particular, for an indeterminate form $\frac{0}{0}$, one can construct specific functions $f(x, y)$ and $g(x, y)$ such that the limit of their ratio will be an arbitrary chosen a priory constant (including zero) or infinity, or this limit will not exist.

Example 4. "If $\lim_{(x,y)\to(a,b)} f(x, y) = A$, $\lim_{(x,y)\to(a,b)} g(x, y) = B$ and there is a deleted neighborhood of (a, b) where $f(x, y) < g(x, y)$, then $A < B$."

(compare with Example 4, section 2.3)

Solution.

Let $f(x, y) = -x^2 - y^2$, $g(x, y) = x^2 + y^2$ and $(a, b) = (0, 0)$. Then both limits exist and in any deleted neighborhood of the origin $f(x, y) < g(x, y)$, but $\lim_{(x,y)\to(0,0)} f(x, y) = \lim_{(x,y)\to(0,0)} g(x, y) = 0$.

Remark. All Remarks to Example 4 of section 2.3 are applicable here with corresponding adjustments to the case of $\mathbb{R}^2$.

Example 5. "If $f(x, y)$ is unbounded in any neighborhood of a point (a, b), then at least one of the path limits of $|f(x, y)|$, as (x, y) approaches (a, b), is infinite."

(compare with Example 9, section 2.3)

Solution.

Let us consider the function $f(x) = \frac{1}{\sqrt{x^2+y^2}} \cos \frac{1}{\sqrt{x^2+y^2}}$ defined in $\mathbb{R}^2 \setminus \{(0, 0)\}$. First, notice that the restriction of this function to the x-axis coincides (except for the absolute value) with the function in Example 9, section 2.3. Therefore, choosing for this restriction the same points as in Example 9, one can show that the function is unbounded in any neighborhood of the origin: for $(x_k, 0) = \left(\frac{1}{2k\pi}, 0\right)$, $k \in \mathbb{Z} \setminus \{0\}$ one gets $\lim_{(x_k,y_k)\to(0,0)} |f(x_k, y_k)| = \lim_{k\to\infty} |2k\pi| \cos |2k\pi| = +\infty$. Next, let us show that no path limit of $|f(x, y)|$, as (x, y) approaches the origin, is infinite. For an arbitrary (continuous, one-to-one) curve $(x, y) = (\varphi(t), \psi(t))$, $t \in [\alpha, \beta]$ such that $(\varphi(t_0), \psi(t_0)) = (0, 0)$, $t_0 \in (\alpha, \beta)$, it is possible to find a set of points (x_n, y_n) whose distance to the origin is $\sqrt{x_n^2 + y_n^2} = \frac{1}{\pi/2+n\pi}$, $n \in \mathbb{N}$, $n > N_1$. Indeed, the distance function $d(t) = \sqrt{\varphi^2(t) + \psi^2(t)}$ is continuous on $[\alpha, \beta]$, that implies that $d(t)$ assumes all the values between $d(\alpha) > 0$ and $d(t_0) = 0$. In particular, choosing N_1 large enough for the inequality $\frac{1}{\pi/2+N_1\pi} < d(\alpha)$ to be satisfied, one ensures that for any $n \in \mathbb{N}$, $n > N_1$ there are points $(x_n, y_n) = (\varphi(t_n), \psi(t_n))$ on the chosen curve $(\varphi(t), \psi(t))$ such that $d(t_n) = \sqrt{x_n^2 + y_n^2} = \frac{1}{\pi/2+n\pi} < d(\alpha)$.

The use of these points along the chosen path, leads to the following result:

$$\lim_{(x_n,y_n)\to(0,0)} |f(x_n, y_n)| = \lim_{n\to+\infty} \left(\frac{\pi}{2} + n\pi\right) \left|\cos\left(\frac{\pi}{2} + n\pi\right)\right| = \lim_{n\to+\infty} 0 = 0,$$

which contradicts to the statement.

Remark. This is a refined version of Example 9 in section 2.3, because under the same conditions we have shown here that none of the path limits can be infinite (while in Example 9 it was shown only for the two types of path limits allowable on $\mathbb{R}$ - one-sided limits along x-axis).

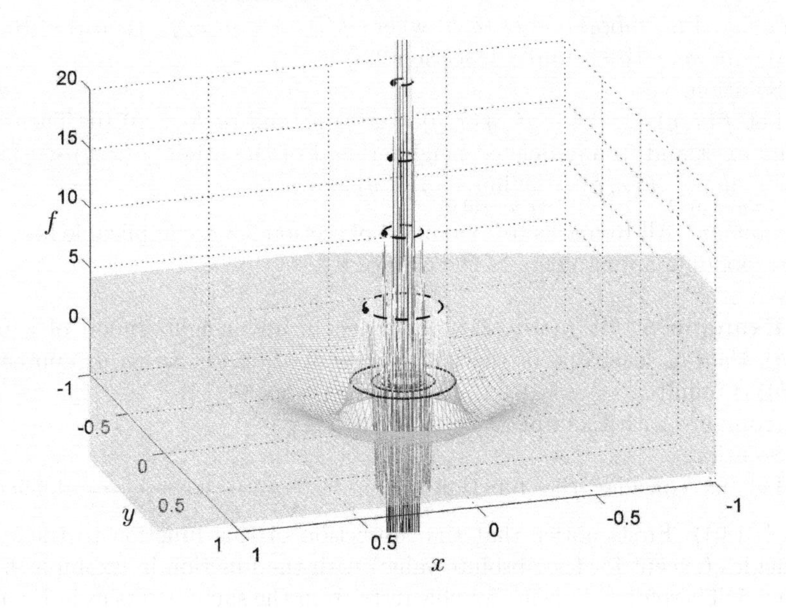

FIGURE 7.2.3: Example 5

Example 6. "If $f(x, y)$ is continuous at the origin and $f(0,0) = 0$, then the inequality $|f(x,y)| \le \sqrt{x^2+y^2}$ holds at least in some neighborhood of the origin."

(compare with Example 6, section 3.2)

Solution.

If $f(x,y) = \sqrt[3]{x^2+y^2}$ then $f(x,y)$ is continuous at the origin (since $\lim_{(x,y)\to(0,0)} f(x,y) = \lim_{(x,y)\to(0,0)} \sqrt[3]{x^2+y^2} = 0 = f(0,0)$), however $\sqrt[3]{x^2+y^2} > \sqrt{x^2+y^2}$ for any (x, y) in any neighborhood of the origin of the radius less than 1.

Remark. The converse is true.

Example 7. "If $f(x,y)$ is defined in a neighborhood of a point (a,b) and is decreasing with respect to the distance to (a,b), then (a,b) is a local maximum point of $f(x,y)$."

(compare with Example 10, section 3.2)

Solution.

The function $f(x,y) = \begin{cases} \frac{1}{x^2+y^2}, & x^2+y^2 \neq 0 \\ 0, & x^2+y^2 = 0 \end{cases}$ satisfies the conditions of the statement in any neighborhood of the origin, but $f(0,0) = 0 < \frac{1}{x^2+y^2} = f(x,y)$ for $\forall (x,y) \neq (0,0)$.

Remark. This statement will be true if the condition of continuity at the point (a,b) would be added.

7.2.2 Global properties

Example 8. "If $f(x,y)$ is continuous on a connected bounded set S and is not anywhere constant, then it can not take its global maximum and minimum values infinitely many times."

(compare with Example 2, section 3.3)

Solution.

Let us consider $f(x,y) = \begin{cases} x\left(\cos\frac{1}{x+y} - 1\right) - y\,, & (x,y) \neq (0,0) \\ 0\,, & (x,y) = (0,0) \end{cases}$ on the unit square $S = [0,1] \times [0,1]$. Due to arithmetic and composition rules this function is continuous on $S \backslash \{(0,0)\}$. It is also continuous at $(0,0)$, because $\lim\limits_{(x,y)\to(0,0)} f(x,y) = 0 = f(0,0)$ according to the squeeze theorem: $|f(x,y)| \leq 2|x| + |y| \underset{(x,y)\to(0,0)}{\to} 0$. Therefore, $f(x,y)$ is continuous on S. Also there is no connected open subset of S where $f(x,y)$ is constant. However, this function takes its global maximum infinitely many times. Indeed, $f(x,y) = x\left(\cos\frac{1}{x+y} - 1\right) - y \leq 0$ for $\forall (x,y) \in S$ and $f(0,0) = 0$, so zero is the global maximum value of $f(x,y)$. Furthermore, the function takes this value at all the points in S with $x_n = \frac{1}{2n\pi}$ and $y_n = 0$, $n \in \mathbb{N}$.

Remark. A similar example can be constructed for a global minimum or for both minimum and maximum.

Example 9. "If a function is defined on $\mathbb{R}^2$, it cannot be continuous at only one point."

(compare with Example 3, section 3.3)

Solution.

The function $f(x,y) = \begin{cases} |x| + |y|\,, & |x| + |y| \in \mathbb{Q} \\ -|x| - |y|\,, & |x| + |y| \in \mathbb{I} \end{cases}$ is discontinuous at any point, except for the origin. In fact, in any neighborhood of every point (x_0, y_0) there are points (x,y) satisfying both the condition $|x| + |y| \in \mathbb{Q}$ and

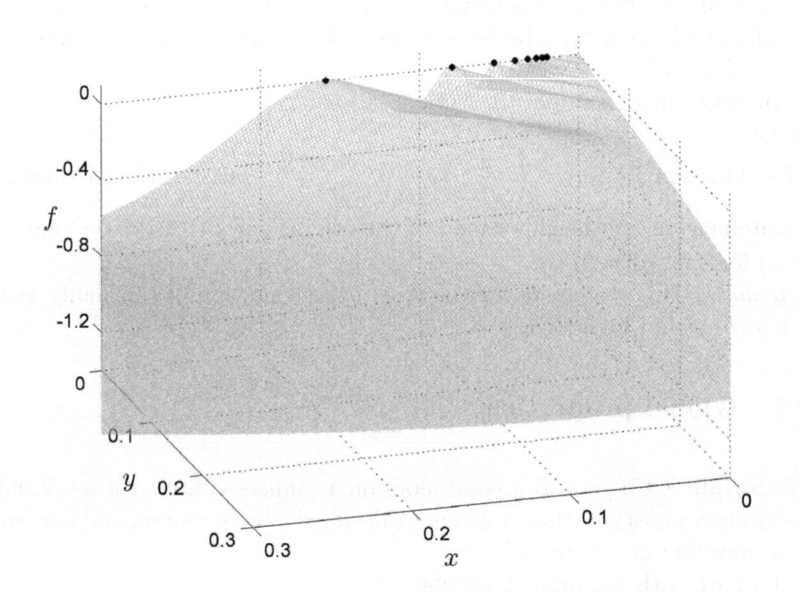

FIGURE 7.2.4: Example 8

$|x|+|y|\in\mathbb{I}$. Hence, using the former points, one gets the partial limit

$$\lim_{(x,y)\to(x_0,y_0)} f(x,y)=\lim_{(x,y)\to(x_0,y_0)} |x|+|y|=|x_0|+|y_0|,$$

while using the latter one arrives to

$$\lim_{(x,y)\to(x_0,y_0)} f(x,y)=\lim_{(x,y)\to(x_0,y_0)} -|x|-|y|=-|x_0|-|y_0|.$$

The two results are different for any $(x_0,y_0)\neq(0,0)$, that means that the general limit does not exist (and the function is not continuous) at any point different from the origin. If $(x_0,y_0)=(0,0)$, then any partial limit gives the same value zero, which coincide with $f(0,0)$. This means that the function is continuous at the origin.

Remark 1. The function $\tilde{d}(x,y)=|x|+|y|$ represents another kind of a distance from the origin (norm) in $\mathbb{R}^2$. For this distance, the points equidistant from the origin are located on the sides of a square centered at the origin and parallel to the bisectrices of the coordinate quadrants. In these terms, the function $f(x,y)$ equals to the distance from the origin, when this distance is a rational number, and to the minus distance, otherwise.

Remark 2. The statement is also false for an arbitrary domain.

Example 10. "If $f(x,y)$ is continuous on $\mathbb{R}^2$ and $\lim\limits_{n\to+\infty} f(n,n)=A$, then $\lim\limits_{(x,y)\to\infty} f(x,y)=A$."

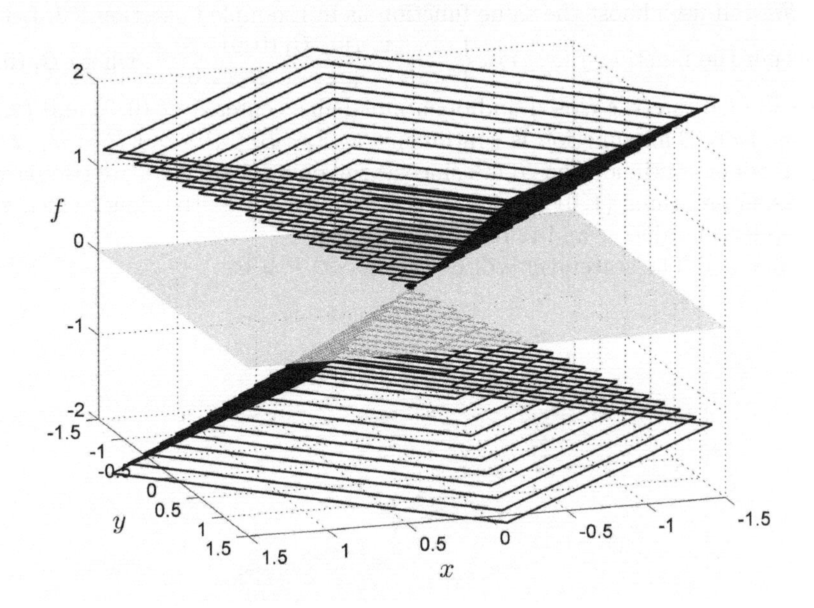

FIGURE 7.2.5: Example 9

(compare with Example 5, section 3.3)
Solution.

We can use almost the same function as in Example 5, section 3.3. The function $f(x,y) = \sin \pi x$ is continuous on $\mathbb{R}^2$ and $\lim\limits_{n\to+\infty} f(n,n) = \lim\limits_{n\to+\infty} \sin \pi n = 0$, but this is only one of the partial limits when x approaches $+\infty$. Another partial limit $\lim\limits_{n\to+\infty} f\left(2n+\frac{1}{2}, n\right) = \lim\limits_{n\to+\infty} \sin\left(2\pi n + \frac{\pi}{2}\right) = 1$ gives a different result, so $\lim\limits_{(x,y)\to\infty} f(x,y)$ does not exist.

Remark. The converse is true: if $\lim\limits_{(x,y)\to\infty} f(x,y) = A$ and $f(x,y)$ is defined on $\mathbb{N}\times\mathbb{N}$, then $\lim\limits_{n\to+\infty} f(n,n) = A$.

General Remark to the next examples on global properties of continuous functions. Each characterization of a set (such as connectedness, closeness, etc.) is related to the space where this set is considered, that is, any characterization of a domain (or its part) of a function is considered in $\mathbb{R}^2$, while a property of its image is considered in $\mathbb{R}$.

Example 11. "If $f(x,y)$ is continuous on connected sets S and P, then it is also continuous on $S\cup P$."

(compare with Example 6, section 3.3)
Solution.

We can use almost the same function as in Example 6, section 3.3. Let us consider the function $f(x,y) = \begin{cases} x, & (x,y) \in \bar{B}_1(0,0) \\ x+1, & (x,y) \in B_1(2,0) \end{cases}$, where $\bar{B}_1(0,0)$ and $B_1(2,0)$ are the closed and open unit balls, centered at $(0,0)$ and $(2,0)$, respectively. This function is continuous on $S = \bar{B}_1(0,0)$ and $P = B_1(2,0)$, but it is not continuous on $S \cup P$, since at the point $(1,0)$ there are two partial limits, approaching $(1,0)$ along x-axis at the left and at the right, which give two different values 1 and 2, respectively.

Remark. The statement will be true if $S \cap P \neq \varnothing$.

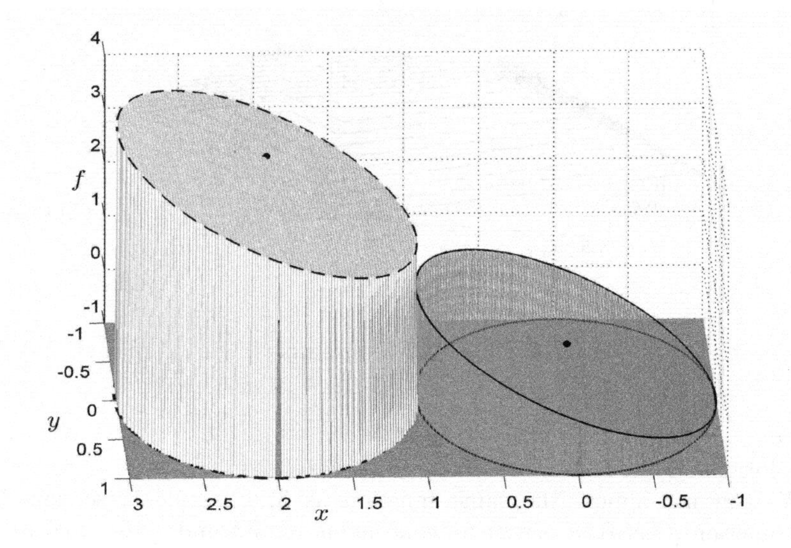

FIGURE 7.2.6: Example 11

Example 12. "If $f(x)$ is continuous on an open set S, then its image is also an open set."

(compare with Example 1, section 3.4)

Solution.

The application of the counterexample in Example 1, Section 3.4 is straightforward: the function $f(x,y) = x^2 + y^2$ is continuous on $(-1,1) \times (-1,1)$, but its image is the half-open interval $[0,2)$.

Another simple counterexample is the function $f(x,y) = \sin(x+y)$: it is continuous on the open set $(0,2\pi) \times (0,2\pi)$, but its image on this set is the closed interval $[-1,1]$.

Remark. The Remarks to Example 1 of section 3.4 are applicable here with corresponding adjustments to the case of $\mathbb{R}^2$.

Example 13. "If $f(x,y)$ attains its global minimum and maximum on a compact S, then it is continuous on S."

(compare with Example 10, section 3.4)
Solution.
The extension of one-dimensional counterexamples is straightforward. For instance, the two-dimensional version of Dirichlet's function $D(x,y) = \begin{cases} 1, & \sqrt{x^2+y^2} \in \mathbb{Q} \\ 0, & \sqrt{x^2+y^2} \in \mathbb{I} \end{cases}$ attains its global minimum 0 and maximum 1 on any compact set, containing interior points, but it does not have a limit at any point.

Another simple counterexample is $f(x,y) = \operatorname{sgn}(x+y)$ considered on any compact set containing the origin as the interior point (for instance, $S = [-1,1] \times [-1,1]$).

Remark. The converse statement is true and represents the second Weierstrass theorem.

Example 14. "If $f(x,y)$ is continuous on S, then $f(x,y)$ assumes any value between $f(x_1,y_1)$ and $f(x_2,y_2)$, (x_1,y_1), $(x_2,y_2) \in S$ on the points of S."
(compare with Example 12, section 3.4)
Solution.
The function from Example 12, Section 3.4 can be easily adapted to the two-dimensional case. Let us consider $f(x,y) = \operatorname{sgn} x$ on the set $S = B_1(-1,0) \cup B_1(1,0)$, where $B_1(-1,0)$ and $B_1(1,0)$ are the open unit balls, centered at $(-1,0)$ and $(1,0)$, respectively. Evidently, this function is continuous on S, but it does not assume any value between $f(-1,0) = -1$ and $f(1,0) = 1$.

Remark. The intermediate value property is satisfied for the functions continuous on a connected sets.

Example 15. "Between two local minima there is a local maximum."
(compare with Example 13, section 3.4)
Solution.
The function in Example 13, section 3.4 is easily adaptable to this case: $f(x,y) = \sec^2 x$ has infinitely many local minima on $\mathbb{R}^2$ and no local maximum.

Remark. This statement is true for continuous functions with domain in $\mathbb{R}$. For multivariate functions the statement is false even when $f(x,y)$ is continuous (see Example 11 in section 7.3).

Example 16. "If $f(x,y)$ is continuous and bounded on a bounded connected set S, then $f(x,y)$ is uniformly continuous on this set."
(compare with Example 1, section 3.5)
Solution.
The function $f(x,y) = \sin\frac{\pi}{x}$ (adapted from Example 1, section 3.5) is bounded and continuous on $S = (0,1) \times (-1,1)$, but it does not possess uniform continuity on S. In fact, for the sequences $(x_n, 0) = \left(\frac{1}{n}, 0\right)$, $\forall n \in$

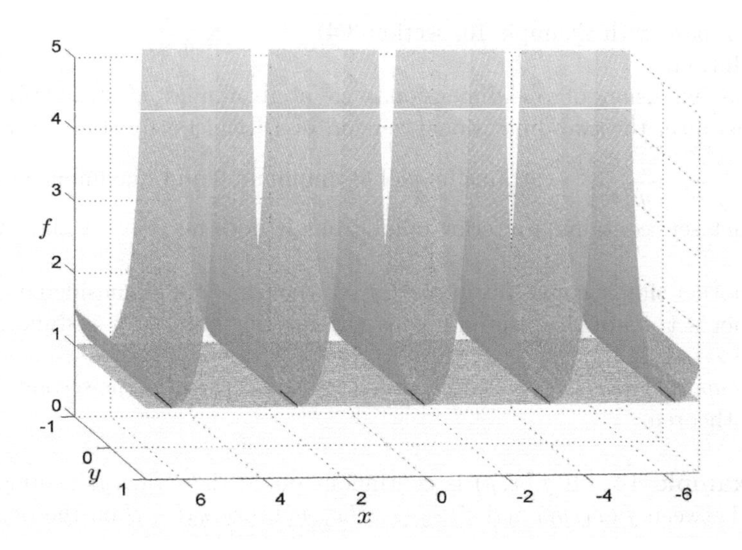

FIGURE 7.2.7: Example 15

$\mathbb{N}$, $n > 1$ and $(\tilde{x}_n, 0) = \left(\frac{2}{1+4n}, 0\right)$, $\forall n \in \mathbb{N}$ we have the property that the distance between the corresponding points of the two sets tends to zero

$$\sqrt{(x_n - \tilde{x}_n)^2 + (0-0)^2} = |x_n - \tilde{x}_n| = \left|\frac{1}{n} - \frac{2}{1+4n}\right| \underset{n\to+\infty}{\to} 0,$$

but

$$|f(x_n, 0) - f(\tilde{x}_n, 0)| = |0 - 1| = 1, \forall n \in \mathbb{N},\ n > 1.$$

Therefore, $f(x, y)$ is not uniformly continuous on S.

Remark. The statement will be correct if S would be compact (according to the Cantor theorem).

Example 17. "If $f(x, y)$ is uniformly continuous on $\mathbb{R}^2$, then $f^2(x, y)$ is also uniformly continuous on $\mathbb{R}^2$."

(compare with Example 4, section 3.5)

Solution.

The function $f(x, y) = x$ (adapted from Example 4, section 3.5) is uniformly continuous on $\mathbb{R}^2$, but $f^2(x, y) = x^2$ is not, since for the two sequences $(x_n, 0) = \left(n + \frac{1}{n}, 0\right)$, $\forall n \in \mathbb{N}$ and $(\tilde{x}_n, 0) = (n, 0)$, $\forall n \in \mathbb{N}$ the distance between the corresponding points of the two sets gets as close as we wish for sufficiently large n: $|x_n - \tilde{x}_n| = \frac{1}{n} \underset{n\to+\infty}{\to} 0$, but

$$\left|f^2(x_n) - f^2(\tilde{x}_n)\right| = 2 + \frac{1}{n^2} > 2, \forall n \in \mathbb{N}.$$

7.3 Multidimensional essentials

Example 1. "If a function approaches zero, as a point approaches the origin along any of the coordinate axis, then its limit is zero at the origin."

Solution.

The function $f(x,y)=\begin{cases}\frac{xy}{x^2+y^2}, & x^2+y^2\neq 0\\ 0, & x^2+y^2=0\end{cases}$ has both coordinate limits equal to zero at the origin: $\lim\limits_{x\to 0} f(x,0)=\lim\limits_{x\to 0}\frac{0}{x^2}=0$ and $\lim\limits_{y\to 0} f(0,y)=\lim\limits_{y\to 0}\frac{0}{y^2}=0$. However, it does not possess the general limit, since the partial limits along straight lines $y=kx$, $k\in\mathbb{R}$, depend on the chosen line: $\lim\limits_{x\to 0} f(x,kx)=\lim\limits_{x\to 0}\frac{kx^2}{x^2+k^2x^2}=\frac{k}{1+k^2}$.

Remark. For the continuity, this false statement can be reformulated as follows: "if a function is continuous with respect to each of variables at some point, then it is continuous at this point". The same function provides counterexample.

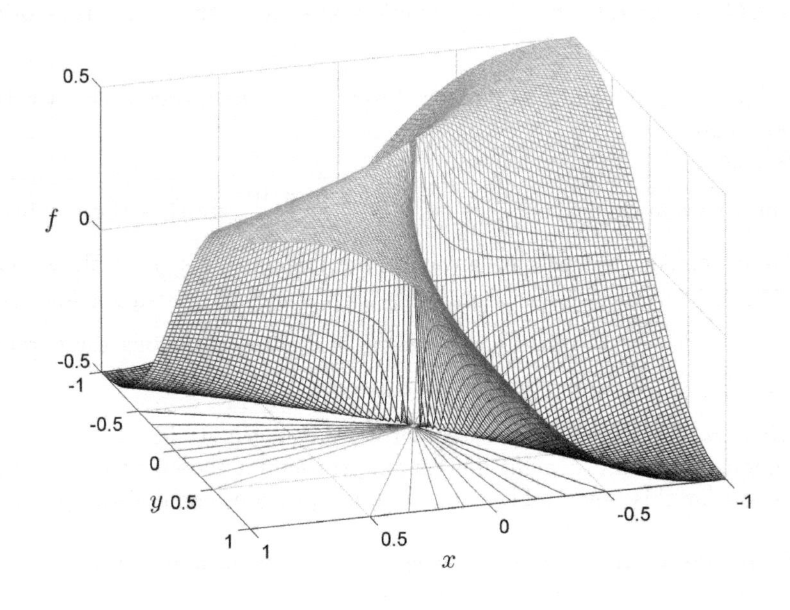

FIGURE 7.3.1: Example 1, graph of the function

Example 2. "If a function approaches zero, as a point approaches the

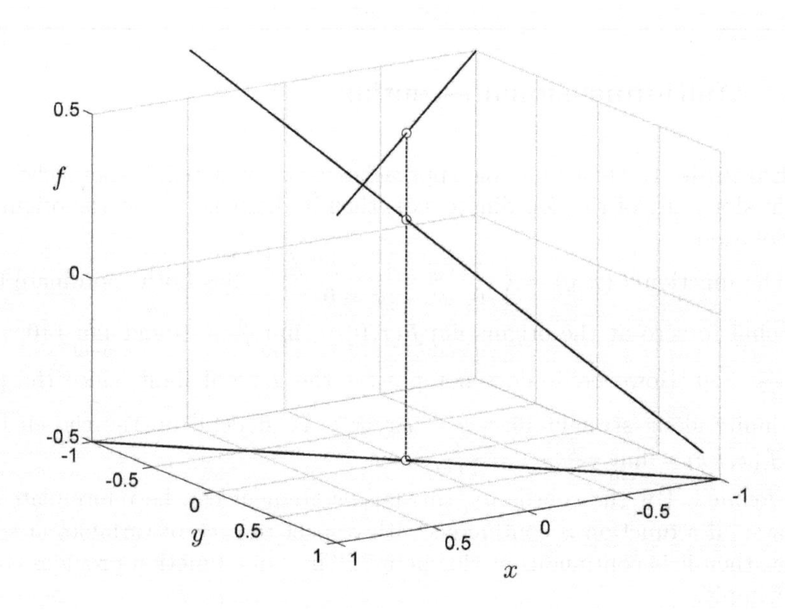

FIGURE 7.3.2: Example 1, the function on some paths through the origin

origin along any straight line passing through the origin, then its limit is zero at the origin."

Solution.

The function $f(x,y) = \begin{cases} \frac{x^2 y}{x^4+y^2}, & x^2+y^2 \neq 0 \\ 0, & x^2+y^2=0 \end{cases}$ satisfies the conditions of the statement. Indeed, for any straight line $y = kx$, $k \in \mathbb{R}$, we have $\lim_{x\to 0} f(x,kx) = \lim_{x\to 0} \frac{kx^3}{x^4+k^2x^2} = \lim_{x\to 0} \frac{kx}{x^2+k^2} = 0$, and for the straight line $x=0$ the limit is also zero: $\lim_{y\to 0} f(0,y) = \lim_{y\to 0} \frac{0}{y^2} = 0$. However, along the parabola $y = x^2$ the result is different: $\lim_{x\to 0} f(x,x^2) = \lim_{x\to 0} \frac{x^4}{x^4+x^4} = \frac{1}{2}$.

Remark. Analogous false statement for the continuity is: "if a function is continuous along any straight line passing through a chosen point, then it is continuous at this point". The same function provides counterexample.

Example 3. "If $f(x,y)$ approaches zero, as a point approaches the origin along any algebraic curve $y = cx^{m/n}$, where $c \in \mathbb{R}$ and $m, n \in \mathbb{N}$ ($x \geq 0$ in case the exponential fraction is in lowest terms and n is even), then the limit of $f(x,y)$ is zero at the origin."

Solution.

Let us show that the function $f(x,y) = \begin{cases} \frac{e^{-1/x^2} y}{e^{-2/x^2}+y^2}, & x \neq 0 \\ 0, & x=0 \end{cases}$ satisfies

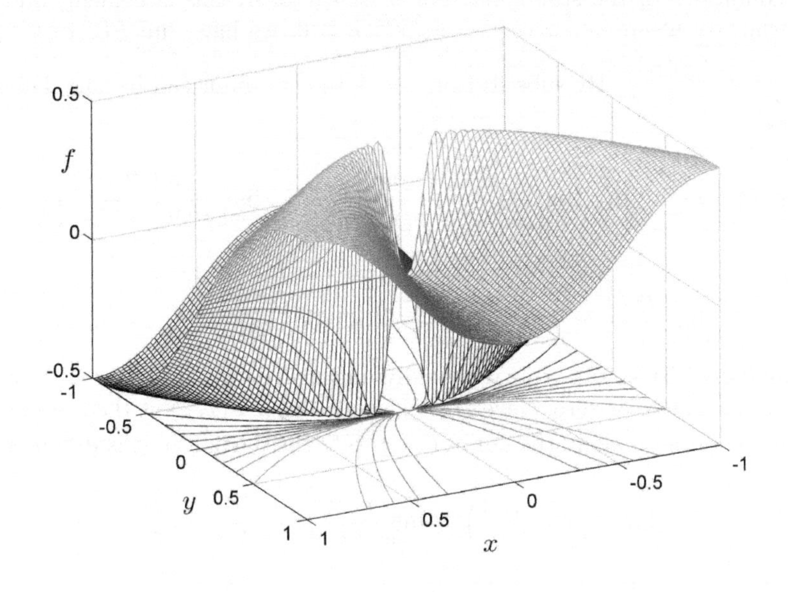

FIGURE 7.3.3: Example 2, graph of the function

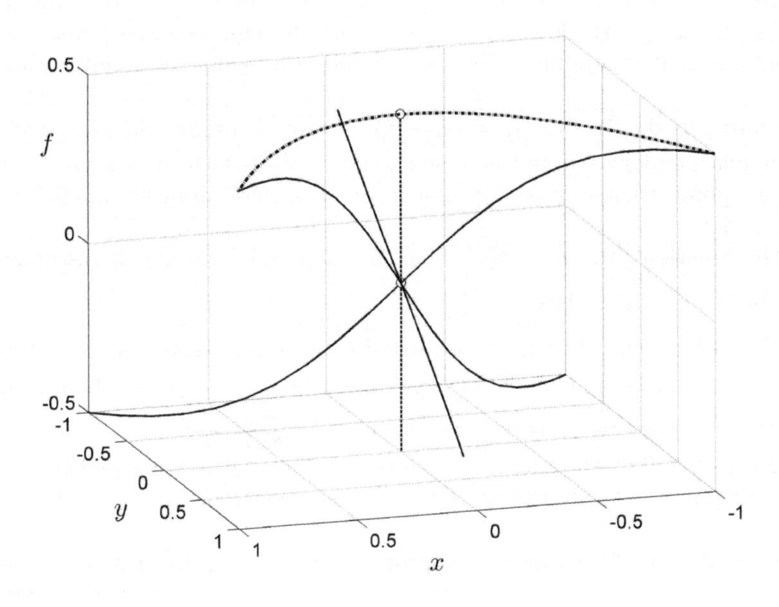

FIGURE 7.3.4: Example 2, the function on some paths through the origin

the conditions of the statement. For $y=0$ $(c=0)$ this is evident, and for an arbitrary algebraic curve $y=cx^{m/n}$, $c\neq 0$, we have $\lim\limits_{x\to 0} f\left(x,cx^{m/n}\right)=\lim\limits_{x\to 0}\frac{e^{-1/x^2}cx^{m/n}}{e^{-2/x^2}+c^2x^{2m/n}}$. By substitution $x=\frac{1}{t}$ the last limit can be calculated as follows:

$$\lim_{x\to 0} f\left(x,cx^{m/n}\right)=\lim_{t\to\infty}\frac{ce^{-t^2}\left(1/t\right)^{m/n}}{e^{-2t^2}+c^2\left(1/t\right)^{2m/n}}=\lim_{t\to\infty}\frac{ct^{m/n}/e^{t^2}}{t^{2m/n}/e^{2t^2}+c^2}=0,$$

since

$$\lim_{t\to\infty}\frac{t^{m/n}}{e^{t^2}}=\lim_{t\to\infty}\frac{m}{n}\frac{t^{m/n-1}}{2te^{t^2}}=\lim_{t\to\infty}\frac{m}{n}\frac{t^{m/n-2}}{2e^{t^2}}=\ldots=0$$

by applying L'Hospital's rule repeatedly until the power of t in the numerator becomes negative or zero. Nevertheless, the limit of $f(x,y)$ at the origin does not exist, because along the curve $y=e^{-1/x^2}$ the partial limit is different:

$$\lim_{x\to 0} f\left(x,e^{-1/x^2}\right)=\lim_{x\to 0}\frac{e^{-1/x^2}\cdot e^{-1/x^2}}{e^{-2/x^2}+e^{-2/x^2}}=\frac{1}{2}.$$

Remark 1. Of course, the family of paths(curves) that assures the same value for the partial limits can be even broader(wider) than the set of algebraic curves, but still the general limit may not exist.

Remark 2. For the continuity, a similar false statement is: "if a function is continuous along any algebraic curve passing through a chosen point, then it is continuous at this point". The same function provides the counterexample.

Example 4. "If $f(x,y)=\frac{P_n(x,y)}{Q_m(x,y)}$, where $P_n(x,y)$ and $Q_m(x,y)$ are polynomials of degree n and m, respectively, and the latter polynomial is zero at some point, then $f(x,y)$ does not have the general limit at this point."

Solution.

The function $f(x,y)=\frac{x^2-y^2}{x+y}$ has the required form and it also has the limit at the origin: $\lim\limits_{(x,y)\to(0,0)}\frac{x^2-y^2}{x+y}=\lim\limits_{(x,y)\to(0,0)}(x-y)=0$.

Remark 1. The statement is also false if the numerator or denominator functions are not polynomials. For example, $f(x,y)=\frac{\sin\left(x^2+y^2\right)}{x^2+y^2}$ has the limit equal to 1 at the origin, as well as $f(x,y)=\frac{x^2+y^2}{\arctan(x^2+y^2)}$.

Remark 2. The statement will be true if $P_n(x,y)$ is not equal to zero at the limit point and only the finite limits are considered.

Example 5. "If a function approaches zero along any ray $x=t\cos\alpha$, $y=t\sin\alpha$, $0\le t<+\infty$, as t approaches infinity, then it approaches zero at infinity."

Solution.

The function $f(x,y)=x^2e^{-\left(x^2-y\right)}$ satisfies the conditions of the statement. Indeed, if $\cos\alpha=0$ $(\alpha=\pm\frac{\pi}{2})$, then $f(t\cos\alpha,t\sin\alpha)=0$. For any other α we have

$$\lim_{t\to+\infty} f(t\cos\alpha, t\sin\alpha) = \lim_{t\to+\infty} \frac{t^2\cos^2\alpha}{e^{t^2\cos^2\alpha - t\sin\alpha}}$$

and applying L'Hospital's rule twice we obtain:

$$\lim_{t\to+\infty} f(t\cos\alpha, t\sin\alpha) = \lim_{t\to+\infty} \frac{2t\cos^2\alpha}{e^{t^2\cos^2\alpha - t\sin\alpha}\cdot(2t\cos^2\alpha - \sin\alpha)}$$

$$= \lim_{t\to+\infty} \frac{2\cos^2\alpha}{e^{t^2\cos^2\alpha - t\sin\alpha}\cdot\left((2t\cos^2\alpha - \sin\alpha)^2 + 2\cos^2\alpha\right)} = 0.$$

However, approaching infinity along the parabola $y = x^2$, we obtain $\lim\limits_{x\to+\infty} f(x, x^2) = \lim\limits_{x\to+\infty} x^2 e^{-\left(x^2 - x^2\right)} = \lim\limits_{x\to+\infty} x^2 = +\infty$, which means that a general limit at infinity does not exist.

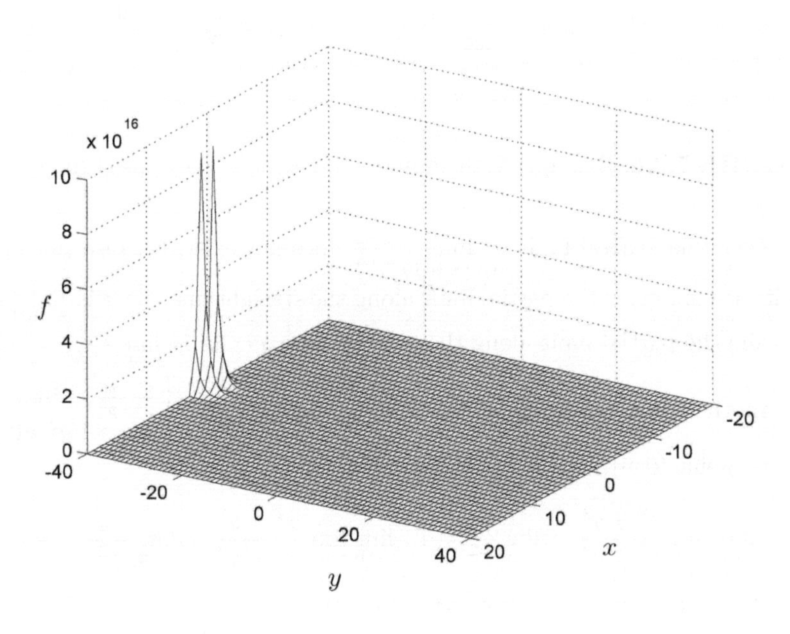

FIGURE 7.3.5: Example 5, graph of the function

Example 6. "If iterated limits of $f(x, y)$ exist, then the general (double) limit also exists."

Solution.

For the function $f(x, y) = \frac{x-y}{x+y}$ both iterated limits exist at the point $(0, 0)$:

$$\lim_{x\to 0}\lim_{y\to 0} \frac{x-y}{x+y} = \lim_{x\to 0} \frac{x}{x} = \lim_{x\to 0} 1 = 1, \quad \lim_{y\to 0}\lim_{x\to 0} \frac{x-y}{x+y} = \lim_{x\to 0} \frac{-y}{y} = \lim_{x\to 0} (-1) = -1.$$

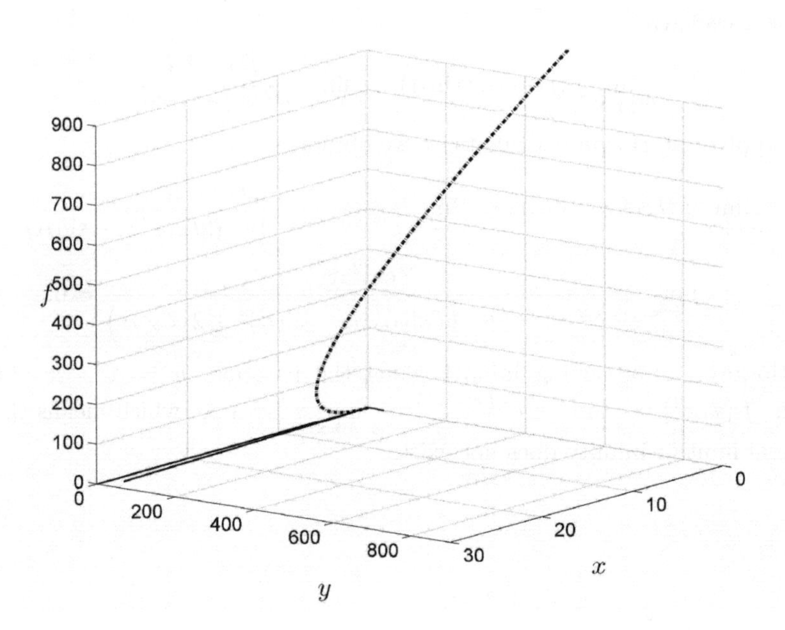

FIGURE 7.3.6: Example 5, the function on some paths approaching infinity

However, the general limit $\lim\limits_{(x,y)\to(0,0)} \frac{x-y}{x+y}$ does not exist, because two partial limits are different: the partial limit along the straight line $y = x$ is $\lim\limits_{x\to 0} \frac{x-x}{x+x} = 0$, while the partial limit along the straight line $y = 2x$ is $\lim\limits_{x\to 0} \frac{x-2x}{x+2x} = -\frac{1}{3}$.

Another counterexample of the same kind is $f(x,y) = \begin{cases} \frac{x^2-y^2}{x^2+y^2}, & x^2+y^2 \neq 0 \\ 0, & x^2+y^2 = 0 \end{cases}$.

At the point $(0,0)$ both iterated limits exist:

$$\lim_{x\to 0}\lim_{y\to 0} \frac{x^2-y^2}{x^2+y^2} = \lim_{x\to 0} \frac{x^2}{x^2} = 1, \ \lim_{y\to 0}\lim_{x\to 0} \frac{x^2-y^2}{x^2+y^2} = \lim_{x\to 0} \frac{-y^2}{y^2} = -1,$$

but the general limit does not exist since the partial limit along the straight line $y = x$, $\lim\limits_{x\to 0} \frac{x^2-x^2}{x^2+x^2} = 0$, is different from the partial limit along the straight line $y = 2x$, $\lim\limits_{x\to 0} \frac{x^2-4x^2}{x^2+4x^2} = -\frac{3}{5}$.

Remark. Both functions are also counterexamples to the following false statement: "if iterated limits exist, then they are equal".

Example 7. "If both iterated limits exist and are equal, then the general (double) limit also exists."

Solution.

The function of the Example 1, $f(x,y) = \begin{cases} \frac{xy}{x^2+y^2}, & x^2+y^2 \neq 0 \\ 0, & x^2+y^2 = 0 \end{cases}$ has both iterated limits at the point $(0,0)$ equal to 0:

$$\lim_{x\to 0}\lim_{y\to 0}\frac{xy}{x^2+y^2}=\lim_{x\to 0}\frac{0}{x^2}=0,\ \lim_{y\to 0}\lim_{x\to 0}\frac{xy}{x^2+y^2}=\lim_{x\to 0}\frac{0}{y^2}=0.$$

However, the general limit $\lim\limits_{(x,y)\to(0,0)}\frac{xy}{x^2+y^2}$ does not exist since the partial limits along the straight lines $y=kx$ are different (see Example 1): for example, along the straight line $y=x$ one has $\lim\limits_{x\to 0}\frac{x^2}{x^2+x^2}=\frac{1}{2}$, while along the straight line $y=2x$ one gets $\lim\limits_{x\to 0}\frac{2x^2}{x^2+4x^2}=\frac{2}{5}$.

Another counterexample of the same kind is the function

$$f(x,y)=\begin{cases}\frac{x^2y^2}{x^2y^2+(x-y)^2}, & x^2+y^2\neq 0\\ 0, & x^2+y^2=0\end{cases}.$$

In fact, both iterated limits are equal to 0:

$$\lim_{x\to 0}\lim_{y\to 0}\frac{x^2y^2}{x^2y^2+(x-y)^2}=\lim_{x\to 0}\frac{0}{x^2}=0,\ \lim_{y\to 0}\lim_{x\to 0}\frac{x^2y^2}{x^2y^2+(x-y)^2}=\lim_{x\to 0}\frac{0}{y^2}=0,$$

but the general limit does not exist, since for $y=x$ one obtains $\lim\limits_{x\to 0}\frac{x^4}{x^4}=1$, while for $y=0$ the result is different $\lim\limits_{x\to 0}\frac{0}{x^2}=0$.

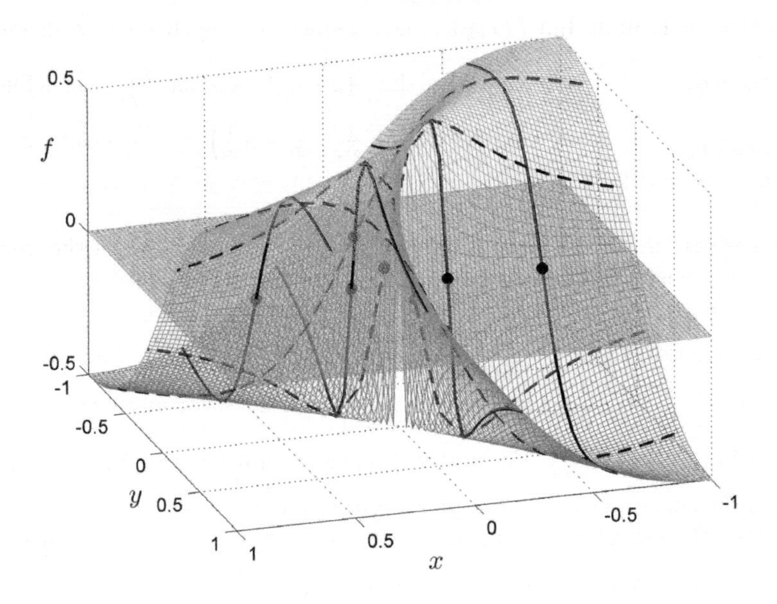

FIGURE 7.3.7: Example 7

Example 8. "If the general limit of $f(x,y)$ exists, then iterated limits also exist."

Solution.

The function $f(x,y) = \begin{cases} (x+y)\sin\frac{1}{xy}, & xy \neq 0 \\ 0, & xy = 0 \end{cases}$ has the general limit at the origin. In fact, the inequality $\left|(x+y)\sin\frac{1}{xy}\right| \leq |x| + |y|$ leads to the evaluation $|f(x,y)| \leq |x| + |y|$ for an arbitrary point (x,y). Therefore, due to the squeeze theorem, the fact that $\lim\limits_{(x,y)\to(0,0)} (|x| + |y|) = 0$ implies that $\lim\limits_{(x,y)\to(0,0)} f(x,y) = 0$. At the same time, $\lim\limits_{y\to 0} (x+y)\sin\frac{1}{xy}$ does not exist for any $x \neq 0$, because for any given $x_0 \neq 0$ there is a sequence of the points $y_n = \frac{1}{2n\pi x_0} \underset{n\to\infty}{\to} 0$ such that $\lim\limits_{y_n\to 0} (x_0 + y_n)\sin\frac{1}{x_0 y_n} = 0$ and another sequence $y_k = \frac{1}{\left(2k\pi + \frac{\pi}{2}\right)x_0} \underset{k\to\infty}{\to} 0$ such that $\lim\limits_{y_k\to 0} (x_0 + y_k)\sin\frac{1}{x_0 y_k} = x_0$. It means, that the iterated limit $\lim\limits_{x\to 0}\lim\limits_{y\to 0} f(x,y)$ does not exist. Due to the same reasons, the second iterated limit $\lim\limits_{y\to 0}\lim\limits_{x\to 0} f(x,y)$ does not exist either.

Similar counterexample is provided with the function $f(x,y) = \begin{cases} x\sin\frac{1}{y} + y\sin\frac{1}{x}, & xy \neq 0 \\ 0, & xy = 0 \end{cases}$. The general limit is equal to 0 due to inequality $\left|x\sin\frac{1}{y} + y\sin\frac{1}{x}\right| \leq |x| + |y|$ and application of the squeeze theorem. Nevertheless, the iterated limit $\lim\limits_{x\to 0}\lim\limits_{y\to 0} f(x,y)$ does not exist, because the interior(internal) limit $\lim\limits_{y\to 0} f(x,y)$ is not defined for any fixed $x \neq 0$: for the sequence $y_n = \frac{1}{n\pi} \underset{n\to\infty}{\to} 0$ one gets $\lim\limits_{y_n\to 0}\left(x\sin\frac{1}{y_n} + y_n\sin\frac{1}{x}\right) = 0$, while for $y_k = \frac{2}{(4k+1)\pi} \underset{k\to\infty}{\to} 0$ one has $\lim\limits_{y_k\to 0}\left(x\sin\frac{1}{y_k} + y_k\sin\frac{1}{x}\right) = x$. Consideration of another iterated limit is analogous.

Example 9. "If the general limit of $f(x,y)$ exists and one of the iterated limits also exists, then the second iterated limit exists as well."

Solution.

The function $f(x,y) = \begin{cases} y + x\sin\frac{1}{y}, & y \neq 0 \\ 0, & y = 0 \end{cases}$ has the zero limit at the origin, because $|f(x,y)| \leq |x| + |y| \underset{(x,y)\to(0,0)}{\to} 0$. Besides, for any fixed $y \neq 0$ we obtain $\lim\limits_{x\to 0}\left(y + x\sin\frac{1}{y}\right) = y$, and consequently $\lim\limits_{y\to 0}\lim\limits_{x\to 0} f(x,y) = \lim\limits_{y\to 0} y = 0$. Hence, the general limit and one of the iterated limits exist and are equal. Nevertheless, the second iterated limit does not exist, since the limit $\lim\limits_{y\to 0} f(x,y)$ is not defined for any fixed $x \neq 0$: for the sequence $y_n = \frac{1}{n\pi} \underset{n\to\infty}{\to} 0$ one gets

$\lim\limits_{y_n \to 0} \left(y_n + x \sin \frac{1}{y_n} \right) = 0$, while for $y_k = \frac{2}{(4k+1)\pi} \underset{k \to \infty}{\to} 0$ the result is different $\lim\limits_{y_k \to 0} \left(y_k + x \sin \frac{1}{y_k} \right) = x$.

Remark. The function $\tilde{f}(x, y) = \begin{cases} x + y \sin \frac{1}{x}, & x \neq 0 \\ 0, & x = 0 \end{cases}$ provides the counterexample with the interchanged variables.

Example 10. "If the general limit and one of the iterated limits of $f(x, y)$ do not exist, then the second iterated limit also does not exist."

Solution.

The function $f(x, y) = \begin{cases} \frac{xy}{x^2+y^2} + x \sin \frac{1}{y}, & y \neq 0 \\ 0, & y = 0 \end{cases}$ does not have a general limit, because two partial limits give different results: along the line $y = x$ one obtains $\lim\limits_{x \to 0} \left(\frac{x^2}{2x^2} + x \sin \frac{1}{x} \right) = \frac{1}{2}$ (since $\left| x \sin \frac{1}{x} \right| \leq |x| \underset{x \to 0}{\to} 0$), while along the line $y = 2x$ one has $\lim\limits_{x \to 0} \left(\frac{2x^2}{x^2+4x^2} + x \sin \frac{1}{2x} \right) = \frac{2}{5}$. Moreover, the limit $\lim\limits_{x \to 0} \lim\limits_{y \to 0} f(x, y)$ also does not exist, because $\lim\limits_{y \to 0} f(x, y)$ is not defined for any fixed $x \neq 0$: if $y_n = \frac{1}{n\pi} \underset{n \to \infty}{\to} 0$ then $\lim\limits_{y_n \to 0} \left(\frac{x y_n}{x^2+y_n^2} + x \sin \frac{1}{y_n} \right) = 0$, but if $y_k = \frac{2}{(4k+1)\pi} \underset{k \to \infty}{\to} 0$ then $\lim\limits_{y_k \to 0} \left(\frac{x y_k}{x^2+y_k^2} + x \sin \frac{1}{y_k} \right) = x$. Nevertheless, the second iterated limit exists:

$$\lim_{y \to 0} \lim_{x \to 0} f(x, y) = \lim_{y \to 0} \lim_{x \to 0} \left(\frac{xy}{x^2+y^2} + x \sin \frac{1}{y} \right) = \lim_{y \to 0} 0 = 0.$$

Remark. The function $\tilde{f}(x, y) = \begin{cases} \frac{xy}{x^2+y^2} + y \sin \frac{1}{x}, & x \neq 0 \\ 0, & x = 0 \end{cases}$ provides the counterexample with the interchanged variables.

Example 11. "If a continuous on a connected set S function $f(x, y)$ possesses two local maxima in S, then there exists a local minimum of $f(x, y)$ in S."

Solution.

The function $f(x, y) = \sin x - y^2$ considered on $S = (-2\pi, 2\pi) \times (-1, 1)$ has two local (and global) maxima at the points $P_1 = \left(-\frac{3\pi}{2}, 0 \right)$ and $P_2 = \left(\frac{\pi}{2}, 0 \right)$. Indeed, the function $\sin x$ assumes its maximum values at the points $-\frac{3\pi}{2}$ and $\frac{\pi}{2}$ in the interval $(-2\pi, 2\pi)$ and the function $-y^2$ has the maximum value at the point 0 of the interval $(-1, 1)$. However, there is no minimum point of $f(x, y)$, since along the line $x = x_0$ the original function is reduced to one-variable function $f(x_0, y) = \sin x_0 - y^2$, which does not have a minimum on all its domain $y \in \mathbb{R}$.

Remark. The same statement is true for one-variable functions.

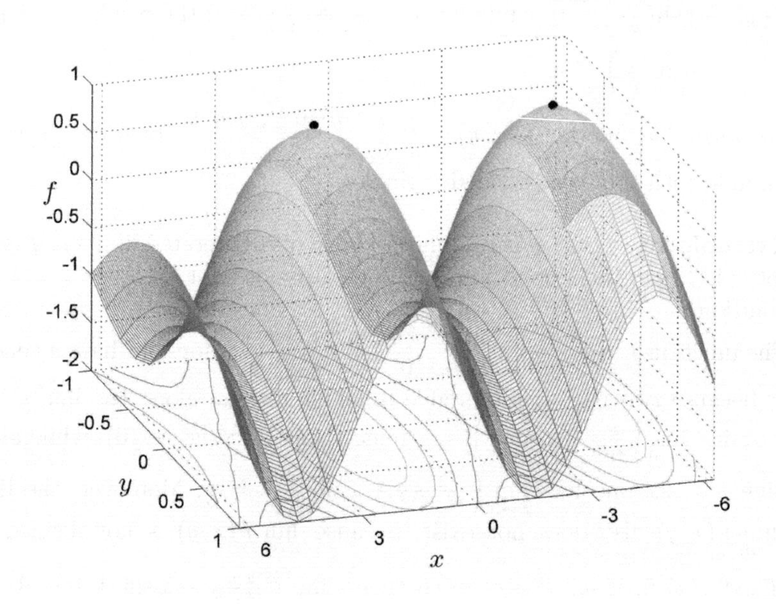

FIGURE 7.3.8: Example 11

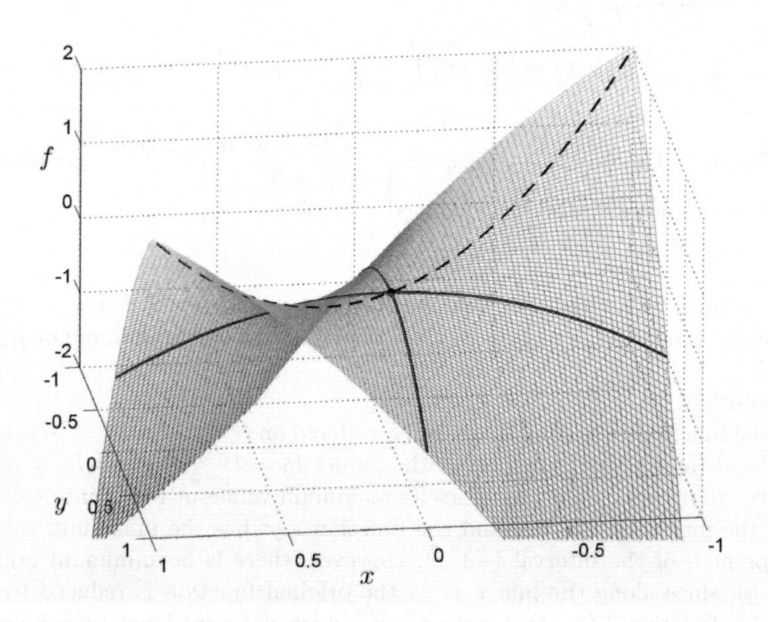

FIGURE 7.3.9: Example 12

Example 12. "If a function $f(x, y)$ is continuous in a neighborhood of the point (x_0, y_0), and both $f(x, y_0)$ and $f(x_0, y)$ have local maxima at this point, then $f(x, y)$ possesses a local maximum at (x_0, y_0)."

Solution.

The function $f(x, y) = -x^2 + 4xy - y^2$ is continuous on $\mathbb{R}^2$, in particular, in any neighborhood of $(0, 0)$. The functions $f(x, 0) = -x^2$ and $f(0, y) = -y^2$ have local (and global) maxima at the origin. However, the function $f(x, y)$ does not have a local maximum at $(0, 0)$, because along the line $y = x$ we get $f(x, x) = 2x^2$ and for this function the origin is the local minimum.

Exercises

1. Show that if a function approaches zero, as a point approaches the origin along any straight line and any parabola, it does not mean that this function has a limit at the origin. Formulate this result as a counterexample. ($f(x,y) = \begin{cases} \frac{2x^3 y}{x^6+y^2}, & x^2+y^2 \neq 0 \\ 0, & x^2+y^2=0 \end{cases}$)

2. Show that the function $f(x,y) = \frac{3x^4-5y^2}{x^4+y^2}$ has the iterated limits (albeit different), but does not have the general limit at the origin. Notice that this is another counterexample to Example 6, section 7.3.

3. Show that the function $f(x,y) = \begin{cases} \frac{x^3 y}{x^4+y^4}, & x^2+y^2 \neq 0 \\ 0, & x^2+y^2=0 \end{cases}$ (see Example 1 section 7.3) has the equal iterated limits, but does not have the general limit. Notice that this is another counterexample to Example 7, section 7.3.

4. Prove that

$$\lim_{x\to\infty}\lim_{y\to\infty}\frac{x^2+y^2}{1+(x-y)^4} = \lim_{y\to\infty}\lim_{x\to\infty}\frac{x^2+y^2}{1+(x-y)^4},$$

but the general limit $\lim\limits_{(x,y)\to\infty}\frac{x^2+y^2}{1+(x-y)^4}$ does not exist. Formulate as a counterexample. Compare with Example 7, section 7.3.

5. Prove that $f(x,y) = \begin{cases} (2x^2+y^2)\cos\frac{1}{xy}, & xy \neq 0 \\ 0, & xy=0 \end{cases}$ is another counterexample to Example 8, section 7.3.

6. Prove that $f(x,y) = \begin{cases} \frac{x^2y^2}{x^4+y^4} + x^2\cos\frac{1}{y^2}, & y \neq 0 \\ 0, & y=0 \end{cases}$ is another counterexample to Example 10, section 7.3.

7. Give an example of a function $f(x,y)$ discontinuous at every point of the square $[0,1]\times[0,1]$, but continuous with respect to x for any fixed $y\in[0,1]$. Formulate as a counterexample. (Hint: use $f(x,y)=D(y)$.)

8. Show that the continuity of a function does not imply that small increments of function values produce small increments of independent variables. Formulate as a counterexample. (Hint: try $f(x,y) = \begin{cases} 0, & x^2+y^2 \leq 1 \\ x^2+y^2-1, & x^2+y^2>1 \end{cases}$)

9. Show that the condition that small increments of function values lead to small increments of independent variables does not guarantee the continuity. Formulate as a counterexample. (Hint: consider $f(x,y) = \begin{cases} x^2+y^2, & x \geq 0 \\ x^2+y^2-1, & x<0 \end{cases}$ and notice that both small and large increments the independent variables correspond to small increments of the function values.)

10. Differently from the theorem on composition of two continuous functions, the composition of two discontinuous functions can result in a continuous one. Formulate as a counterexample. (Hint: use $u = D(x-y)$, $v = D(x+y)$ and $z = D(u+v)$, where $D(t)$ is Dirichlet's function.)

11. "If $u = f(x,y)$ and $v = g(x,y)$ are discontinuous functions and $z = h(u,v)$ is a continuous function, then $z = h(f(x,y), g(x,y))$ is a discontinuous function." Prove by example that this is a false statement. (Hint: use $u = D(x-y)$, $v = D(x+y)$ and $z = (u-\frac{1}{2})^2 + (v-\frac{1}{2})^2$, where $D(t)$ is Dirichlet's function.)

12. Provide a counterexample to the following false statement: "if $f(x,y)$ is continuous on a compact set S, except for a one point, then $f(x,y)$ is bounded on S". What if additionally the function is defined on S?

13. Provide a counterexample to the following false statement: "if $f(x,y)$ is bounded on a compact set S and is continuous on S, except for a one point, then $f(x,y)$ attains its global extrema on S".

14. Prove by example that the following statement is false: "if $f(x,y)$ is continuous on a closed set S, then its image is also a closed set".

15. Prove by example that the following statement is false: "if $f(x,y)$ is continuous on a bounded set S and $f(S)$ is a compact, then S is also a compact set".

16. Show by example that the following statement is false: "if $f(x,y)$ satisfies the intermediate value property on a connected set S, then $f(x,y)$ is continuous on S".

17. Provide a counterexample to the following false statement: "if $f(x,y)$ is uniformly continuous on S and Q, then $f(x,y)$ is uniformly continuous on $S \cup Q$".

18. Provide a counterexample to the following false statement: "if $f(x,y)$ and $g(x,y)$ are uniformly continuous on S, then $f(x,y) \cdot g(x,y)$ is uniformly continuous on S". What about two non-uniformly continuous function? Should their product be non-uniformly continuous also?

19. Provide a counterexample to the following false statement: "if $f(x,y)$ is uniformly continuous on any compact set then it is uniformly continuous on $\mathbb{R}^2$".

20. Show by example that the following statement is false "if $f(x,y)$ is uniformly continuous and positive on S, then $\frac{1}{f(x,y)}$ is also uniformly continuous on S".

Chapter 8

Differentiability

8.1 Elements of Theory

Concepts

Partial derivatives. Suppose function $f(x, y)$ is defined on $X \subset \mathbb{R}^2$. Let us keep one of the independent coordinates, let say y, fixed and consider the function of one variable $g(x) = f(x, y_0)$. If this function is differentiable at x_0, then the derivative $g'(x_0)$ is called the *partial derivative* of $f(x, y)$ with respect to x at the point (x_0, y_0). The following notations are common: $f_x(x_0, y_0) \equiv \partial_x f(x_0, y_0) \equiv \partial_1 f(x_0, y_0) \equiv \frac{\partial f}{\partial x}(x_0, y_0)$. The specification of the point is frequently omitted if it is clear from the context. In the same way can be defined the partial derivative with respect to y.

Geometrically the partial derivative $f_x(x_0, y_0)$ represents the slope of the tangent line at (x_0, y_0) to the trace of the function graph on the plane $x = x_0$ in $\mathbb{R}^3$.

Partial derivatives on a set. A function $f(x, y)$ defined on X has partial derivative on a set $S \subset X$, if f_x exists at every point of S.

Higher (order) partial derivatives. If there exists a partial derivative of the function $h(x, y) = f_x(x, y)$ with respect to x at the point (x_0, y_0), then this partial derivative is called the *second order (or simply the second) partial derivative* of the original function $f(x, y)$ with respect to x, x at the point (x_0, y_0). The usual notations are (omitting the point indication): $f_{xx} \equiv \partial_{xx} f \equiv \partial_{11} f \equiv \frac{\partial^2 f}{\partial x^2}$. In the same way can be defined other three partial derivative of the second order: f_{xy}, f_{yx} and f_{yy}.

Similarly, if there exists the partial derivative with respect to x of the function $h(x, y) = f_{x^m}(x, y)$ representing the m-th partial derivative of $f(x, y)$, then this partial derivative is called the $(m+1)$-*th partial derivative* of the original function $f(x)$ with respect to x, ..., x. The standard notations are: $f_{x^{m+1}} \equiv f_{x\ldots x} \equiv \partial_{1\ldots1} f \equiv \frac{\partial^{m+1} f}{\partial x^{m+1}}$. In the same way can be defined all others $2^{m+1}-1$ partial derivative of the $(m+1)$-th order.

Clairaut's Theorem. If the mixed derivatives f_{xy} and f_{yx} are continuous at (x_0, y_0), then they are equal at (x_0, y_0).

Remark. A similar result holds for the mixed m-th partial derivatives: if the two mixed corresponding m-th partial derivatives are continuous at a point (x_0, y_0), then they are equal at (x_0, y_0).

Differentiation at a point. A function $f(x,y)$ defined on X is called *differentiable* at a point $(x_0, y_0) \in X$, if for any point (x,y) in a neighborhood of (x_0, y_0) the following representation holds: $f(x,y) = f(x_0,y_0) + a \cdot (x-x_0) + b \cdot (y-y_0) + \alpha \cdot (x-x_0) + \beta \cdot (y-y_0)$, where a, b are constants, and α, β are the functions of x, y such that $\alpha, \beta \underset{(x,y)\to(x_0,y_0)}{\to} 0$.

Remark 1. The same definition is frequently written in the terms of the increments of the arguments $\Delta x = x - x_0$, $\Delta y = y - y_0$ and the function $\Delta f = f(x,y) - f(x_0,y_0)$ as follows: $\Delta f = a \cdot \Delta x + b \cdot \Delta y + \alpha \cdot \Delta x + \beta \cdot \Delta y$, where a, b are constants and α, β are the functions of Δx, Δy such that $\alpha, \beta \underset{(\Delta x, \Delta y)\to(0,0)}{\to} 0$.

Remark 2. The same definition can be formulated in the following equivalent form: a function $f(x,y)$ is called differentiable at a point (x_0, y_0), if for any point in a neighborhood of (x_0, y_0) the following representation holds: $f(x,y) = f(x_0,y_0) + a \cdot (x-x_0) + b \cdot (y-y_0) + \gamma(x,y)$, where a, b are constants, and γ is the function of x, y that approaches zero faster than $\rho(x, y) = \sqrt{(x-x_0)^2 + (y-y_0)^2}$, that is $\frac{\gamma(x,y)}{\rho(x,y)} \underset{\rho\to0}{\to} 0$.

Differential and remainder. The expression $a \cdot \Delta x + b \cdot \Delta y \equiv df$ is called the *differential* of $f(x,y)$ at a point (x_0, y_0) and the difference $\gamma = \Delta f - df$ is called the remainder. Note that the term df is the linear (with respect to Δx, Δy) part of Δf, and this is also the main part of Δf in a small neighborhood of (x_0, y_0) if $a^2 + b^2 \neq 0$. Thus, the differential represents a linear approximation to the function increment in a neighborhood of (x_0, y_0).

Theorem. If $f(x,y)$ is differentiable at (x_0, y_0), then there exist both first partial derivatives f_x, f_y at (x_0, y_0) and $a = f_x$, $b = f_y$.

Gradient. The vector of the first partial derivatives (f_x, f_y) is called the *gradient*. The common notations are ∇f and grad$f$.

Remark. According to the last theorem and the gradient definition, the definition of the differentiation can be rewritten using the dot product in the form $\Delta f = \nabla f \cdot \Delta r + (\alpha, \beta) \cdot \Delta r$, where $\Delta r = (\Delta x, \Delta y)$ and α, β satisfy the required for differentiation properties. Sometimes the last formula is used as a definition of the differentiation. Accordingly, the definition of the differential can be rewritten as $df = \nabla f \cdot \Delta r$.

Differentiation on a set. If $f(x,y)$ is differentiable at every point of a set S, then $f(x,y)$ is differentiable on S.

Continuous differentiability. If the differential $g(x,y) \equiv df = f_x(x,y)\Delta x + f_y(x,y)\Delta y$ is a continuous function at (x_0, y_0) (keeping the increments $\Delta x, \Delta y$ fixed), then the original function $f(x,y)$ is continuously differentiable at (x_0, y_0). If the differential df is continuous on a set S, then $f(x,y)$ is continuously differentiable on S.

Higher (order) differentiation at a point. Let a function $f(x,y)$ be differentiable in a neighborhood of a point (x_0, y_0). If the differential $g(x,y) \equiv df = f_x(x,y)\Delta x + f_y(x,y)\Delta y$ is a differentiable function at (x_0, y_0) (keeping the increments $\Delta x, \Delta y$ fixed), then the original function $f(x,y)$ is *twice differentiable* and its *second (order) differential* is (by definition) equal to the differential of $g(x,y)$:

$$d^2 f \equiv dg = g_x \Delta x + g_y \Delta y = f_{xx}\Delta x \Delta x + f_{xy}\Delta x \Delta y + f_{yx}\Delta x \Delta y + f_{yy}\Delta y \Delta y.$$

In the case when the mixed partial derivatives are equal, the second differential can be simplified to the form

$$d^2 f = f_{xx}\Delta x \Delta x + 2 f_{xy}\Delta x \Delta y + f_{yy}\Delta y \Delta y.$$

The following symbolic form is helpful: $d^2 f = \left(\Delta x \partial_x + \Delta y \partial_y\right)^2 f$.

Similarly, if a function $f(x,y)$ is m times differentiable in a neighborhood of a point (x_0, y_0) and its m-th differential $h(x) \equiv d^m f = \left(\Delta x \partial_x + \Delta y \partial_y\right)^m f$ is a differentiable function at (x_0, y_0) (keeping the increments $\Delta x, \Delta y$ fixed), then the original function $f(x,y)$ is $m+1$ *times differentiable* at (x_0, y_0) and its *(m + 1)-th differential* can be found by the formula: $d^{m+1} f = \left(\Delta x \partial_x + \Delta y \partial_y\right)^{m+1} f$.

Directional derivative. Let us consider a half-straight line defined by the initial point (x_0, y_0) and the direction (non-zero) vector $v = (a, b)$: $x = x_0 + at$, $y = y_0 + bt$, $t \ge 0$ (t is the parameter). Along this ray a function $f(x,y)$ is the function of the one variable t: $h(t) \equiv f(x_0 + at, y_0 + bt)$. If the function $h(t)$ is differentiable at $t_0 = 0$, then its derivative is the directional derivative $D_v f(x_0, y_0)$ of the original function $f(x,y)$, that is, $D_v f(x_0, y_0) = h'(0)$.

Tangent plane. If $f(x,y)$ is differentiable at a point (x_0, y_0), then the following tangent plane to the graph of $f(x)$ at (x_0, y_0) can be defined: $z - f(x_0, y_0) = f_x(x_0, y_0) \cdot (x - x_0) + f_y(x_0, y_0) \cdot (y - y_0)$.

It follows immediately from the definition that the tangent plane contains the tangent lines at (x_0, y_0) to the two traces of the function graph - on the plane $x = x_0$ and on $y = y_0$.

Notice, that the linear approximation of a function through its tangent plane corresponds to the linear approximation of its increment through differential: $z - f(x_0, y_0) = df(x_0, y_0)$.

Hessian. The Hessian matrix (Hessian) is the matrix of the second partial derivatives $H(f) = \begin{pmatrix} f_{xx} & f_{xy} \\ f_{xy} & f_{yy} \end{pmatrix}$.

Basic properties

Theorem. If $f(x, y)$ is differentiable at (x_0, y_0), then it is continuous at (x_0, y_0).

Similarly, if $f(x, y)$ is differentiable on a set, then it is continuous on this set.

Theorem. $f(x, y)$ is continuously differentiable at (x_0, y_0) if, and only if, the first partial derivatives f_x, f_y are continuous at (x_0, y_0).

Theorem. $f(x, y)$ is continuously differentiable on S if, and only if, the partial derivatives f_x, f_y are continuous on S.

Arithmetic (algebraic) properties

If $f(x, y)$ and $g(x, y)$ are differentiable at a point (x, y), then

1) $f(x, y) + g(x, y)$ is differentiable at (x, y) and $d(f + g) = df + dg$
2) $f(x, y) - g(x, y)$ is differentiable at (x, y) and $d(f - g) = df - dg$
3) $f(x, y) \cdot g(x, y)$ is differentiable at (x, y) and $d(f \cdot g) = df \cdot g + f \cdot dg$
4) $\frac{f(x,y)}{g(x,y)}$ is differentiable at (x, y) and $d\left(\frac{f}{g}\right) = \frac{df \cdot g - f \cdot dg}{g^2}$ (under the additional condition $g(x, y) \neq 0$)

Chain rule. If $p = f(x, y)$ and $q = g(x, y)$ are differentiable functions at a point (x_0, y_0), and $h(p, q)$ is a differentiable function at the point $(p_0, q_0) = (f(x_0, y_0), g(x_0, y_0))$, then the composite function $\varphi(x, y) \equiv h(f(x, y), g(x, y))$ is differentiable at the point (x_0, y_0) and its partial derivatives can be calculated as follows: $\varphi_x = h_p p_x + h_q q_x$, $\varphi_y = h_p p_y + h_q q_y$.

Theorem. If $f(x, y)$ is differentiable at a point (x_0, y_0), then $f(x, y)$ has a directional derivative at this point in the direction of any vector $v = (a, b)$ and the following formula holds $D_v f = \nabla f \cdot \frac{v}{|v|} = \frac{f_x a + f_y b}{\sqrt{a^2+b^2}}$.

Extremum properties of the gradient:

1) If $f(x, y)$ is differentiable and $\nabla f \neq 0$, then the maximum and minimum values of $D_v f$ are $|\nabla f|$ and $-|\nabla f|$, that occur in the directions ∇f and $-\nabla f$, respectively.
2) If $f(x, y)$ is differentiable and $\nabla f \neq 0$, then ∇f is the normal vector to

level curves of $f(x,y)$, that is, ∇f is the normal vector to the tangent lines of the level curves.

Theorem. If $f(x,y)$ is differentiable at a point (x_0,y_0), then the tangent plane at (x_0,y_0) contains all the tangent lines passing through (x_0,y_0).

Mean Value Theorem. If $f(x,y)$ is continuously differentiable in a neighborhood of a point (x_0,y_0), then for any point (x,y) in this neighborhood there exists a point $(\tilde{x},\tilde{y}) = (x_0+\theta(x-x_0), y_0+\theta(y-y_0))$, $\theta \in (0,1)$ in the same neighborhood such that $\Delta f(x_0) = df(\tilde{x})$.

Remark. The condition of continuous differentiability can be substituted by the condition of continuity of the first order partial derivatives.

Taylor's Theorem. If $f(x,y)$ is $n+1$ continuously differentiable in a neighborhood of a point (x_0,y_0), then for any point (x,y) in this neighborhood the following formula holds:

$$f(x,y) = f(x_0,y_0) + \frac{1}{1!}df(x_0,y_0) + \frac{1}{2!}d^2 f(x_0,y_0) + \dots$$

$$+\frac{1}{n!}d^n f(x_0,y_0) + \frac{1}{(n+1)!}d^{n+1} f(\tilde{x},\tilde{y}),$$

where $(\tilde{x},\tilde{y})$ is the point located on a straight-line segment between (x_0,y_0) and (x,y), that is, there exists $\theta \in (0,1)$ such that $(\tilde{x},\tilde{y}) = (x_0+\theta(x-x_0), y_0+\theta(y-y_0))$.

Remark 1. The continuous differentiability can be substituted by the existence of continuous partial derivatives of $n+1$-th order.

Remark 2. Taylor's theorem is a generalization of the Mean Value Theorem.

Applications

Criterion of a constant function. A differentiable on a connected set S function $f(x,y)$ is constant on S if, and only if, $df(x,y) = 0$, $\forall (x,y) \in S$.

Critical (stationary) point. If $df(x_0,y_0) = 0$, then (x_0,y_0) is called a *critical (stationary)* point.

Theorem (necessary condition of a local extremum). If $f(x,y)$ is differentiable at a local extremum point (x_0,y_0), then $df(x_0,y_0) = 0$.

Second differential test (sufficient condition of a local extremum). Let $f(x,y)$ be twice differentiable at a critical point (x_0,y_0). Then

1) if $d^2 f(x_0, y_0)$ maintains the same sign for small increments $\Delta x, \Delta y$, then (x_0, y_0) is a strict local extremum; more specifically, if $d^2 f(x_0, y_0) > 0$, then (x_0, y_0) is a strict local minimum, and if $d^2 f(x_0, y_0) < 0$, then (x_0, y_0) is a strict local maximum;
2) if $d^2 f(x_0, y_0)$ changes its sign for small increments $\Delta x, \Delta y$, then (x_0, y_0) is not a local extremum.

Second derivatives test (sufficient condition of a local extremum). Let $f(x, y)$ be twice continuously differentiable at a critical point (x_0, y_0). Then
1) if the Hessian matrix $H(f) = \begin{pmatrix} f_{xx} & f_{xy} \\ f_{xy} & f_{yy} \end{pmatrix}$ is definite at (x_0, y_0), that is, if the determinant of the Hessian matrix is positive at (x_0, y_0), then (x_0, y_0) is a strict local extremum; more specifically, if $H(f)$ is positive definite, that is, if $f_{xx} > 0$ and $f_{xx} f_{yy} - f_{xy}^2 > 0$, then (x_0, y_0) is a strict local minimum; if $H(f)$ is negative definite, that is, if $f_{xx} < 0$ and $f_{xx} f_{yy} - f_{xy}^2 > 0$, then (x_0, y_0) is a strict local maximum;
2) if the Hessian matrix is indefinite at (x_0, y_0), that is, if $f_{xx} f_{yy} - f_{xy}^2 < 0$ at (x_0, y_0), then (x_0, y_0) is not a local extremum.

8.2 One-dimensional links

General Remark. Like in section 7.2, the examples in this section are modified versions of some examples for one-variable functions presented in Part I and, correspondingly, the majority of the provided below counterexamples use the functions of the chapter 4 adapted to the case of $\mathbb{R}^2$. Each example in this section has the reference to the corresponding one-dimensional counterpart.

8.2.1 Concepts and local properties

Example 1. "If $f(x, y)$ is not differentiable at a point (a, b), then it is not continuous at this point."

(compare with Example 1, section 4.2)

Solution.

The function $f(x, y) = \sqrt{x^2 + y^2}$ is continuous at the origin, since $\lim_{(x,y)\to(0,0)} f(x, y) = 0 = f(0, 0)$. However, the restriction of $f(x, y)$ to x-axis is the absolute value function $f(x, 0) = |x|$, which is not differentiable at

0. Therefore, $f(x,y)$ does not possess the partial derivative in x at the origin, that implies that $f(x,y)$ is not differentiable at the origin.

Remark. The Remarks to Example 1 in Section 4.2 are applicable here with corresponding adjustments to the case of $\mathbb{R}^2$.

Example 2. "If a function $f(x,y)$ is not differentiable at a point (a,b), then $f^2(x,y)$ also is not differentiable at (a,b)."

(compare with Example 4, section 4.2)

Solution.

The function $f(x,y)=\sqrt{x^2+y^2}$ of Example 1 is not differentiable at the origin, but its square $g(x,y)=f^2(x,y)=x^2+y^2$ is differentiable function on $\mathbb{R}^2$, since the partial derivatives $g_x(x,y)=2x$ and $g_y(x,y)=2y$ are the continuous function on $\mathbb{R}^2$.

Remark. The Remarks to Example 4 in section 4.2 are applicable here with corresponding adjustments to the case of $\mathbb{R}^2$.

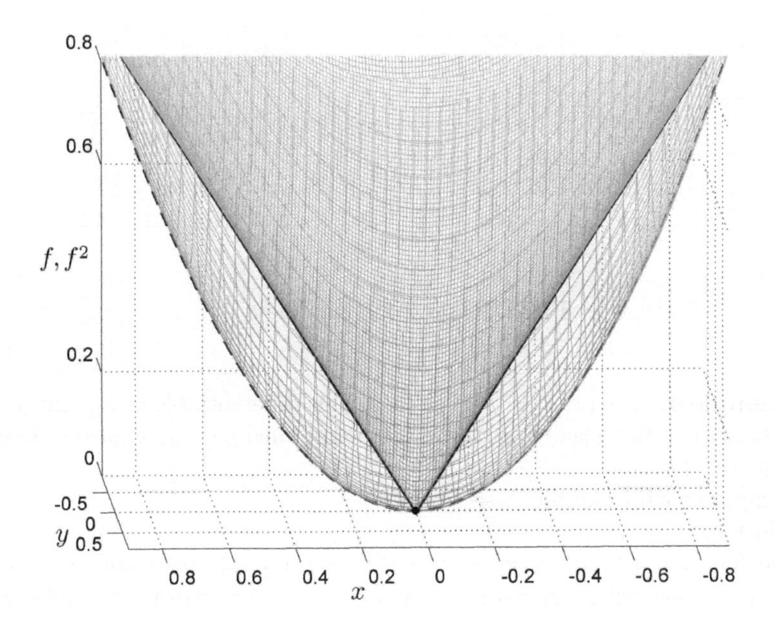

FIGURE 8.2.1: Example 1 and Example 2

Example 3. "If $h(x,y)=f(x,y)+g(x,y)$ is differentiable at a point (a,b), then both $f(x,y)$ and $g(x,y)$ are differentiable at (a,b)."

(compare with Example 1, section 4.3)

Solution.

The functions $f(x,y)=\sqrt{x^2+y^2}$ and $g(x,y)=-\sqrt{x^2+y^2}$ are not differentiable at $(0,0)$ (see Example 1), but their sum $h(x,y) = f(x,y) + g(x,y) \equiv 0$ is differentiable on $\mathbb{R}^2$.

Remark. The Remarks to Example 1 in section 4.3 are applicable here with corresponding adjustments to the case of $\mathbb{R}^2$.

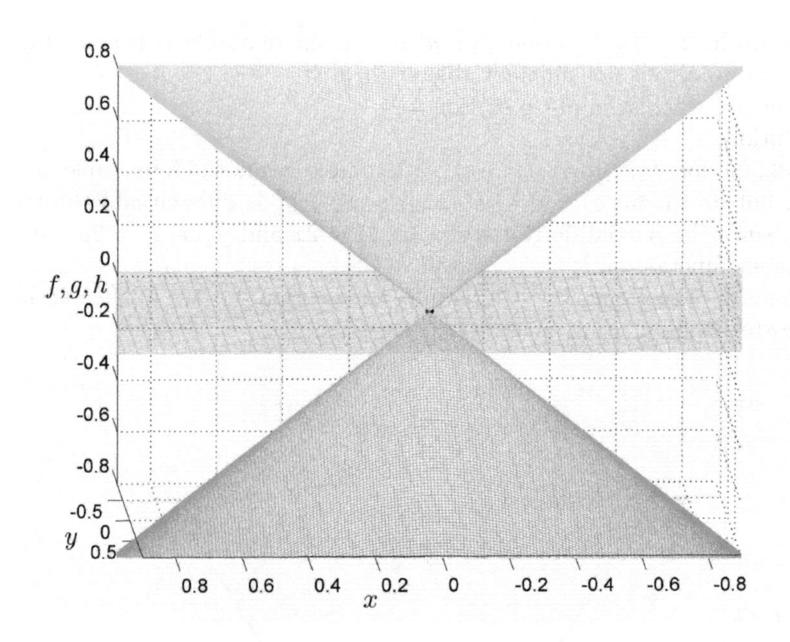

FIGURE 8.2.2: Example 3

Example 4. "If $h(x,y) = f(x,y) \cdot g(x,y)$ is differentiable at a point (a,b) and $f(x,y)$ is differentiable at the same point, then $g(x,y)$ is also differentiable at (a,b)."

(compare with Example 2, section 4.3)

Solution.

The function $f(x,y) = x^2 + y^2$ is differentiable at $(0,0)$ and $g(x,y) = \sqrt[3]{x^2+y^2}$ is not differentiable at $(0,0)$ (for the same reason as $\sqrt{x^2+y^2}$). However, their product $h(x,y) = f(x,y) \cdot g(x,y) = (x^2+y^2)^{4/3}$ is differentiable at the origin, because both partial derivatives exist at the origin

$$f_x(0,0) = \lim_{x\to 0} \frac{f(x,0) - f(0,0)}{x-0} = \lim_{x\to 0} \frac{x^{8/3}}{x} = 0,$$

$$f_y(0,0) = \lim_{y\to 0} \frac{f(0,y) - f(0,0)}{y-0} = \lim_{y\to 0} \frac{y^{8/3}}{y} = 0$$

and the remainder in the definition of differentiability

$$\alpha(x,y) = f(x,y) - (f(0,0) + f_x(0,0) \cdot x + f_y(0,0) \cdot y) = (x^2+y^2)^{4/3}$$

satisfies the required evaluation

$$\lim_{(x,y)\to(0,0)} \frac{\alpha(x,y)}{\rho(x,y)} = \lim_{(x,y)\to(0,0)} \left(x^2+y^2\right)^{5/6} = 0.$$

Remark. The Remarks to Example 2 in section 4.3 are applicable here with corresponding adjustments to the case of $\mathbb{R}^2$.

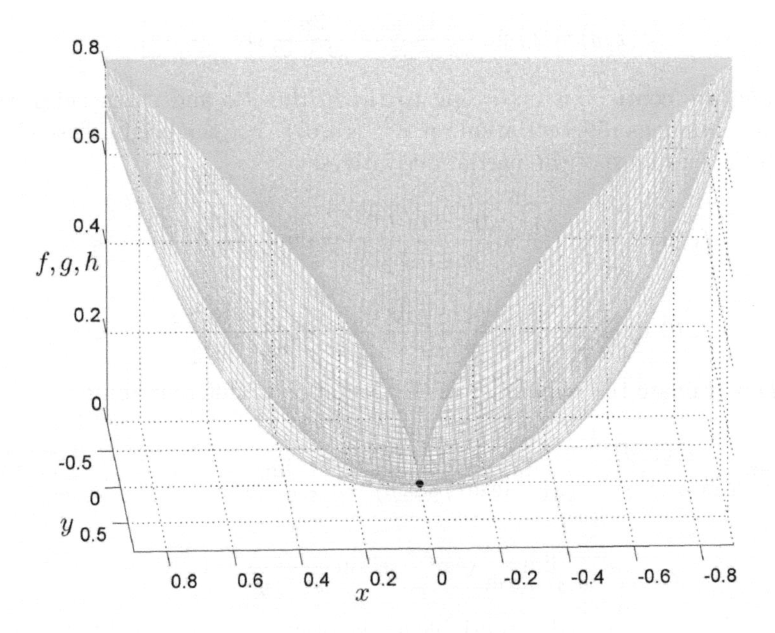

FIGURE 8.2.3: Example 4

Example 5. "If $f(x,y)$ is not differentiable at a point (a,b) and $g(t)$ is not differentiable at the point $f(a,b)$, then $g(f(x,y))$ is not differentiable at (a,b)."

(compare with Example 4, section 4.3)

Solution.

The function $f(x,y) = \begin{cases} \frac{1}{x^2+y^2}, & (x,y) \neq (0,0) \\ 0, & (x,y) = (0,0) \end{cases}$ is not differentiable at $(0,0)$ and the function $g(t) = \begin{cases} \frac{1}{t}, & t \neq 0 \\ 0, & t = 0 \end{cases}$ is not differentiable at 0, but the function $g(f(x,y)) = x^2 + y^2$ is differentiable at $(0,0)$.

Example 6. "If $f(x,y)$ is differentiable on $\mathbb{R}^2$, then at least one of the partial derivatives is bounded in a neighborhood of each point."

(compare with Example 5, section 4.3)

Solution.

The function $f(x,y) = \begin{cases} (x^2+y^2)\sin\frac{1}{x^2+y^2}, & (x,y) \neq (0,0) \\ 0, & (x,y) = (0,0) \end{cases}$ is differentiable on $\mathbb{R}^2$. Indeed, for any $(x,y) \neq (0,0)$, both partial derivatives

$$f_x(x,y) = 2x\sin\frac{1}{x^2+y^2} - \frac{2x}{x^2+y^2}\cos\frac{1}{x^2+y^2},$$

and

$$f_y(x,y) = 2y\sin\frac{1}{x^2+y^2} - \frac{2y}{x^2+y^2}\cos\frac{1}{x^2+y^2}$$

exist and are continuous according to the arithmetic and chain rules, that implies continuous differentiation on $\mathbb{R}^2 \setminus \{(0,0)\}$. For the origin, first we establish the existence of the partial derivatives:

$$f_x(0,0) = \lim_{x\to 0}\frac{f(x,0)-f(0,0)}{x-0} = \lim_{x\to 0}\frac{x^2\sin 1/x^2}{x} = 0,$$

$$f_y(0,0) = \lim_{y\to 0}\frac{f(0,y)-f(0,0)}{y-0} = \lim_{y\to 0}\frac{y^2\sin 1/y^2}{y} = 0,$$

and then evaluate the remainder in the definition of differentiability:

$$\lim_{(x,y)\to(0,0)}\frac{\alpha(x,y)}{\rho(x,y)} = \lim_{(x,y)\to(0,0)}\frac{\Delta f - df}{\rho(x,y)} = \lim_{(x,y)\to(0,0)}\frac{(x^2+y^2)\sin\frac{1}{x^2+y^2}}{\sqrt{x^2+y^2}}$$

$$= \lim_{(x,y)\to(0,0)}\sqrt{x^2+y^2}\sin\frac{1}{x^2+y^2} = 0.$$

Therefore, $f(x,y)$ is differentiable at the origin as well, and the statement condition is satisfied. However, by choosing the sequence of the points $(x_k,0) = \left(\frac{1}{\sqrt{2k\pi}},0\right)$, $k \in \mathbb{N}$ which approach the origin, we obtain $\lim_{x_k\to 0} f_x(x_k,0) = \lim_{k\to+\infty}\left(-2\sqrt{2k\pi}\cos 2k\pi\right) = -\infty$, that is, $f_x(x,y)$ is unbounded in any neighborhood of the origin. In a similar way, approaching the origin by the points $(0,y_n) = \left(0,\frac{1}{\sqrt{2n\pi}}\right)$, $n \in \mathbb{N}$, it can be shown that $f_y(x,y)$ is also unbounded in any neighborhood of the origin.

Example 7. "There is no function differentiable only at one point of its domain."

(compare with Example 6, section 4.3)

Solution.

Using the same reasoning as in Example 9, section 7.2, one can prove that the function $f(x,y) = \begin{cases} x^2+y^2, & \sqrt{x^2+y^2} \in \mathbb{Q} \\ 0, & \sqrt{x^2+y^2} \in \mathbb{I} \end{cases}$ is not continuous (and consequently not differentiable) at any point $(x,y) \neq (0,0)$. However, at the origin, both partial derivatives exist $f_x(0,0) = 0$ and $f_y(0,0) = 0$, that means that the remainder assumes the form $\alpha(x,y) = \Delta f - df = \begin{cases} x^2+y^2, \ \sqrt{x^2+y^2} \in \mathbb{Q} \\ 0, \ \sqrt{x^2+y^2} \in \mathbb{I} \end{cases}$ and for all points in a neighborhood of the origin it can be evaluated as follows: $|\alpha(x,y)| \le x^2+y^2$. Therefore, $\frac{|\alpha(x,y)|}{\rho(x,y)} \le \sqrt{x^2+y^2}$ and according to the squeese theorem $\lim\limits_{(x,y)\to(0,0)} \frac{\alpha(x,y)}{\rho(x,y)} = 0$, that is, $f(x,y)$ is differentiable at the origin.

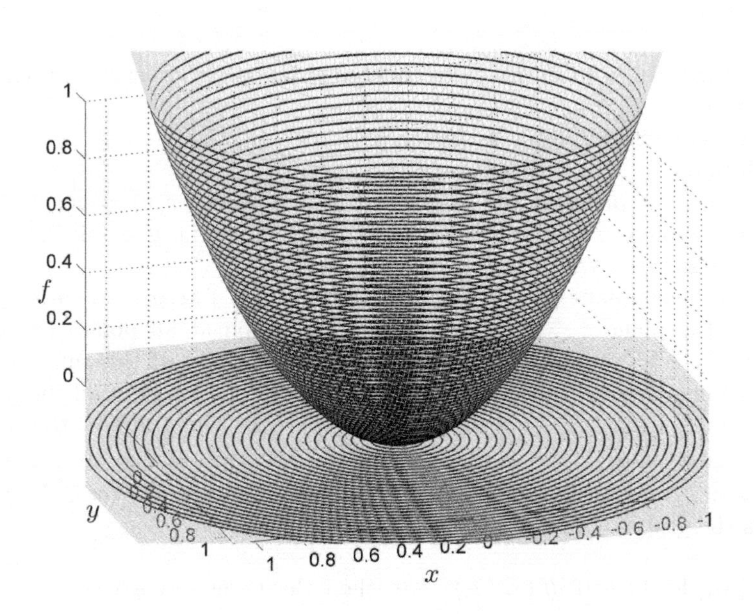

FIGURE 8.2.4: Example 7

8.2.2 Global properties and applications

Example 8. "If both functions $f(x,y)$ and $g(x,y)$ are differentiable on $\mathbb{R}^2$ and $f(x,y) > g(x,y)$, $\forall (x,y) \in \mathbb{R}^2$, then there exists at least one point in $\mathbb{R}^2$ where $f_x(x,y) > g_x(x,y)$ or $f_y(x,y) > g_y(x,y)$."

(compare with Example 1, section 4.4)

Solution.

Both functions $f(x,y) = e^{-x} + e^{-y}$ and $g(x,y) = -e^{-x} - e^{-y}$ are differentiable on $\mathbb{R}^2$ and $f(x,y) > g(x,y)$, $\forall (x,y) \in \mathbb{R}^2$. However, $f_x(x,y) = -e^{-x} < e^{-x} = g_x(x,y)$ and $f_y(x,y) = -e^{-y} < e^{-y} = g_y(x,y)$, $\forall (x,y) \in \mathbb{R}^2$.

Example 9. "If $df(x,y) = 0$ on a set S then $f(x,y) = const$ on this set."

(compare with Example 10, section 4.4)

Solution.

Consider $f(x,y) = \begin{cases} 1, & (x,y) \in B_1(0,0) \\ 2, & (x,y) \in B_1(2,0) \end{cases}$ with the set $S = B_1(0,0) \cup B_1(2,0)$, where $B_1(0,0)$ and $B_1(2,0)$ are the open unit balls, centered at $(0,0)$ and $(2,0)$, respectively. This function is differentiable at each point of S and $df(x,y) = 0$ on S, but the function is not a constant.

Remark 1. The statement will be correct if the set S is connected.

Remark 2. A generalization of this Example can be found in Example 17, section 8.3.

Example 10. "A tangent plane touches or intersects the surface at exactly one point."

(compare with Example 2, section 4.5)

Solution.

The tangent plane to the graph of the function $f(x,y) = \cos x$ at the origin is the plane $z = 1$, which touches the graph of $f(x,y)$ at all the points $(x_n, y) = (2n\pi, y)$, $\forall n \in \mathbb{Z}$, $\forall y$.

The tangent plane to the graph of the function $f(x,y) = x \sin x$ at the origin is the plane $z = 0$, which crosses the graph at all points $(x_n, y) = (n\pi, y)$, $\forall n \in \mathbb{Z}$, $\forall y$. At the same time, at the point $\left(\frac{\pi}{2}, 0\right)$ this function has the tangent plane $z = x$, which touches the graph of $f(x,y)$ at all the points $(x_k, y) = \left(\frac{\pi}{2} + 2k\pi, y\right)$, $\forall k \in \mathbb{Z}$, $\forall y$ and also crosses the graph at the origin. The tangent plane $z = -x$ at the point $\left(-\frac{\pi}{2}, 0\right)$ has similar properties: it touches the graph at all the points $(x_m, y) = \left(-\frac{\pi}{2} + 2m\pi, y\right)$, $\forall m \in \mathbb{Z}, \forall y$ and crosses the graph at the origin.

Example 11. "If $df(x_0, y_0) = 0$, then the point (x_0, y_0) is a local extremum of the function $f(x,y)$."

(compare with Example 7, section 4.5)

Solution.

The function $f(x,y) = x^3$ is differentiable on $\mathbb{R}^2$ and $df(0,0) = 0$, but $(0,0)$ is not a local extremum, since in any neighborhood of the origin $f(x,y)$ assumes both negative (for $x < 0$) and positive (for $x > 0$) values.

Another counterexample, classical for two-variable functions, is the example of the hyperbolic paraboloid $f(x,y) = x^2 - y^2$ that has a saddle point at the origin. Again, $f(x,y)$ is differentiable on $\mathbb{R}^2$ and $df(0,0) = 0$, but $(0,0)$ is not a local extremum, since in any neighborhood of the origin $f(x,y)$ assumes both negative (on y-axis) and positive (on x-axis) values. Furthermore, $f(x,0) = x^2$ is upward parabola with the minimum point at $(0,0)$, while $f(0,y) = -y^2$ is downward parabola with the maximum point at $(0,0)$. Such points are called saddle points.

Example 12. "If $f(x,y)$ has a strict local extremum at (x_0, y_0), then $f_x(x_0, y_0) = f_y(x_0, y_0) = 0$."

(compare with Example 8, section 4.5)

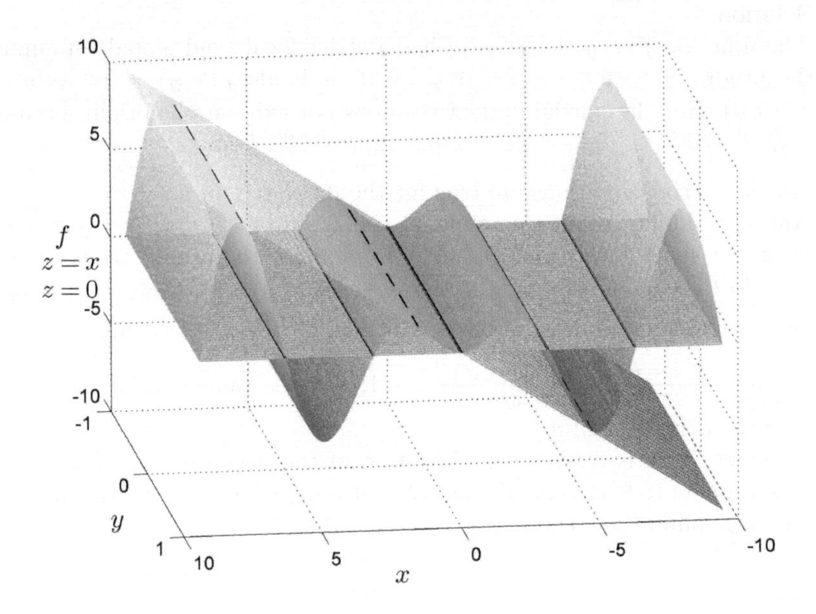

FIGURE 8.2.5: Example 10, graph of the second function

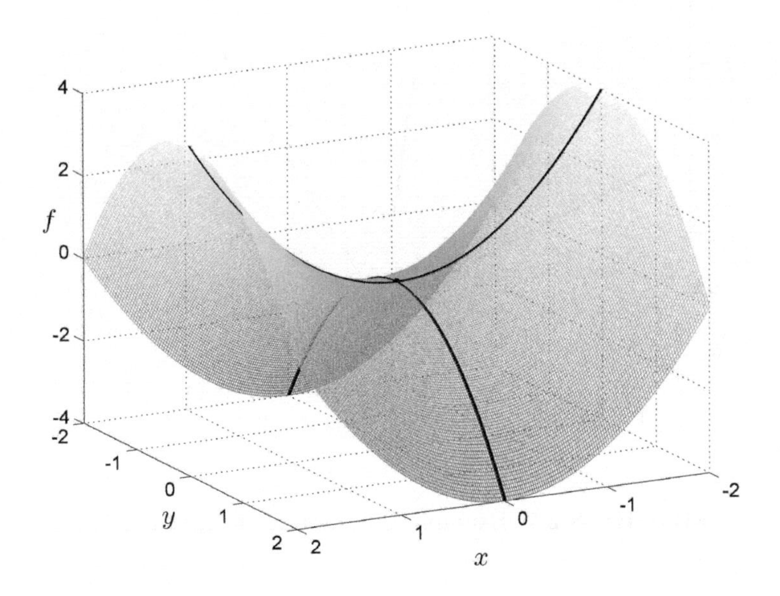

FIGURE 8.2.6: Example 11, graph of the second function

Solution.

The function $f(x,y) = |x| + |y|$ has a strict local (and global) minimum at the origin, since for any $(x,y) \neq (0,0)$ it holds $f(x,y) = |x| + |y| > 0 = f(0,0)$. But the partial derivatives does not exist at the origin, because $\lim\limits_{x\to 0_+} \frac{f(x,0)-f(0,0)}{x-0} = \lim\limits_{x\to 0_+} \frac{x}{x} = 1$ while $\lim\limits_{x\to 0_-} \frac{f(x,0)-f(0,0)}{x-0} = \lim\limits_{x\to 0_-} \frac{-x}{x} = -1$; and the same considerations are true for the derivative in y.

Another similar counterexample is the function $f(x,y) = \sqrt{x^2+y^2}$, which has a strict local (and global) minimum at the origin, since for any $(x,y) \neq (0,0)$ it holds $f(x,y) = \sqrt{x^2+y^2} > 0 = f(0,0)$. However, the partial derivatives does not exist at the origin: $\lim\limits_{x\to 0_+} \frac{f(x,0)-f(0,0)}{x-0} = \lim\limits_{x\to 0_+} \frac{\sqrt{x^2}}{x} = 1$, while $\lim\limits_{x\to 0_-} \frac{f(x,0)-f(0,0)}{x-0} = \lim\limits_{x\to 0_-} \frac{\sqrt{x^2}}{x} = -1$; and the same considerations are true for the derivative in y.

Remark. This is a wrong modification of the necessary condition for a local extremum: if $f(x,y)$ is differentiable at a point (x_0,y_0) and this point is a local extremum, then $f_x(x_0,y_0) = f_y(x_0,y_0) = 0$.

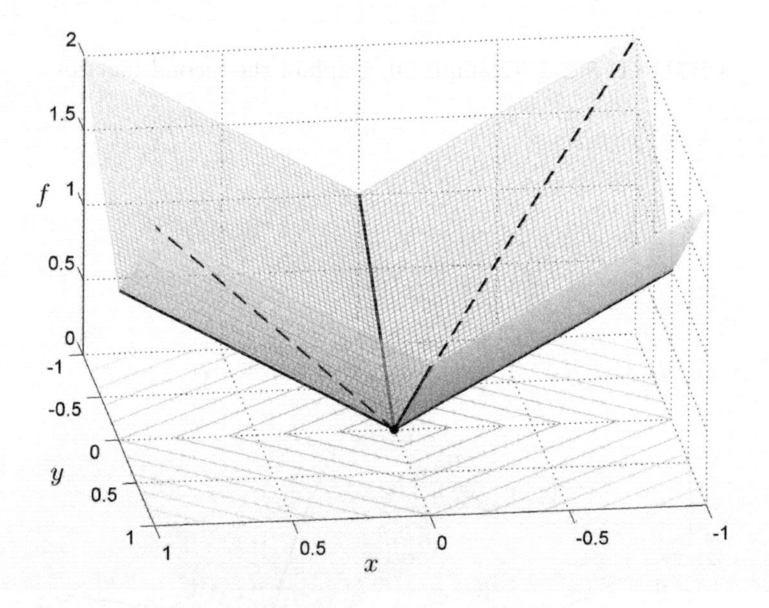

FIGURE 8.2.7: Example 12, graph of the first function

Example 13. "If (x_0,y_0) is a local extremum of $f(x,y)$ and $f(x,y)$ is twice differentiable at (x_0,y_0), then $d^2 f(x_0,y_0) \neq 0$."
(compare with Example 9, section 4.5)
Solution.

The function $f(x,y) = x^4 + y^4$ is twice differentiable at the origin, which is its local (and global) minimum. The second differential has the form

$$d^2 f = f_{xx}\Delta x^2 + 2f_{xy}\Delta x\Delta y + f_{yy}\Delta y^2 = 12x^2\Delta x^2 + 12y^2\Delta y^2$$

and it vanishes at the origin, that contradicts the above statement.

Remark. This is the false converse to the Second Differential test: if $f(x,y)$ is twice differentiable at a critical point (x_0, y_0) and $d^2 f(x_0, y_0) \neq 0$, then (x_0, y_0) is a local extremum of $f(x,y)$.

8.3 Multidimensional essentials

Example 1. "If a function has partial derivatives at some point, then it is continuous at this point."

Solution.

The function of Example 1, section 7.3, $f(x,y) = \begin{cases} \frac{xy}{x^2+y^2}, & x^2+y^2 \neq 0 \\ 0, & x^2+y^2 = 0 \end{cases}$ has partial derivatives at the origin:

$$f_x(0,0) = \lim_{x\to 0}\frac{f(x,0) - f(0,0)}{x - 0} = \lim_{x\to 0}\frac{0-0}{x} = 0$$

and

$$f_y(0,0) = \lim_{y\to 0}\frac{f(0,y) - f(0,0)}{y - 0} = \lim_{y\to 0}\frac{0-0}{y} = 0.$$

Moreover, both partial derivatives exist at each point in $\mathbb{R}^2$ and for $(x,y) \neq (0,0)$ can be calculated by applying the arithmetic rules: $f_x(x,y) = \frac{y(y^2-x^2)}{(x^2+y^2)^2}$ and $f_y(x,y) = \frac{x(x^2-y^2)}{(x^2+y^2)^2}$. Nevertheless, $f(x,y)$ is not continuous at the origin (see proof in Example 1, section 7.3).

A similar example can be provided with the function of Example 2, section 7.3, $f(x,y) = \begin{cases} \frac{x^2 y}{x^4+y^2}, & x^2+y^2 \neq 0 \\ 0, & x^2+y^2 = 0 \end{cases}$. Again, the partial derivatives at the origin should be evaluated by definition:

$$f_x(0,0) = \lim_{x\to 0}\frac{f(x,0) - f(0,0)}{x - 0} = \lim_{x\to 0}\frac{0-0}{x} = 0,$$

$$f_y(0,0) = \lim_{y\to 0}\frac{f(0,y) - f(0,0)}{y - 0} = \lim_{y\to 0}\frac{0-0}{y} = 0,$$

and at any other point one can use the arithmetic rules: $f_x(x,y) = \frac{2xy(y^2-x^4)}{(x^4+y^2)^2}$,

$f_y(x,y) = \frac{x^2(x^4-y^2)}{(x^4+y^2)^2}$. However, it was shown in Example 2, section 7.3 that $f(x,y)$ is not continuous at the origin.

Remark 1. In both examples, although the partial derivatives exist on $\mathbb{R}^2$, it can be shown that they are discontinuous at the origin. For example, the partial limit of $f_x(x,y)$ of the first function along the path $y = 2x$ is $\lim\limits_{x\to 0} f_x(x,2x) = \lim\limits_{x\to 0} \frac{2x \cdot 3x^2}{(5x^2)^2} = \infty$.

Remark 2. For functions of one variable, the existence of derivative implies the continuity (at a point or on a set). The situation is different for functions of several variables.

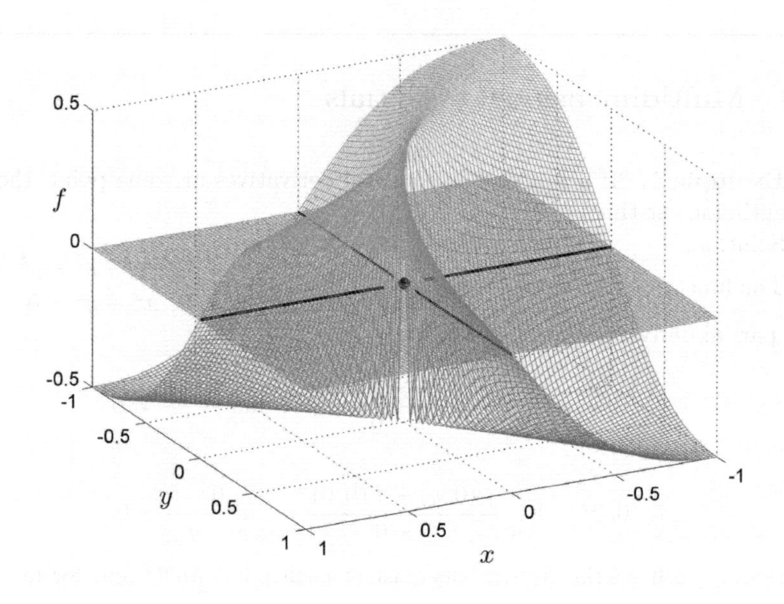

FIGURE 8.3.1: Example 1

Example 2. "If all the directional derivatives and partial derivatives of $f(x,y)$ exist and are equal at some point, then $f(x,y)$ is continuous at the same point."

Solution.

The function $f(x,y) = \begin{cases} \frac{x^6}{(y-x^2)^2+x^6}, & x^2+y^2 \neq 0 \\ 0, & x^2+y^2 = 0 \end{cases}$ has zero partial derivatives at the origin:

$$f_x(0,0) = \lim_{x\to 0} \frac{f(x,0) - f(0,0)}{x-0} = \lim_{x\to 0} \frac{x^6}{(x^4+x^6)\,x} = \lim_{x\to 0} \frac{x}{1+x^2} = 0$$

and

$$f_y(0,0) = \lim_{y\to 0} \frac{f(0,y) - f(0,0)}{y-0} = \lim_{y\to 0} \frac{0}{y} = 0.$$

The directional derivatives are also zero:

$$D_u f(0,0) = \lim_{t\to 0_+} \frac{f(\alpha t, \beta t) - f(0,0)}{t-0} = \lim_{t\to 0_+} \frac{\alpha^6 t^6}{t^2 (\beta - \alpha^2 t)^2 + \alpha^6 t^6} \frac{1}{t}$$

$$= \lim_{t\to 0_+} \frac{\alpha^6 t^3}{(\beta - \alpha^2 t)^2 + \alpha^6 t^4} = 0$$

in any direction $u = (\alpha, \beta)$. Nevertheless, $f(x,y)$ is not continuous at the origin, because along the path $y = x^2$ one gets

$$f(x, x^2) = \frac{x^6}{0 + x^6} = 1 \underset{x\to 0}{\not\to} 0 = f(0,0).$$

Remark 1. Consequently, this function is not differentiable at the origin.
Remark 2. This is a strengthened version of Example 1.

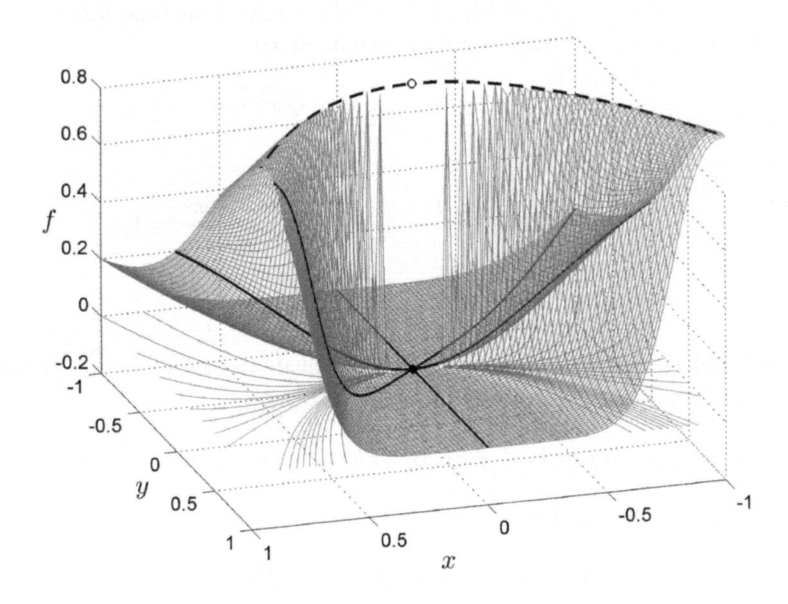

FIGURE 8.3.2: Example 2

Example 3. "If a function is continuous at a point (x_0, y_0) and has partial derivatives at this point, then it is differentiable at (x_0, y_0)."
Solution.

The function $f(x,y) = \sqrt{|xy|}$ satisfies the conditions of the statement at the origin. In fact, $\lim\limits_{(x,y)\to(0,0)} f(x,y) = \lim\limits_{(x,y)\to(0,0)} \sqrt{|xy|} = 0$ and

$$f_x(0,0) = \lim_{x\to 0} \frac{f(x,0) - f(0,0)}{x-0} = \lim_{x\to 0} \frac{0-0}{x} = 0,$$

$$f_y(0,0) = \lim_{y\to 0} \frac{f(0,y) - f(0,0)}{y-0} = \lim_{y\to 0} \frac{0-0}{y} = 0.$$

However, $f(x,y)$ is not differentiable at the origin, because the remainder in the definition of differentiability

$$\alpha(x,y) = f(x,y) - (f(0,0) + f_x(0,0)\cdot x + f_y(0,0)\cdot y) = \sqrt{|xy|}$$

does not approach zero faster than $\rho = \sqrt{x^2+y^2}$: along the path $y = x$ one gets

$$\lim_{\rho\to 0} \frac{\alpha}{\rho} = \lim_{x\to 0} \frac{|x|}{\sqrt{2}\,|x|} = \frac{1}{\sqrt{2}} \neq 0.$$

Another counterexample can be given with the function $f(x,y) = \sqrt[3]{x^3+y^3}$ considered in a neighborhood of the origin. This function is continuous and has partial derivatives at the origin, since

$$\lim_{(x,y)\to(0,0)} f(x,y) = \lim_{(x,y)\to(0,0)} \sqrt[3]{x^3+y^3} = 0$$

and

$$f_x(0,0) = \lim_{x\to 0} \frac{f(x,0) - f(0,0)}{x-0} = \lim_{x\to 0} \frac{\sqrt[3]{x^3}}{x} = 1,$$

$$f_y(0,0) = \lim_{y\to 0} \frac{f(0,y) - f(0,0)}{y-0} = \lim_{x\to 0} \frac{\sqrt[3]{y^3}}{x} = 1.$$

To check differentiability at the origin, let us evaluate the difference between $f(x,y)$ and its linear approximation near the origin:

$$\alpha(x,y) = f(x,y) - (f(0,0) + f_x(0,0)\cdot x + f_y(0,0)\cdot y) = \sqrt[3]{x^3+y^3} - x - y.$$

Along the line $y = x$ we have

$$\lim_{x\to 0} \frac{\alpha(x,x)}{\rho(x,x)} = \lim_{x\to 0} \frac{\sqrt[3]{2x^3} - 2x}{\sqrt{2}\,|x|} = \lim_{x\to 0} \frac{\sqrt[3]{2} - 2}{\sqrt{2}} \operatorname{sgn} x \neq 0$$

(the last limit does not exist), therefore $\lim\limits_{\rho\to 0} \frac{\alpha(\rho)}{\rho} \neq 0$ as it should be for a differentiable function. Hence, $f(x,y)$ is not differentiable at the origin.

Example 4. "If $f(x,y)$ is continuous and has bounded partial derivatives on $\mathbb{R}^2$, then $f(x,y)$ is differentiable on $\mathbb{R}^2$."

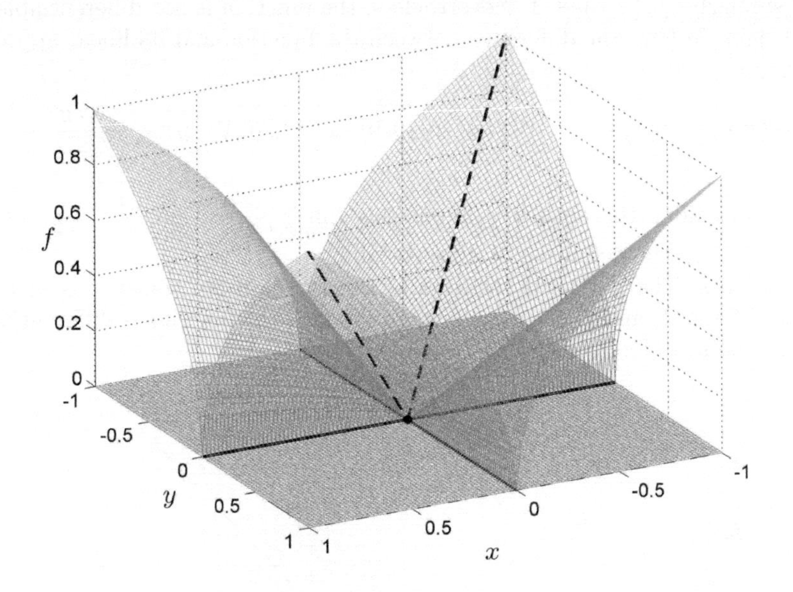

FIGURE 8.3.3: Example 3, graph of the first function

Solution.

The function $f(x,y) = \begin{cases} \frac{xy}{\sqrt{x^2+y^2}}, & x^2+y^2 \neq 0 \\ 0, & x^2+y^2 = 0 \end{cases}$ is continuous at any point $(x,y) \neq (0,0)$ due to arithmetic rules, and at the origin we should appeal to the definition:

$$|f(x,y)-f(0,0)| = \frac{|xy|}{\sqrt{x^2+y^2}} = \frac{2|xy|}{x^2+y^2} \cdot \frac{1}{2}\sqrt{x^2+y^2} \leq \frac{1}{2}\sqrt{x^2+y^2} \underset{(x,y)\to(0,0)}{\to} 0.$$

In the same way, the partial derivatives can be calculated at any point $(x,y) \neq (0,0)$ according to the arithmetic rules:

$$f_x(x,y) = \frac{y^3}{(x^2+y^2)^{3/2}}, \; f_y(x,y) = \frac{x^3}{(x^2+y^2)^{3/2}}$$

and at the origin by using the definition:

$$f_x(0,0) = \lim_{x\to 0} \frac{f(x,0)-f(0,0)}{x-0} = \lim_{x\to 0} \frac{0-0}{x} = 0,$$

$$f_y(0,0) = \lim_{y\to 0} \frac{f(0,y)-f(0,0)}{y-0} = \lim_{y\to 0} \frac{0-0}{y} = 0.$$

Additionally, the partial derivatives are bounded on $\mathbb{R}^2$:

$$|f_x(x,y)| = \frac{|y^3|}{(x^2+y^2)^{3/2}} = \frac{y^2}{x^2+y^2} \cdot \frac{|y|}{(x^2+y^2)^{1/2}} \leq 1,$$

and similarly $|f_y(x,y)| \le 1$. Nevertheless, the function is not differentiable at the origin. In fact, the difference between the function and its linear approximation is

$$\alpha(x,y) = f(x,y) - (f(0,0) + f_x(0,0)\cdot x + f_y(0,0)\cdot y) = \frac{xy}{\sqrt{x^2+y^2}}.$$

Therefore, along the path $y = x$ one has $\lim\limits_{x\to 0} \frac{\alpha(x,x)}{\rho(x,x)} = \lim\limits_{x\to 0} \frac{x^2}{2x^2} = \frac{1}{2}$, which contradicts to the definition of differentiability.

Remark. The partial derivatives of the function are continuous at each point different from the origin, which implies that the function is differentiable at each point different from the origin.

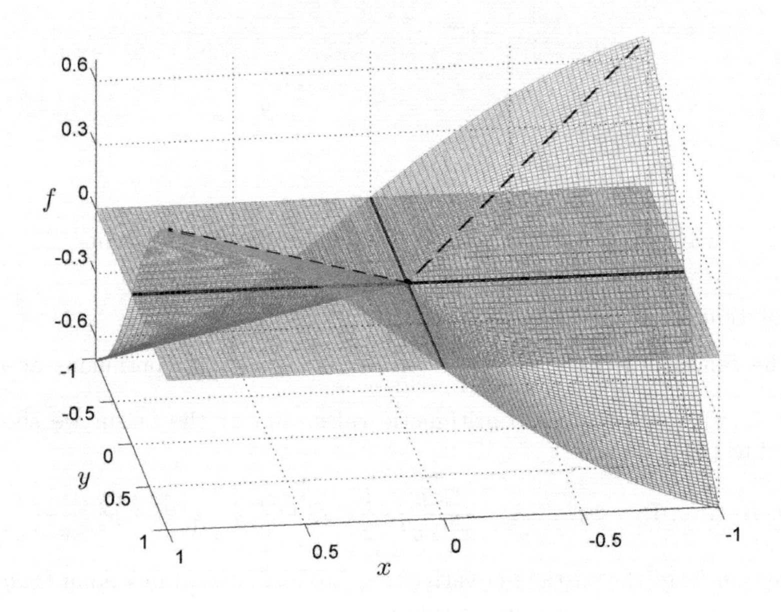

FIGURE 8.3.4: Example 4

Example 5. "If $f(x,y)$ is continuous and possesses partial derivatives and all the directional derivatives at some point, then $f(x,y)$ is differentiable at this point."

Solution.

It was shown in Example 3 that the function $f(x,y) = \sqrt{|xy|}$ is continuous and possesses partial derivatives at the origin, but it is not differentiable at the origin. It remains to prove that $f(x,y)$ has all the directional derivative. Indeed, choosing an arbitrary unitary direction vector $u = (\alpha, \beta)$, it follows by definition that

$$D_u f(0,0) = \lim_{t\to 0_+} \frac{f(\alpha t, \beta t) - f(0,0)}{t-0} = \lim_{t\to 0_+} \frac{\sqrt{|\alpha\beta|\, t^2} - 0}{t} = \sqrt{|\alpha\beta|}.$$

One can show that a similar example is provided by the function $f(x,y) = \begin{cases} \frac{xy}{\sqrt{x^2+y^2}}, & x^2+y^2 \neq 0 \\ 0, & x^2+y^2 = 0 \end{cases}$ from Example 4.

Example 6. "If $f(x,y)$ is continuous and has bounded partial derivatives on $\mathbb{R}^2$, and for any differentiable curve $x(t)$, $y(t)$ passing through the origin and such that $(x'(0))^2 + (y'(0))^2 > 0$ the composite function $g(t) = f(x(t), y(t))$ is differentiable for $\forall t \in \mathbb{R}$, then $f(x,y)$ is differentiable at the origin."

Solution.

Let us show that the function $f(x,y) = \begin{cases} \frac{x^3}{x^2+y^2}, & x^2+y^2 \neq 0 \\ 0, & x^2+y^2 = 0 \end{cases}$ satisfies the conditions of the statement. First, this function is continuous at any point $(x,y) \neq (0,0)$ due to arithmetic rules. At the origin we should appeal to the definition:

$$|f(x,y) - f(0,0)| = |x| \frac{x^2}{x^2+y^2} \leq |x| \underset{(x,y)\to(0,0)}{\to} 0,$$

so $\lim_{(x,y)\to(0,0)} f(x,y) = 0 = f(0,0)$. Then the partial derivatives can be calculated at any point but the origin by applying the arithmetic rules:

$$f_x(x,y) = \frac{x^2(x^2+3y^2)}{(x^2+y^2)^2}, \quad f_y(x,y) = \frac{-2x^3 y}{(x^2+y^2)^2}.$$

At the origin we apply the definition:

$$f_x(0,0) = \lim_{x\to 0} \frac{f(x,0) - f(0,0)}{x-0} = \lim_{x\to 0} \frac{x}{x} = 1,$$

$$f_y(0,0) = \lim_{y\to 0} \frac{f(0,y) - f(0,0)}{y-0} = \lim_{y\to 0} \frac{0-0}{y} = 0.$$

Moreover, the partial derivatives are bounded on $\mathbb{R}^2$:

$$|f_x(x,y)| = \frac{x^2(x^2+3y^2)}{(x^2+y^2)^2} = \frac{x^2(x^2+y^2)}{(x^2+y^2)^2} + \frac{2x^2y^2}{(x^2+y^2)^2} \leq 1+1 = 2$$

and

$$|f_y(x,y)| = \frac{2|x^3 y|}{(x^2+y^2)^2} = \frac{x^2}{(x^2+y^2)} \cdot \frac{2|xy|}{(x^2+y^2)} \leq 1 \cdot 1 = 1.$$

Let us prove the additional condition of this example. Since the given curve $x(t), y(t)$ is differentiable and the partial derivatives of $f(x,y)$ exist, the composite function $g(t) = f(x(t), y(t))$ can be differentiated by applying the chain rule: for any $t \neq 0$ we obtain

$$g'(t) = f_x(x, y) \cdot x'(t) + f_y(x, y) \cdot y'(t) = \frac{x^4 + 3x^2y^2}{(x^2+y^2)^2} \cdot x'(t) - \frac{2x^3y}{(x^2+y^2)^2} \cdot y'(t),$$

and for $t = 0$ we have

$$g'(0) = f_x(0,0) \cdot x'(0) + f_y(0,0) \cdot y'(0) = 1 \cdot x'(0) + 0 \cdot y'(0) = x'(0).$$

Hence, all the conditions of the statement hold. Nevertheless, $f(x,y)$ is not differentiable at the origin. Indeed, the difference between the function and its linear approximation in a neighborhood of the origin is as follows:

$$\alpha(x,y) = f(x,y) - (f(0,0) + f_x(0,0) \cdot x + f_y(0,0) \cdot y) = \frac{x^3}{x^2+y^2} - x = -\frac{xy^2}{x^2+y^2}.$$

Then, along the path $y = x$ the ratio $\frac{\alpha(x,y)}{\sqrt{x^2+y^2}}$ does not tend to zero:

$$\lim_{x \to 0} \frac{\alpha(x,x)}{\rho(x,x)} = \lim_{x \to 0} \frac{-x^3}{2\sqrt{2}\,|x|\,x^2} = \lim_{x \to 0} \frac{-1}{2\sqrt{2}} \operatorname{sgn} x \neq 0.$$

Therefore, $f(x,y)$ is not differentiable at the origin.

Remark to Examples 3-6. These examples represent a sequence of refined, but still insufficient, conditions for differentiation.

Remark to Examples 1-6. These examples show that when the partial derivatives are discontinuous at some point, then the function may be non-differentiable at this point. Continuity of the partial derivatives guarantees the differentiability, but this is a sufficient condition as it is shown in the next Example 7.

Example 7. "If $f(x,y)$ is differentiable at a point (x_0, y_0), then its partial derivatives are continuous at this point."

Solution.

The function $f(x,y) = \begin{cases} (x^2+y^2)\sin\frac{1}{x^2+y^2}, & x^2+y^2 \neq 0 \\ 0, & x^2+y^2 = 0 \end{cases}$ satisfies the conditions of the statement at the origin. First, let us find its partial derivatives. For any point $(x,y) \neq (0,0)$, applying the arithmetic rules we obtain:

$$f_x(x,y) = 2x\sin\frac{1}{x^2+y^2} - \frac{2x}{x^2+y^2}\cos\frac{1}{x^2+y^2},$$

$$f_y(x,y) = 2y\sin\frac{1}{x^2+y^2} - \frac{2y}{x^2+y^2}\cos\frac{1}{x^2+y^2}.$$

At the origin we should appeal to the definition:

$$f_x(0,0) = \lim_{x\to 0} \frac{f(x,0) - f(0,0)}{x - 0} = \lim_{x\to 0} \frac{x^2 \sin \frac{1}{x^2}}{x} = 0,$$

$$f_y(0,0) = \lim_{y\to 0} \frac{f(0,y) - f(0,0)}{y - 0} = \lim_{y\to 0} \frac{y^2 \sin \frac{1}{y^2}}{y} = 0.$$

Second, the remainder in the definition of differentiability at the origin takes the form

$$\alpha(x,y) = f(x,y) - (f(0,0) + f_x(0,0)\cdot x + f_y(0,0)\cdot y) = (x^2+y^2)\sin\frac{1}{x^2+y^2},$$

and consequently,

$$\lim_{(x,y)\to(0,0)} \frac{\alpha(x,y)}{\rho(x,y)} = \lim_{(x,y)\to(0,0)} \sqrt{x^2+y^2}\sin\frac{1}{x^2+y^2} = 0$$

due to the following evaluation

$$\left|\sqrt{x^2+y^2}\sin\frac{1}{x^2+y^2}\right| \le \sqrt{x^2+y^2} \underset{(x,y)\to(0,0)}{\to} 0.$$

Therefore, $f(x,y)$ is differentiable at the origin. (Notice, that $f(x,y)$ is also differentiable at each point different from the origin, because its partial derivatives are continuous there.) Now let us show that the partial derivatives are discontinuous at the origin. Indeed, choosing the sequence of the points (x_n, y_n), $x_n = \frac{1}{\sqrt{2n\pi}} \underset{n\to\infty}{\to} 0$, $y_n = 0 \underset{n\to\infty}{\to} 0$, we obtain

$$f_x(x_n, y_n) = 2x_n \sin 2n\pi - \frac{2}{x_n}\cos 2n\pi = -\frac{2}{x_n} \underset{n\to\infty}{\to} -\infty,$$

which means that $f_x(x,y)$ is discontinuous (and even unbounded) at the origin. Similarly for (x_k, y_k), $x_k = 0 \underset{k\to\infty}{\to} 0$, $y_k = \frac{1}{\sqrt{2k\pi}} \underset{k\to\infty}{\to} 0$, one gets $f_y(x_k, y_k) \underset{k\to\infty}{\to} -\infty$, that is, $f_y(x,y)$ is also discontinuous and unbounded at the origin.

Another counterexample is provided by the function

$$f(x,y) = \begin{cases} x^2 \sin\frac{1}{x} + y^2 \sin\frac{1}{y}, & xy \ne 0 \\ x^2 \sin\frac{1}{x}, & x \ne 0,\ y = 0 \\ y^2 \sin\frac{1}{y}, & x = 0,\ y \ne 0 \\ 0, & x = y = 0 \end{cases}$$

Remark. The considered counterexample shows that the above statement can be reformulated in the following (still false) form: “if $f(x,y)$ is differentiable on $\mathbb{R}^2$, then its partial derivatives are continuous at the origin”.

Example 8. "If $f(x,y)$ is differentiable at a point (x_0,y_0) and one of its partial derivatives is continuous at this point, then another partial derivative is also continuous at (x_0,y_0)."

Solution.

Let us consider the function $f(x,y) = \begin{cases} \frac{4x^6y^2}{(x^4+y^2)^2}, & x^2+y^2 \neq 0 \\ 0, & x^2+y^2=0 \end{cases}$ in a neighborhood of the origin. The partial derivatives can be found by applying the arithmetic rules for any point different from the origin:

$$f_x(x,y) = \frac{24x^5y^4 - 8x^9y^2}{(x^4+y^2)^3}, \quad f_y(x,y) = \frac{8x^{10}y - 8x^6y^3}{(x^4+y^2)^3};$$

and by using the definition at the origin:

$$f_x(0,0) = \lim_{x\to 0} \frac{f(x,0)-f(0,0)}{x-0} = \lim_{x\to 0} \frac{0-0}{x} = 0,$$

$$f_y(0,0) = \lim_{y\to 0} \frac{f(0,y)-f(0,0)}{y-0} = \lim_{y\to 0} \frac{0-0}{y} = 0.$$

Since both partial derivatives are zero at the origin, the remainder in the definition of differentiability takes the form $\alpha(x,y) = f(x,y) = \frac{4x^6y^2}{(x^4+y^2)^2}$, and consequently,

$$\frac{\alpha(x,y)}{\rho(x,y)} = \frac{4x^6y^2}{(x^4+y^2)^2\sqrt{x^2+y^2}} = \left(\frac{2x^2y}{x^4+y^2}\right)^2 \cdot \frac{|x|}{\sqrt{x^2+y^2}} \cdot |x| \leq |x| \underset{(x,y)\to(0,0)}{\to} 0.$$

Therefore, $f(x,y)$ is differentiable at the origin. Notice that due to the arithmetic rules of continuity both partial derivatives are continuous at each point different from the origin. It guarantees differentiability of $f(x,y)$ on $\mathbb{R}^2$. Finally, the continuity of $f_x(x,y)$ at the origin follows from the evaluation:

$$|f_x(x,y) - f_x(0,0)| = \frac{|24x^5y^4 - 8x^9y^2|}{(x^4+y^2)^3}$$

$$\leq 6\left(\frac{2x^2y}{x^4+y^2}\right)^2 \cdot \frac{y^2}{x^4+y^2} \cdot |x| + 2\left(\frac{2x^2y}{x^4+y^2}\right)^2 \cdot \frac{x^4}{x^4+y^2} \cdot |x|$$

$$\leq 6|x| + 2|x| = 8|x| \underset{(x,y)\to(0,0)}{\to} 0.$$

Hence, all the conditions of the statement hold. However, $f_y(x,y)$ is not continuous at the origin, because along the path $y=2x^2$ we obtain

$$\lim_{x\to 0} f_y(x,2x^2) = \lim_{x\to 0} \frac{16x^{12} - 64x^{12}}{(5x^4)^3} = -\frac{48}{125} \neq 0 = f_y(0,0).$$

Another counterexample can be given with the function $f(x,y) = \begin{cases} x^2 \sin\frac{1}{x} + y^2, \ x \neq 0 \\ y^2, \ x = 0 \end{cases}$. The partial derivatives are readily calculated by applying the arithmetic rules at each point out of the y-axis:

$$f_x(x,y) = 2x \sin\frac{1}{x} - \cos\frac{1}{x}, \ f_y(x,y) = 2y.$$

For the points of the y-axis we appeal to the definition: at the origin we have

$$f_x(0,0) = \lim_{x\to 0} \frac{f(x,0) - f(0,0)}{x - 0} = \lim_{x\to 0} \frac{x^2 \sin 1/x - 0}{x} = \lim_{x\to 0} x \sin\frac{1}{x} = 0,$$

$$f_y(0,0) = \lim_{y\to 0} \frac{f(0,y) - f(0,0)}{y - 0} = \lim_{y\to 0} \frac{y^2 - 0}{y} = 0,$$

and for any other point $(0, y_0)$, $y_0 \neq 0$ we obtain

$$f_x(0,y_0) = \lim_{x\to 0} \frac{f(x,y_0) - f(0,y_0)}{x - 0} = \lim_{x\to 0} \frac{x^2 \sin\frac{1}{x} + y_0^2 - y_0^2}{x} = \lim_{x\to 0} x \sin\frac{1}{x} = 0,$$

$$f_y(0,y_0) = \lim_{y\to y_0} \frac{f(0,y) - f(0,y_0)}{y - y_0} = \lim_{y\to y_0} \frac{y^2 - y_0^2}{y - y_0} = 2y_0.$$

Since both partial derivatives are continuous on $\mathbb{R}^2 \backslash (0,y)$, $f(x,y)$ is differentiable on $\mathbb{R}^2 \backslash (0,y)$. Let us verify differentiability of $f(x,y)$ on the y-axis. At the origin the remainder of the linear approximation is $\alpha(x,y) = f(x,y) = \begin{cases} x^2 \sin\frac{1}{x} + y^2, \ x \neq 0 \\ y^2, \ x = 0 \end{cases}$. For the points of the y-axis we have $\frac{\alpha(0,y)}{\rho(0,y)} = \frac{y^2}{\sqrt{y^2}} = |y| \underset{(x,y)\to(0,0)}{\to} 0$ and for the points outside of the y-axis we obtain

$$\left|\frac{\alpha(x,y)}{\rho(x,y)}\right| = \left|\frac{x^2 \sin 1/x + y^2}{\sqrt{x^2+y^2}}\right| \le \frac{x^2+y^2}{\sqrt{x^2+y^2}} = \sqrt{x^2+y^2} \underset{(x,y)\to(0,0)}{\to} 0.$$

Therefore, $f(x,y)$ is differentiable at the origin. For any other point $(0, y_0)$, $y_0 \neq 0$ the remainder is

$$\alpha(x,y) = f(x,y) - (f(0,y_0) + f_x(0,y_0) \cdot x + f_y(0,y_0) \cdot (y - y_0))$$

$$= \begin{cases} x^2 \sin\frac{1}{x} + y^2, \ x \neq 0 \\ y^2, \ x = 0 \end{cases} - \left(y_0^2 + 2y_0(y - y_0)\right).$$

Therefore, for the points of the y-axis we have

$$\frac{\alpha(0,y)}{\rho(0,y)} = \frac{y^2 + y_0^2 - 2y_0 y}{\sqrt{(y-y_0)^2}} = \frac{(y-y_0)^2}{\sqrt{(y-y_0)^2}} \underset{(x,y)\to(0,0)}{\to} 0,$$

and for the points outside of the y-axis we obtain

$$\left|\frac{\alpha(x,y)}{\rho(x,y)}\right| = \left|\frac{x^2 \sin 1/x + (y-y_0)^2}{\sqrt{x^2+(y-y_0)^2}}\right|$$

$$\leq \frac{x^2+(y-y_0)^2}{\sqrt{x^2+(y-y_0)^2}} = \sqrt{x^2+(y-y_0)^2} \underset{(x,y)\to(0,y_0)}{\to} 0.$$

It implies that $f(x,y)$ is differentiable at any point $(0,y_0)$, $y_0 \neq 0$. Summarizing, $f(x,y)$ is differentiable on $\mathbb{R}^2$. Notice that the formulas found for the partial derivative in y can be rewritten in the form $f_y(x,y) = 2y$ on $\mathbb{R}^2$, so this derivative is continuous on $\mathbb{R}^2$. Hence, all the conditions of the statement hold. However, $f_x(x,y)$ is discontinuous at each point of the y-axis, because approaching a point $(0,y)$ by the sequence of the points (x_n,y), $x_n = \frac{1}{2n\pi}$, $\forall n \in \mathbb{Z}\setminus\{0\}$, we obtain

$$\lim_{x_n\to 0} f_x(x_n,y) = \lim_{n\to\infty}\left(\frac{1}{n\pi}\sin 2n\pi - \cos 2n\pi\right) = -1 \neq 0 = f_x(0,y).$$

Remark. Notice that both functions are also counterexamples to the following version of the false statement: "if $f(x,y)$ is differentiable on $\mathbb{R}^2$ and one of its partial derivatives is continuous on $\mathbb{R}^2$, then another partial derivative is also continuous on $\mathbb{R}^2$".

Example 9. "If $f(x,y)$ has partial derivatives at some point, then $f(x,y)$ has all the directional derivatives at the same point."

Solution.

It was shown in Example 1 that the function $f(x,y) = \begin{cases} \frac{xy}{x^2+y^2}, & x^2+y^2 \neq 0 \\ 0, & x^2+y^2=0 \end{cases}$ has zero partial derivatives at the origin (notice that its partial derivatives exist at every point in $\mathbb{R}^2$). However, $f(x,y)$ has no directional derivative at the origin, except for the directions along the coordinate axes. In fact, let $u = (\alpha,\beta)$ be an arbitrary unitary direction vector not parallel to coordinate axes, that is, $\alpha\beta \neq 0$. Then evaluating the directional derivative by definition one obtains

$$D_u f(0,0) = \lim_{t\to 0_+}\frac{f(\alpha t,\beta t) - f(0,0)}{t-0} = \lim_{t\to 0_+}\frac{\frac{\alpha\beta t^2}{(\alpha^2+\beta^2)t^2} - 0}{t} = \lim_{t\to 0_+}\frac{\alpha\beta}{t} = \infty,$$

that is, these derivatives do not exist.

Example 10. "If $f(x,y)$ is continuous and possesses all the directional derivatives at some point, then there exist a partial derivative of $f(x,y)$ at this point."

Solution.

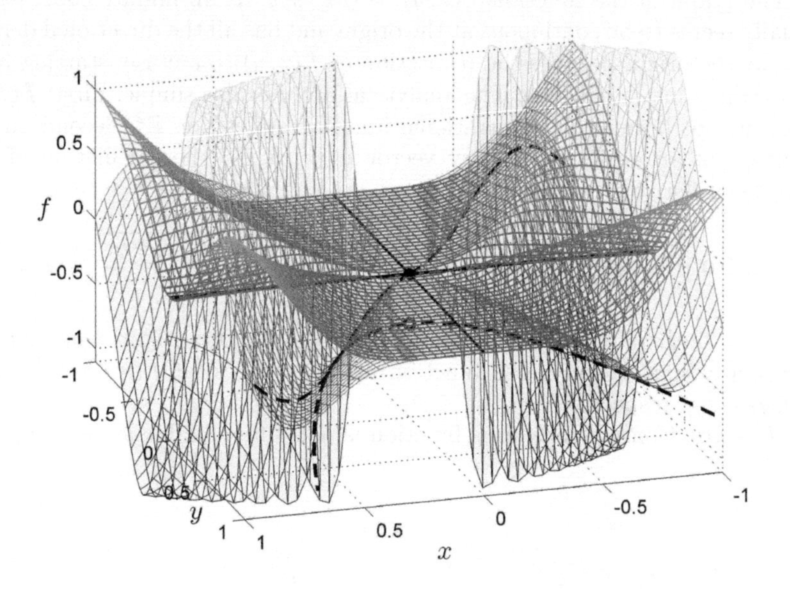

FIGURE 8.3.5: Example 8, graph of the first function

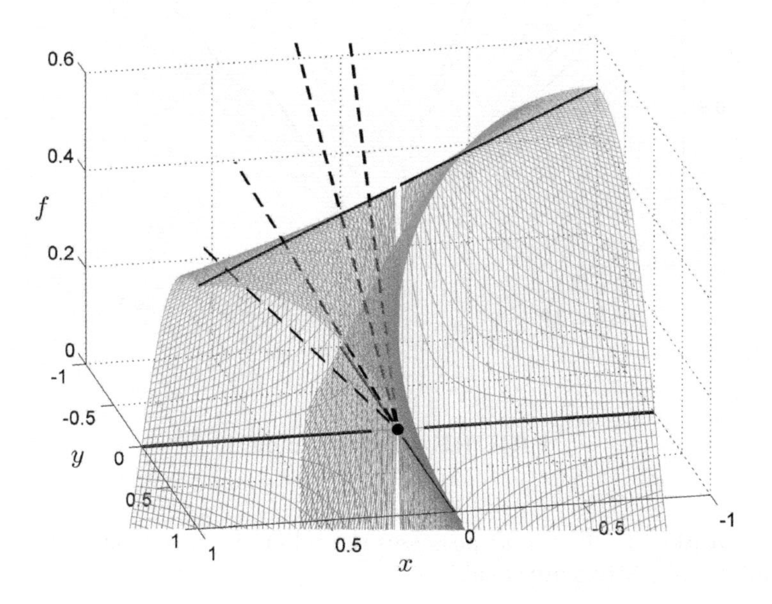

FIGURE 8.3.6: Example 9

The graph of the function $f(x,y) = \sqrt{x^2+y^2}$ is an infinite cone, which visually seems to be continuous at the origin and has all the directional derivatives at the origin because the restriction of $f(x,y)$ to any ray starting from the origin is just a half-line. The analytical proof is also simple. First, $f(x,y)$ is continuous because it is a irrational function defined on $\mathbb{R}^2$. Second, in the direction of an arbitrary unitary vector $u = (\alpha, \beta)$, the calculation of the directional derivatives gives:

$$D_u f(0,0) = \lim_{t \to 0_+} \frac{f(\alpha t, \beta t) - f(0,0)}{t-0} = \lim_{t \to 0_+} \frac{\sqrt{\alpha^2 t^2 + \beta^2 t^2} - 0}{t} = 1.$$

However, both partial derivatives do not exist at the origin, since on the coordinates axis x and y the function assumes the form $f(x,0) = |x|$ and $f(0,y) = |y|$, respectively.

Remark. Consequently, this function is not differentiable at the origin.

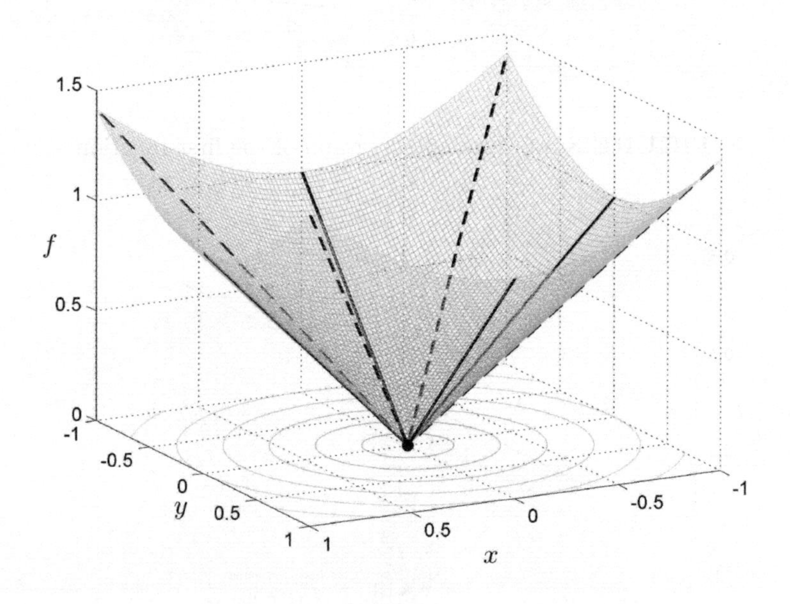

FIGURE 8.3.7: Example 10

Example 11. "If $f(x,y)$ possesses second-order partial derivatives, then the mixed derivatives are equal."

Solution.

Let us consider the function $f(x,y) = \begin{cases} xy\frac{x^2-y^2}{x^2+y^2}, & x^2+y^2 \neq 0 \\ 0, & x^2+y^2 = 0 \end{cases}$. Applying arithmetic rules for partial derivative calculation and simple algebra for simplification we can obtain the following expressions for partial derivatives at any point different from the origin:

$$f_x(x,y) = \frac{x^4y + 4x^2y^3 - y^5}{(x^2+y^2)^2}, \quad f_y(x,y) = \frac{-xy^4 - 4x^3y^2 + x^5}{(x^2+y^2)^2}.$$

At the origin both partial derivatives can be calculated by the definition:

$$f_x(0,0) = \lim_{x\to 0} \frac{f(x,0) - f(0,0)}{x-0} = \lim_{x\to 0} \frac{0-0}{x} = 0,$$

$$f_y(0,0) = \lim_{y\to 0} \frac{f(0,y) - f(0,0)}{y-0} = \lim_{y\to 0} \frac{0-0}{y} = 0.$$

Therefore, $f_x(x,y)$ and $f_y(x,y)$ are defined on $\mathbb{R}^2$. To calculate the mixed second-order derivative at the origin we should again apply the definition:

$$f_{xy}(0,0) = \lim_{y\to 0} \frac{f_x(0,y) - f_x(0,0)}{y-0} = \lim_{y\to 0} \frac{-y-0}{y} = -1,$$

$$f_{yx}(0,0) = \lim_{x\to 0} \frac{f_y(x,0) - f_y(0,0)}{x-0} = \lim_{x\to 0} \frac{x-0}{x} = 1.$$

Hence, both mixed derivatives exist, but they are different. It is simple to show that the second-order mixed derivatives exist on $\mathbb{R}^2$ and coincide in all points but the origin. Other two second order derivatives also exist on $\mathbb{R}^2$.

Another counterexample is the function $f(x,y) = \begin{cases} xy\,, & |y| \le |x| \\ -xy, & |y| > |x| \end{cases}$. For $|y| < |x|$ we readily get $f_x(x,y) = y$, $f_y(x,y) = x$, and for $|y| > |x|$: $f_x(x,y) = -y$, $f_y(x,y) = -x$. For the points with $|y| = |x| \neq 0$ the first-order derivatives do not exist, because approaching any fixed point (x_0, y_0) $|y_0| = |x_0| \neq 0$ from the side where $|y| < |x|$ we obtain

$$\lim_{x\to x_0} \frac{f(x,y_0) - f(x_0,y_0)}{x-x_0} = \lim_{x\to x_0} \frac{xy_0 - x_0y_0}{x-x_0} = y_0,$$

while approaching the same point from the side with $|y| > |x|$ we get

$$\lim_{x\to x_0} \frac{f(x,y_0) - f(x_0,y_0)}{x-x_0} = \lim_{x\to x_0} \frac{-xy_0 - x_0y_0}{x-x_0} = \infty,$$

and similarly for derivative with respect to y. Finally, at the origin both derivatives can be obtained by the definition:

$$f_x(0,0) = \lim_{x\to 0} \frac{f(x,0) - f(0,0)}{x-0} = \lim_{x\to 0} \frac{0-0}{x} = 0,$$

$$f_y(0,0) = \lim_{y\to 0} \frac{f(0,y) - f(0,0)}{y-0} = \lim_{y\to 0} \frac{0-0}{y} = 0.$$

Hence, the first-order derivatives are defined on $\mathbb{R}^2$, including the origin, but excluding the two lines $y = x$ and $y = -x$ (without the origin). Since both partial derivatives are defined at the origin, and the last is a limit point for their domains, we can consider the second-order partial derivatives at the origin. In particular, for the mixed derivatives we obtain:

$$f_{xy}(0,0) = \lim_{y\to 0} \frac{f_x(0,y) - f_x(0,0)}{y-0} = \lim_{y\to 0} \frac{-y-0}{y} = -1,$$

$$f_{yx}(0,0) = \lim_{x\to 0} \frac{f_y(x,0) - f_y(0,0)}{x-0} = \lim_{x\to 0} \frac{x-0}{x} = 1.$$

Therefore, the mixed derivatives are different at the origin.

Remark 1. The condition that guarantees the equality of the mixed derivatives is their continuity. So in this Example the mixed derivatives are discontinuous at the origin. Let us show this explicitly for the first function. Applying arithmetic rules for derivatives and simple algebra we can obtain the following results for any $(x,y) \neq (0,0)$:

$$f_{xy}(x,y) = \frac{x^6 + 9x^4y^2 - 9x^2y^4 - y^6}{(x^2+y^2)^3}, \quad f_{yx}(x,y) = \frac{x^6 + 9x^4y^2 - 9x^2y^4 - y^6}{(x^2+y^2)^3},$$

that is, the mixed derivatives are continuous and coincide on $\mathbb{R}^2 \backslash (0,0)$. However, at the origin these derivatives are discontinuous. Indeed, approaching the origin along the line $y = x$ we obtain:

$$\lim_{x\to 0} f_{xy}(x,x) = 0 \neq -1 = f_{xy}(0,0) \text{ and } \lim_{x\to 0} f_{yx}(x,x) = 0 \neq 1 = f_{yx}(0,0).$$

Remark 2. The continuity of the mixed derivatives is a sufficient, but not necessary, condition for their equality, as it is shown in the next Examples 12, 13 and 14.

Example 12. "If $f(x,y)$ possesses first- and second-order partial derivatives at some point (or on $\mathbb{R}^2$) and the mixed derivatives coincide, then $f(x,y)$ is continuous at this point (or on $\mathbb{R}^2$)."

Solution.

Let us consider the function $f(x,y) = \begin{cases} \frac{x^2y^2}{x^4+y^4}, & x^2+y^2 \neq 0 \\ 0, & x^2+y^2 = 0 \end{cases}$. Its first-order derivatives exist on $\mathbb{R}^2$ and can be found by using arithmetic rules at any point different from the origin:

$$f_x(x,y) = \frac{2xy^6 - 2x^5y^2}{(x^4+y^4)^2}, \quad f_y(x,y) = \frac{2x^6y - 2x^2y^5}{(x^4+y^4)^2},$$

and by applying definition at the origin:

$$f_x(0,0) = \lim_{x\to 0} \frac{f(x,0) - f(0,0)}{x-0} = \lim_{x\to 0} \frac{0-0}{x} = 0,$$

$$f_y(0,0) = \lim_{y\to 0} \frac{f(0,y) - f(0,0)}{y-0} = \lim_{y\to 0} \frac{0-0}{y} = 0.$$

In the same mode we can calculate the second-order derivatives on $\mathbb{R}^2$. In particular, for the mixed derivatives we obtain

$$f_{xy}(x,y) = f_{yx}(x,y) = \frac{24x^5y^5 - 4x^9y - 4xy^9}{(x^4+y^4)^3}$$

for $(x,y) \neq (0,0)$, and $f_{xy}(0,0) = f_{yx}(0,0) = 0$. Hence, the conditions of the statement are satisfied. However, the function is not continuous at the origin, since along the line $y = x$ we get

$$\lim_{x\to 0} f(x,x) = \lim_{x\to 0} \frac{x^4}{2x^4} = \frac{1}{2} \neq 0 = f(0,0).$$

Actually, the function even does not have a limit at the origin, since along another line $y = 2x$ we obtain different result: $\lim\limits_{x\to 0} f(x,2x) = \lim\limits_{x\to 0} \frac{4x^4}{17x^4} = \frac{4}{17}$.

Remark. Evidently, at least one of the first-order derivatives is discontinuous at the origin, because otherwise the function would be differentiable (and therefore, continuous) at the origin. It is simple to show explicitly: along $y = x$ we get $\lim\limits_{x\to 0} f_x(x,x) = \lim\limits_{x\to 0} \frac{2x^7-2x^7}{4x^8} = 0$, while along $y = 2x$ we obtain: $\lim\limits_{x\to 0} f_x(x,2x) = \lim\limits_{x\to 0} \frac{128x^7-8x^7}{289x^8} = \infty$.

Example 13. "If $f(x,y)$ possesses partial derivatives of all orders at some point (or on $\mathbb{R}^2$) and the mixed derivatives do not depend on the order of differentiation, then $f(x,y)$ is continuous at this point (or on $\mathbb{R}^2$)."

Solution.

Let us consider the function $f(x,y) = \begin{cases} \exp\left(-\frac{x^2}{y^2} - \frac{y^2}{x^2}\right), & xy \neq 0 \\ 0, \ xy = 0 \end{cases}$ and show that it is discontinuous at the origin. Indeed, along the line $x = y$ one gets

$$\lim_{x\to 0} f(x,x) = \lim_{x\to 0} \exp\left(-\frac{x^2}{x^2} - \frac{x^2}{x^2}\right) = e^{-2} \neq 0 = f(0,0).$$

Even being discontinuous this function possesses all partial derivatives. We start with an arbitrary point out of the coordinate axes. In this case, applying the arithmetic and chain rules we readily obtain

$$f_x(x,y) = \exp\left(-\frac{x^2}{y^2} - \frac{y^2}{x^2}\right) \cdot \left(-\frac{2x}{y^2} + \frac{2y^2}{x^3}\right),$$

$$f_y(x,y) = \exp\left(-\frac{x^2}{y^2} - \frac{y^2}{x^2}\right) \cdot \left(-\frac{2y}{x^2} + \frac{2x^2}{y^3}\right).$$

Apart from the form of these derivatives, we can see that any derivative has the form

$$f_{x^m y^n}(x,y) = \exp\left(-\frac{x^2}{y^2} - \frac{y^2}{x^2}\right) \cdot P(x,y),$$

where $m+n=k$ is the order of the derivative, and $P(x,y)$ is a polynomial with respect to $x^i y^j$, $i,j \in \mathbb{Z}$. Indeed, the partial derivatives of each term $h(x,y) = \exp\left(-\frac{x^2}{y^2} - \frac{y^2}{x^2}\right) \cdot x^i y^j$ are

$$h_x(x,y) = \exp\left(-\frac{x^2}{y^2} - \frac{y^2}{x^2}\right) \cdot \left(-\frac{2x}{y^2} x^i y^j + \frac{2y^2}{x^3} x^i y^j + i x^{i-1} y^j\right)$$

and

$$h_y(x,y) = \exp\left(-\frac{x^2}{y^2} - \frac{y^2}{x^2}\right) \cdot \left(-\frac{2y}{x^2} x^i y^j + \frac{2x^2}{y^3} x^i y^j + j x^i y^{j-1}\right).$$

So their sum will give again the conjectured form. Notice also that all these derivatives are continuous when $xy \neq 0$ and it implies that the corresponding mixed derivatives are equal.

Now let us consider the remaining points for which $xy = 0$. Using the method of mathematical induction we will prove that at each of these point the partial derivative of any order is zero. First, we check this for the first-order derivatives. For f_x we obtain the following results at the origin:

$$f_x(0,0) = \lim_{x\to 0} \frac{f(x,0) - f(0,0)}{x-0} = \lim_{x\to 0} \frac{0-0}{x} = 0,$$

at the points $(x_0, 0)$, $x_0 \neq 0$:

$$f_x(x_0,0) = \lim_{x\to x_0} \frac{f(x,0) - f(x_0,0)}{x-x_0} = \lim_{x\to x_0} \frac{0-0}{x-x_0} = 0,$$

and at the points $(0, y_0)$, $y_0 \neq 0$:

$$f_x(0,y_0) = \lim_{x\to 0} \frac{f(x,y_0) - f(0,y_0)}{x-0} = \lim_{x\to 0} \frac{1}{x} \exp\left(-\frac{x^2}{y_0^2} - \frac{y_0^2}{x^2}\right)$$

$$= \lim_{x\to 0} \exp\left(-\frac{x^2}{y_0^2}\right) \cdot \frac{1}{x} \exp\left(-\frac{y_0^2}{x^2}\right)$$

$$= \lim_{x\to 0} \exp\left(-\frac{x^2}{y_0^2}\right) \cdot \lim_{t\to\infty} t \exp\left(-y_0^2 t^2\right) = 1 \cdot 0 = 0$$

(the substitution $t = \frac{1}{x}$ has been used on the penultimate step). Summarizing, $f_x(x,y) = 0$ at each point such that $xy = 0$. Absolutely the same reasoning shows that $f_y(x,y) = 0$ at the same points. Therefore, the base of the induction is proved.

Let us suppose now that at each point of the coordinate axes $f_{x^m y^n}(x,y) = 0$ for all the derivatives of the order $m+n=k$ and show that $f_{x^m y^n x}(x,y) = 0$ and $f_{x^m y^n y}(x,y) = 0$ at these points. Again, it is sufficient to analyze only the former derivative, because all considerations for the latter are the same. At the origin we have

$$f_{x^m y^n x}(0,0) = \lim_{x\to 0} \frac{f_{x^m y^n}(x,0) - f_{x^m y^n}(0,0)}{x-0} = \lim_{x\to 0} \frac{0-0}{x} = 0.$$

At the points $(x_0, 0)$, $x_0 \neq 0$ the situation is also simple:

$$f_{x^m y^n x}(x_0, 0) = \lim_{x \to x_0} \frac{f_{x^m y^n}(x, 0) - f_{x^m y^n}(x_0, 0)}{x - x_0} = \lim_{x \to x_0} \frac{0 - 0}{x - x_0} = 0.$$

For the points $(0, y_0)$, $y_0 \neq 0$ analysis is more involved. First, we transform the definition to the following form:

$$f_{x^m y^n x}(0, y_0) = \lim_{x \to 0} \frac{f_{x^m y^n}(x, y_0) - f_{x^m y^n}(0, y_0)}{x - 0}$$

$$= \lim_{x \to 0} \exp\left(-\frac{x^2}{y_0^2}\right) \cdot \frac{1}{x} P(x, y_0) \exp\left(-\frac{y_0^2}{x^2}\right),$$

where $P(x, y)$ is the polynomial from the definition of $f_{x^m y^n}(x, y)$, $xy \neq 0$. Since the limit of the first multiplier is simple, $\lim\limits_{x \to 0} \exp\left(-\frac{x^2}{y_0^2}\right) = 1$, we should analyze the remaining terms. Notice that just like $P(x, y_0)$, the function $\frac{1}{x} P(x, y_0)$ is also a polynomial with the terms $x^i y_0^j$, $i, j \in \mathbb{Z}$. If $i \in \mathbb{N}$, then such term tends to zero as $x \to 0$. If $i = 0$, then such term is constant, but it is multiplied by $\exp\left(-\frac{y_0^2}{x^2}\right)$ inside the limit, and the last function tends to zero as $x \to 0$. Finally, for $i < 0$ we can apply the following calculations:

$$\lim_{x \to 0} x^i \exp\left(-\frac{y_0^2}{x^2}\right) = \lim_{x \to 0} \frac{\exp\left(-\frac{y_0^2}{x^2}\right)}{x^{-i}} = \lim_{t \to \infty} \frac{t^{-i}}{\exp\left(y_0^2 t^2\right)}$$

$$= \lim_{t \to \infty} \frac{-i \cdot t^{-i-1}}{2t y_0^2 \exp\left(y_0^2 t^2\right)} = \lim_{t \to \infty} \frac{-i \cdot t^{-i-2}}{2 y_0^2 \exp\left(y_0^2 t^2\right)} = \ldots = 0.$$

Here, we used the substitution $x = \frac{1}{t}$ and then applied L'Hospital's rule repeatedly until the power of t in the numerator becomes negative or zero. Based on this analysis, we can conclude that $\lim\limits_{x \to 0} \frac{1}{x} P(x, y_0) \exp\left(-\frac{y_0^2}{x^2}\right) = 0$, and consequently, $f_{x^m y^n x}(0, y_0) = 0$ for each point $(0, y_0)$, $y_0 \neq 0$. Concluding, we have shown that at each point of the coordinate axes the partial derivatives of any order are zero: $f_{x^m y^n}(x, y) = 0$. In particular it means that the mixed derivatives do not depend on the order of differentiation.

Hence, we have proved, that partial derivatives of an arbitrary order can be expressed in the form:

$$f_{x^m y^n}(x, y) = \begin{cases} \exp\left(-\frac{x^2}{y^2} - \frac{y^2}{x^2}\right) \cdot P(x, y), & xy \neq 0 \\ 0, & xy = 0 \end{cases},$$

with the polynomial $P(x, y)$ specified above, and the corresponding mixed derivatives are equal.

Example 14. "If $f(x, y)$ is continuous and possesses first- and second-order partial derivatives on $\mathbb{R}^2$, and the mixed derivatives coincide on $\mathbb{R}^2$, then $f(x, y)$ is differentiable on $\mathbb{R}^2$."

Solution.

Let us consider the function $f(x,y)=\begin{cases}\frac{x^3y^2}{x^4+y^4}, & x^2+y^2\neq 0\\ 0, & x^2+y^2=0\end{cases}$. First, the continuity of $f(x,y)$ at any $(x,y)\neq(0,0)$ follows directly from the arithmetic rules for continuous functions and at the origin it can be shown by the following evaluation:

$$|f(x,y)-f(0,0)|=\left|\frac{x^3y^2}{x^4+y^4}\right|=|x|\frac{x^2y^2}{x^4+y^4}\leq\frac{1}{2}|x|\underset{(x,y)\to(0,0)}{\longrightarrow}0.$$

Second, the first-order derivatives can be found by using the arithmetic rules at any point $(x,y)\neq(0,0)$:

$$f_x(x,y)=\frac{3x^2y^6-x^6y^2}{(x^4+y^4)^2},\quad f_y(x,y)=\frac{2x^7y-2x^3y^5}{(x^4+y^4)^2},$$

and by applying the definition at the origin:

$$f_x(0,0)=\lim_{x\to0}\frac{f(x,0)-f(0,0)}{x-0}=\lim_{x\to0}\frac{0-0}{x}=0,$$

$$f_y(0,0)=\lim_{y\to0}\frac{f(0,y)-f(0,0)}{y-0}=\lim_{y\to0}\frac{0-0}{y}=0.$$

In the same way one can calculate the second-order derivatives: at any $(x,y)\neq(0,0)$ one obtains

$$f_{xx}(x,y)=\frac{-24x^5y^6+2x^9y^2+6xy^{10}}{(x^4+y^4)^3},\quad f_{yy}(x,y)=\frac{2x^{11}-24x^7y^4+6x^3y^8}{(x^4+y^4)^3},$$

$$f_{xy}(x,y)=f_{yx}(x,y)=\frac{24x^6y^5-6x^2y^9-2x^{10}y}{(x^4+y^4)^3}$$

and at the origin

$$f_{xx}(0,0)=0,\ f_{yy}(0,0)=0,\ f_{xy}(0,0)=f_{yx}(0,0)=0.$$

Hence, the conditions of the statement hold. Nevertheless, the function is not differentiable at the origin. In fact, the difference between the function and its linear approximation at the origin has the form:

$$\alpha(x,y)=f(x,y)-(f(0,0)+f_x(0,0)\cdot x+f_y(0,0)\cdot y)=\frac{x^3y^2}{x^4+y^4}.$$

Along the path $y=x$, $x>0$ the ratio $\frac{\alpha(x,y)}{\rho(x,y)}=\frac{x^3y^2}{(x^4+y^4)\sqrt{x^2+y^2}}$ does not approach zero:

$$\lim_{x\to0}\frac{\alpha(x,x)}{\rho(x,x)}=\lim_{x\to0}\frac{x^5}{2\sqrt{2}x^5}=\frac{1}{2\sqrt{2}}\neq0.$$

Therefore, $f(x,y)$ is not differentiable at the origin.

Remark 1. For any $(x,y) \neq (0,0)$ the first-order derivatives are continuous, which ensures differentiability of the function.

Remark 2. Since $f(x,y)$ is not differentiable at the origin, at least one of the first-order derivatives is not continuous there, and consequently at least one of the mixed derivatives is also discontinuous at the origin. It can be shown explicitly that both $f_x(x,y)$ and $f_y(x,y)$ are discontinuous at the origin, and so are $f_{xy}(x,y)$ and $f_{yx}(x,y)$.

Example 15. "If $f(x,y)$ is differentiable at some point, then its partial derivatives exist in a neighborhood of this point."

Solution.

Let us consider the function $f(x,y) = \begin{cases} x^2+y^2, & x,y \in \mathbb{Q} \\ 0, & otherwise \end{cases}$. First, we show that $f(x,y)$ is continuous at the origin. Approaching the origin by the points with rational coordinates, we obtain

$$\lim_{\substack{(x,y)\to(0,0) \\ x,y\in\mathbb{Q}}} f(x,y) = \lim_{(x,y)\to(0,0)} \left(x^2+y^2\right) = 0 = f(0,0).$$

Approaching the origin by other points we have

$$\lim_{\substack{(x,y)\to(0,0) \\ x\notin\mathbb{Q} \text{ or } y\notin\mathbb{Q}}} f(x,y) = \lim_{(x,y)\to(0,0)} 0 = 0 = f(0,0).$$

Since the points used in these two partial limits cover all the points in a neighborhood of the origin, it follows that $\lim\limits_{(x,y)\to(0,0)} f(x,y) = f(0,0)$, i.e., $f(x,y)$ is continuous at the origin. Second, let us consider the partial derivatives at the origin. If $x \in \mathbb{Q}$, then

$$\lim_{x\to 0,\, x\in\mathbb{Q}} \frac{f(x,0)-f(0,0)}{x-0} = \lim_{x\to 0} \frac{x^2-0}{x} = 0,$$

and for $x \in \mathbb{I}$ we obtain the same result:

$$\lim_{x\to 0,\, x\in\mathbb{I}} \frac{f(x,0)-f(0,0)}{x-0} = \lim_{x\to 0} \frac{0-0}{x} = 0.$$

Since the union of the rational and irrational points is equal to all the real points, it follows that $f_x(0,0) = \lim\limits_{x\to 0} \frac{f(x,0)-f(0,0)}{x-0} = 0$. For any point $(x_0,y_0) \neq (0,0)$ with $y_0 \in \mathbb{I}$, f_x also exists:

$$f_x(x_0,y_0) = \lim_{x\to x_0,\, y_0\in\mathbb{I}} \frac{f(x,y_0)-f(x_0,y_0)}{x-x_0} = \lim_{x\to x_0} \frac{0-0}{x-x_0} = 0.$$

On the other hand, for $(x_0,y_0) \neq (0,0)$ with $y_0 \in \mathbb{Q}$, the derivative f_x does not exist:

1) if $x_0 \in \mathbb{Q}$, then for $x \in \mathbb{Q}$ one has

$$\lim_{x\to x_0,\, x\in\mathbb{Q}} \frac{f(x,y_0)-f(x_0,y_0)}{x-x_0} = \lim_{x\to x_0} \frac{x^2+y_0^2-x_0^2-y_0^2}{x-x_0} = 2x_0,$$

while for $x \in \mathbb{I}$ one obtains

$$\lim_{x\to x_0,\, x\in\mathbb{I}} \frac{f(x,y_0)-f(x_0,y_0)}{x-x_0} = \lim_{x\to x_0} \frac{0-x_0^2-y_0^2}{x-x_0} = \infty,$$

and consequently, $\lim\limits_{x\to x_0} \frac{f(x,y_0)-f(x_0,y_0)}{x-x_0}$ does not exist;
2) if $x_0 \in \mathbb{I}$, then again two approaches give different results -

$$\lim_{x\to x_0,\, x\in\mathbb{Q}} \frac{f(x,y_0)-f(x_0,y_0)}{x-x_0} = \lim_{x\to x_0} \frac{x^2+y_0^2-0}{x-x_0} = \infty$$

and

$$\lim_{x\to x_0,\, x\in\mathbb{I}} \frac{f(x,y_0)-f(x_0,y_0)}{x-x_0} = \lim_{x\to x_0} \frac{0-0}{x-x_0} = 0$$

showing non-existence of $\lim\limits_{x\to x_0} \frac{f(x,y_0)-f(x_0,y_0)}{x-x_0}$.

Summarizing, $f_x(x_0,y_0) = 0$ at $(x_0,y_0) = (0,0)$ and at (x_0,y_0) with $y_0 \in \mathbb{I}$. For all other points f_x does not exist. Due to exactly the same reasoning, $f_y(x_0,y_0) = 0$ if $(x_0,y_0) = (0,0)$ or (x_0,y_0) has $x_0 \in \mathbb{I}$, and f_y does not exist at all other points. Consequently, at each point $(x,y) \neq (0,0)$ with rational coordinates both partial derivatives do not exist, and at each point with irrational coordinates both partial derivatives exist. Since any neighborhood of the origin contains both the points with rational and irrational coordinates, it follows that in any neighborhood of the origin there points where both partial derivatives do not exist.

Finally, let us prove that $f(x,y)$ is differentiable at the origin. Since $f_x(0,0) = f_y(0,0) = 0$, the difference between the function and its linear approximation is $\alpha(x,y) = f(x,y)$. Using the points with rational coordinates we obtain

$$\lim_{(x,y)\to(0,0)} \frac{\alpha(x,y)}{\rho(x,y)} = \lim_{(x,y)\to(0,0)} \frac{x^2+y^2}{\sqrt{x^2+y^2}} = 0,$$

and for the rest of points the limit is the same:

$$\lim_{(x,y)\to(0,0)} \frac{\alpha(x,y)}{\rho(x,y)} = \lim_{(x,y)\to(0,0)} \frac{0}{\sqrt{x^2+y^2}} = 0.$$

Therefore, the general limit exists and is equal to zero: $\lim\limits_{(x,y)\to(0,0)} \frac{\alpha(x,y)}{\rho(x,y)} = 0$, which means that the function is differentiable at the origin.

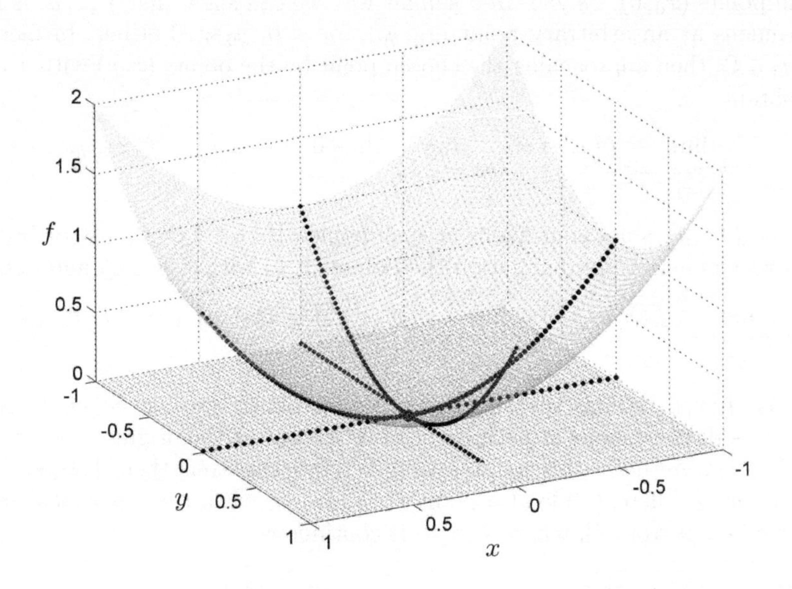

FIGURE 8.3.8: Example 15

Example 16. "If $f(x,y)$ is differentiable at some point, then it is continuous in a neighborhood of this point."

Solution.

We can use the function $f(x,y) = \begin{cases} x^2+y^2, & x,y \in \mathbb{Q} \\ 0, & otherwise \end{cases}$ of the previous Example. It was already shown above that this function is differentiable (and therefore, continuous) at the origin. Let us check if $f(x,y)$ is continuous at $(x,y) \neq (0,0)$. First, we consider the points of the y-axis, that is, $(0,y_0)$, $y_0 \neq 0$. If $y_0 \in \mathbb{Q}$, then approaching the chosen point by the points (x,y) with $x \in \mathbb{I}$, we obtain

$$\lim_{\substack{(x,y)\to(0,y_0) \\ x\in\mathbb{I}}} f(x,y) = \lim_{(x,y)\to(0,y_0)} 0 = 0 \neq y_0^2 = f(0,y_0),$$

that is, $f(x,y)$ is not continuous at such points. If $y_0 \in \mathbb{I}$, then we choose for approximation the points (x,y) with $x,y \in \mathbb{Q}$, and obtain

$$\lim_{\substack{(x,y)\to(0,y_0) \\ x,y\in\mathbb{Q}}} f(x,y) = \lim_{(x,y)\to(0,y_0)} \left(x^2+y^2\right) = y_0^2 \neq 0 = f(0,y_0),$$

that is, $f(x,y)$ is not continuous at these points. Therefore, $f(x,y)$ is discontinuous at all points $(0,y_0)$, $y_0 \neq 0$.

Using the same reasoning we can show that $f(x,y)$ is also discontinuous at all points $(x_0, 0)$, $x_0 \neq 0$. In a similar way we can show that $f(x,y)$ is not continuous at an arbitrary point (x_0, y_0), $x_0 \neq 0$, $y_0 \neq 0$ either. In fact, if $x_0, y_0 \in \mathbb{Q}$, then approaching the chosen point by the points (x,y) with $x \in \mathbb{I}$, we obtain

$$\lim_{\substack{(x,y)\to(x_0,y_0)\\ x\in\mathbb{I}}} f(x,y) = \lim_{(x,y)\to(x_0,y_0)} 0 = 0 \neq x_0^2 + y_0^2 = f(x_0, y_0),$$

that is, $f(x,y)$ is not continuous at such points. If $x_0 \in \mathbb{I}$ or $y_0 \in \mathbb{I}$ (or both), then we choose for approximation the points (x,y) with $x, y \in \mathbb{Q}$, and obtain

$$\lim_{\substack{(x,y)\to(x_0,y_0)\\ x,y\in\mathbb{Q}}} f(x,y) = \lim_{(x,y)\to(x_0,y_0)} \left(x^2+y^2\right) = x_0^2 + y_0^2 \neq 0 = f(x_0, y_0),$$

that is, $f(x,y)$ is also discontinuous. Summarizing, we can conclude that $f(x,y)$ is discontinuous at each point of $\mathbb{R}^2$ except at the origin.

Remark. Actually, this counterexample refute the strengthened statement: "if $f(x,y)$ is differentiable at a point $P_0 = (x_0, y_0)$, then there is at least one point different from P_0 where $f(x,y)$ is continuous".

Example 17. "If $f(x,y)$ is continuously differentiable in a domain $D \subset \mathbb{R}^2$ and $f_y(x,y) = 0$, $\forall (x,y) \in D$, then $f(x,y)$ does not depend on y in D."

Solution.

Let us consider the first quadrant $D_1 = \{(x,y) : x > 0, y > 0\}$ and the domain $D = \mathbb{R}^2 \backslash L$, where $L = \{(x,y) : x \geq 0, y = 0\}$ is the ray, and let us define on D the function $f(x,y) = \begin{cases} x^3, & (x,y) \in D_1 \\ 0, & (x,y) \in D_2 = D \backslash D_1 \end{cases}$. For $\forall (x,y) \in D_1$ we have the continuous partial derivatives: $f_x(x,y) = 3x^2$, $f_y(x,y) = 0$. For $\forall (x,y) \in D_2 \backslash L_1$, where $L_1 = \{(x,y) : x = 0, y > 0\}$, the derivatives are also continuous: $f_x(x,y) = 0$, $f_y(x,y) = 0$. Finally, in the points of L_1 we should use the definitions both for existence and continuity of the partial derivatives. For $f_x(0, y_0)$ there are two different ways of approaching a point $(0, y_0)$, $y_0 > 0$: using the points in D_1, we obtain

$$f_x(0, y_0) = \lim_{\substack{x\to 0\\ (x,y)\in D_1}} \frac{f(x, y_0) - f(0, y_0)}{x - 0} = \lim_{x\to 0} \frac{x^3 - 0}{x - 0} = 0,$$

and for the points in D_2 we have

$$f_x(0, y_0) = \lim_{\substack{x\to 0\\ (x,y)\in D_2}} \frac{f(x, y_0) - f(0, y_0)}{x - 0} = \lim_{x\to 0} \frac{0 - 0}{x - 0} = 0.$$

Since both limits are zero, we conclude that $f_x(0, y_0) = 0$.

For $f_y(0, y_0)$ all the points in the definition lie on the positive part of the y-axis, and the calculation is straightforward:

$$f_y(0, y_0) = \lim_{y\to y_0} \frac{f(0, y) - f(0, y_0)}{y - y_0} = \lim_{y\to y_0} \frac{0 - 0}{y - y_0} = 0.$$

Next, we verify the continuity of the partial derivatives on L_1. Again, there are two different ways of approaching a point $(0, y_0)$, $y_0 > 0$: for the paths in D_1, we obtain

$$\lim_{\substack{(x,y)\to(0,y_0)\\(x,y)\in D_1}} f_x(x,y) = \lim_{x\to 0} 3x^2 = 0,$$

and for the paths in D_2 we get

$$\lim_{\substack{(x,y)\to(0,y_0)\\(x,y)\in D_2}} f_x(x,y) = \lim_{x\to 0} 0 = 0.$$

Therefore, $\lim_{(x,y)\to(0,y_0)} f_x(x,y) = 0 = f_x(0,y_0)$, that is $f_x(0,y_0)$ is continuous on L_1. Finally, since $f_y(x,y) = 0$ at each point of D (including L_1), it follows that this derivative is continuous on D. Hence, all the conditions of the statement hold. Nevertheless, $f(1,1) = 1$, while $f(1,-1) = 0$, that is, $f(x,y)$ does depend on y.

Remark 1. If the intersection of a region D with any line parallel to the y-axis is an interval, then the statement will be true.

Remark 2. Recall that for a function $g(x)$ of a single variable there is the following result: if $g'(x) = 0$ on a connected set (i.e. interval), then $g(x) = const$ on this set, that is, $g(x)$ does not actually depend on x. We see that a direct generalization of this result for functions of several variables is not true.

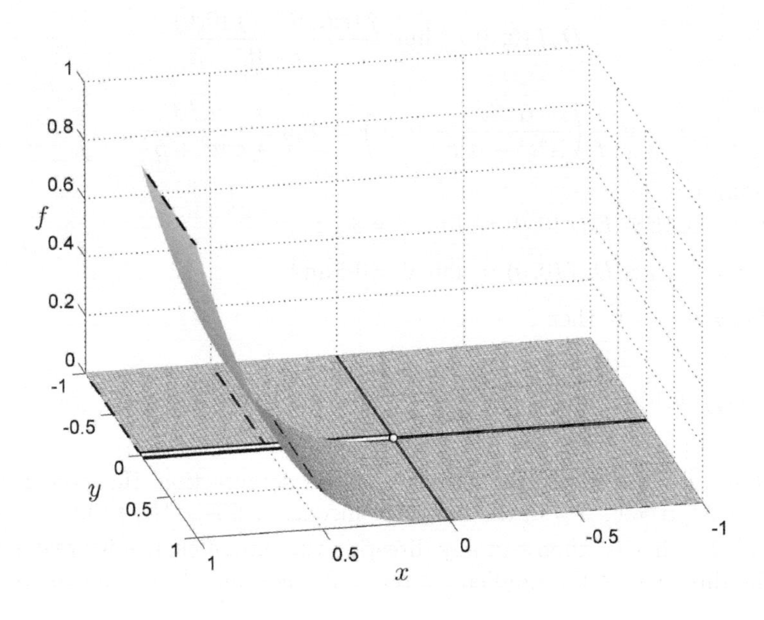

FIGURE 8.3.9: Example 17

Example 18. "If ∇f exists and does not vanish at (x_0, y_0), then the maximum rate of change of $f(x,y)$ occurs in the direction of ∇f."

Solution.

Let us consider the function $f(x,y) = \begin{cases} \frac{x^2 y}{x^4+y^2} + y, \ x^2+y^2 \neq 0 \\ 0, \ x^2+y^2 = 0 \end{cases}$. In Example 2, section 7.3, it was shown that a similar function is not continuous at the origin. Using the same arguments, we can arrive to the same conclusion regarding the above function: the path limits along $y = 0$ and $y = x^2$ give different results -

$$\lim_{x\to 0} f(x,0) = \lim_{x\to 0} 0 = 0, \text{ while } \lim_{x\to 0} f\left(x,x^2\right) = \lim_{x\to 0} \frac{x^4}{x^4+x^4} + x^2 = \frac{1}{2}.$$

At the same time, both partial derivatives exist at the origin:

$$f_x(0,0) = \lim_{x\to 0} \frac{f(x,0) - f(0,0)}{x-0} = 0,$$

$$f_y(0,0) = \lim_{y\to 0} \frac{f(0,y) - f(0,0)}{y-0} = \lim_{y\to 0} \frac{y}{y} = 1,$$

and $\nabla f(0,0) = (0,1) \neq (0,0)$. Since the function is not differentiable at the origin, we appeal to the definition of the directional derivative: choosing an arbitrary unitary direction vector $u = (\alpha, \beta)$, we get

$$D_u f(0,0) = \lim_{t\to 0_+} \frac{f(\alpha t, \beta t) - f(0,0)}{t-0}$$

$$= \lim_{t\to 0_+} \frac{1}{t}\left(\frac{\alpha^2\beta t^3}{\alpha^4 t^4 + \beta^2 t^2} + \beta t\right) = \lim_{t\to 0_+}\left(\frac{\alpha^2\beta}{\alpha^4 t^2 + \beta^2} + \beta\right).$$

Therefore:

1) if $\alpha = 0$, then $D_u f(0,0) = \lim\limits_{t\to 0_+} \beta = \pm 1$;

2) if $\beta = 0$, then $D_u f(0,0) = \lim\limits_{t\to 0_+} 0 = 0$; and

3) if $\alpha \neq 0$, $\beta \neq 0$, then

$$D_u f(0,0) = \lim_{t\to 0_+}\left(\frac{\left(1-\beta^2\right)\beta}{\left(1-\beta^2\right)^2 t^2 + \beta^2} + \beta\right) = \frac{\left(1-\beta^2\right)\beta}{\beta^2} + \beta = \frac{1}{\beta}.$$

Thus, the directional derivative exists in any direction. However, for any $u = (\alpha, \beta)$, $\alpha \neq 0$, $\beta \neq 0$, the following inequality holds: $|D_u f(0,0)| = \frac{1}{|\beta|} > |\nabla f(0,0)|$, that is, almost in any direction the rate of change is greater than in the direction of the gradient. Moreover, since $\lim\limits_{\beta\to 0} \frac{1}{|\beta|} = +\infty$ there is no direction of the maximum rate of change.

Remark. This result is a consequence of discontinuity (and therefore non-differentiability) of $f(x,y)$ at the origin. So the theorem on the extremum properties of the gradient is not applicable here.

Example 19. "If $\nabla f(x_0, y_0) = 0$, then all the directional derivatives of $f(x,y)$ vanishes at (x_0, y_0)."

Solution.

It was shown in Examples 3 and 5 that the function $f(x,y) = \sqrt{|xy|}$ possesses zero partial derivatives and all the directional derivatives at the origin, which have the form $D_u f(0,0) = \sqrt{|\alpha\beta|}$ in the direction of an arbitrary unitary vector $u = (\alpha, \beta)$. Therefore, these derivatives are not zero in any direction different from those of the coordinate axes.

One can prove that another kind of the counterexample is the function $f(x,y) = \begin{cases} \frac{xy}{x^2+y^2}, & x^2+y^2 \neq 0 \\ 0, & x^2+y^2 = 0 \end{cases}$. In Examples 1 and 9 it was shown that $\nabla f(0,0) = 0$ and, at the same time, there is no directional derivative, except in the direction of the coordinate axes.

Remark. The cause of the statement incorrectness is non-differentiability of $f(x,y)$ at the origin.

Example 20. "If $f(x,y)$ is differentiable on $\mathbb{R}^2$ and has two or more local maxima, then it has also a local minimum."

Solution.

Let us show that the infinitely differentiable on $\mathbb{R}^2$ function $f(x,y) = (1+e^y)\cos x - ye^y$ has infinitely many local maxima and none local minimum. First, solving the system for critical points

$$\begin{cases} f_x(x,y) = -\sin x\,(1+e^y) = 0 \\ f_y(x,y) = e^y(\cos x - 1 - y) = 0 \end{cases},$$

we obtain $\sin x = 0$, i.e., $x_n = n\pi$, $\forall n \in \mathbb{Z}$ from the first equation, and, substituting these values of x in the second equation, we get $y_n = \cos n\pi - 1$, i.e., $y_{2n} = 0$ and $y_{2n+1} = -2$. Hence, there are the following critical points: $P_{2n} = (2n\pi, 0)$ and $P_{2n+1} = ((2n+1)\pi, -2)$, $\forall n \in \mathbb{Z}$. To check the sufficient conditions of local extrema at the found critical points we need the second-order derivatives:

$$f_{xx}(x,y) = -(1+e^y)\cos x, \;\; f_{xy}(x,y) = -e^y \sin x,$$

$$f_{yy}(x,y) = e^y(\cos x - 2 - y).$$

If the Hessian matrix $\begin{pmatrix} f_{xx} & f_{xy} \\ f_{xy} & f_{yy} \end{pmatrix}$ is positive/negative definite at a critical point then that point is a local minimum/maximum; if the determinant of the matrix is negative at a critical point then that point is not a local extremum. For the points $P_{2n} = (2n\pi, 0)$ the Hessian matrix assumes the form $\begin{pmatrix} -2 & 0 \\ 0 & -1 \end{pmatrix}$, that is the matrix is negative definite, and consequently, all the points $P_{2n} = (2n\pi, 0)$ are local maxima. On the other hand, for the points $P_{2n+1} = ((2n+1)\pi, -2)$ the corresponding Hessian matrix $\begin{pmatrix} 1+e^{-2} & 0 \\ 0 & -e^{-2} \end{pmatrix}$ is not definite (its determinant is negative) and therefore, all the points $P_{2n+1} = ((2n+1)\pi, -2)$ are not local extrema.

The function $f(x,y) = \left(x^2-1\right)^2 + \left(x^2y - x - 1\right)^2$ gives another counterexample.

The function $f(x,y) = x^4 + y^4 - (x+y)^2$ gives one more counterexample.

Remark 1. A similar example for continuous functions is given in Example 11, section 7.3.

Remark 2. Similar statement for one-variable(single-variable) functions is true.

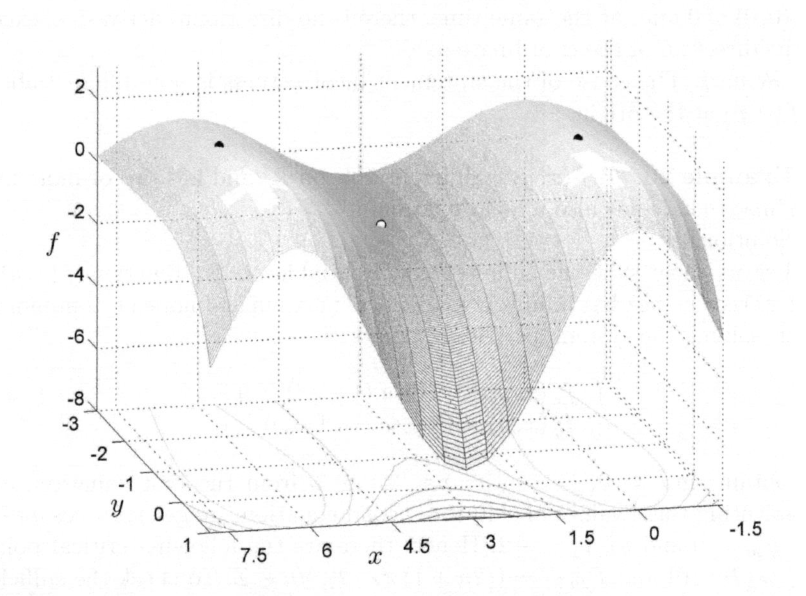

FIGURE 8.3.10: Example 20

Example 21. "If $f(x,y)$ is differentiable on $\mathbb{R}^2$ and its restriction to an arbitrary straight line passing through the origin has a strict local minimum at the origin, then $f(x,y)$ has a local minimum at the origin."

Solution.

Let us consider the infinitely differentiable on $\mathbb{R}^2$ function $f(x,y) = \left(y-x^2\right)\left(y-3x^2\right)$. Its restriction to the x-axis is the function $f(x,0) = 3x^4$, which has the strict local minimum at the origin. The restriction to the y-axis,

$f(0,y)=y^2$, also possesses the strict local minimum at the origin. Finally, the restriction to an arbitrary line $y=kx$, $k=const\neq 0$, results in the function $g(x)=f(x,kx)=k^2x^2-4kx^3+3x^4$. The critical points of the last function, obtained as solutions of the equation $g'(x)=2k^2x-12kx^2+12x^3=0$, are $x_1=0$, $x_{2,3}=\frac{3\pm\sqrt{3}}{6}k$. We are interested in only the first point, because our analysis is for the origin. Since the second derivative $g''(x)=2k^2-24kx+36x^2$ is positive at $x_1=0$, we conclude that $x_1=0$ is the local minimum of $f(x,kx)$. Hence, $f(x,y)$ satisfies the statement conditions. However, it does not have local extrema at the origin, since in any neighborhood of the origin $f(x,y)$ assumes both positive and negative values:

$$f(x,0)=3x^4>0=f(0,0)\,,\ f\left(x,2x^2\right)=-x^4<0=f(0,0)\,.$$

Another similar counterexample is the infinitely differentiable on $\mathbb{R}^2$ function $f(x,y)=\left(x-y^2\right)\left(2x-y^2\right)$. Its restriction to the x-axis, $f(x,0)=2x^2$, has the strict local minimum at the origin. The restriction to the y-axis, $f(0,y)=y^4$, also possesses the strict local minimum at the origin. Finally, the restriction to an arbitrary line $y=kx$, $k=const\neq 0$, gives $g(x)=f(x,kx)=k^4x^4-3k^2x^3+2x^2$. Solving the equation $g'(x)=4k^4x^3-9k^2x^2+4x=0$, we find the critical points: $x_1=0$, $x_{2,3}=\frac{9\pm\sqrt{17}}{8k^2}$. The only point of our concern is $x_1=0$, which is the strict local minimum of $f(x,kx)$, because $g''(x)=12k^4x^2-18k^2x+4$ is positive there. Hence, $f(x,y)$ satisfies the statement conditions. However, it does not have local extrema at the origin, since in any neighborhood of the origin $f(x,y)$ assumes both positive and negative values:

$$f(x,0)=2x^2>0=f(0,0)\,,\ f\left(\frac{2}{3}y^2,y\right)=-\frac{1}{9}y^4<0=f(0,0)\,.$$

Remark. This example is a strengthened form of the statement in Example 12, section 7.3 for continuous functions.

Example 22. "If $f(x,y)$ is differentiable on $\mathbb{R}^2$ and its restriction to an arbitrary algebraic curve $y=cx^{m/n}$, where $c\in\mathbb{R}$ and $m,n\in\mathbb{N}$ ($x\geq 0$ in case the exponential fraction is in lowest terms and n is even), has a strict local minimum at the origin, then $f(x,y)$ has a local minimum at the origin."

Solution.

Let us consider the infinitely differentiable on $\mathbb{R}^2$ function

$$f(x,y)=\begin{cases}\left(y-e^{-1/x^2}\right)\left(y-3e^{-1/x^2}\right), & x\neq 0\\ y^2,\ x=0\end{cases}.$$

For the restriction to the x-axis we have $f(x,0)=3e^{-2/x^2}>0=f(0,0)$, that is, $x=0$ is the strict local minimum. For the restriction to the y-axis, $f(0,y)=y^2$, $y=0$ is the strict local minimum. Finally, the restriction to

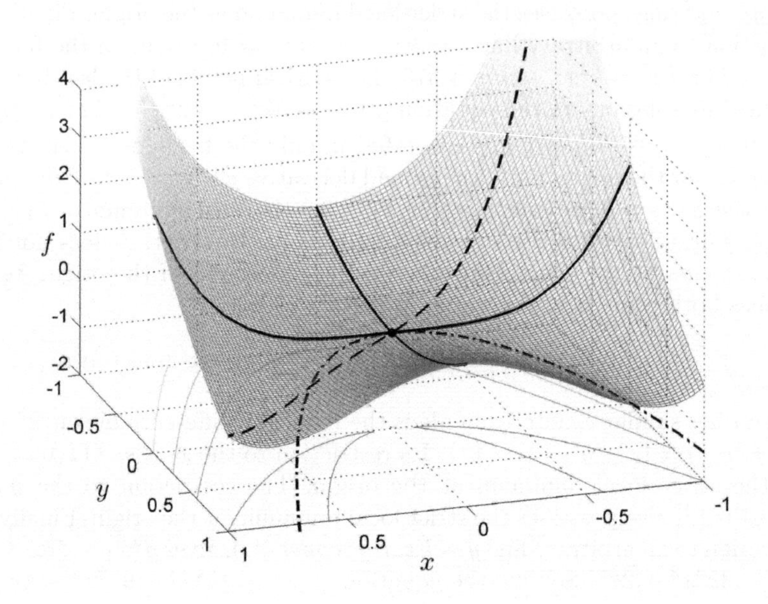

FIGURE 8.3.11: Example 21, graph of the first function

$y = cx^{m/n}$, $c \neq 0$, leads to the function

$$g(x) = f\left(x, cx^{m/n}\right) = \left(cx^{m/n} - e^{-1/x^2}\right)\left(cx^{m/n} - 3e^{-1/x^2}\right) = x^{2m/n} \cdot h(x),$$

where

$$h(x) = c^2 - 4ce^{-1/x^2} \cdot x^{-m/n} + 3e^{-2/x^2} \cdot x^{-2m/n}.$$

To evaluate the behavior of $g(x)$, we need to calculate the auxiliary limit:

$$\lim_{x \to 0} e^{-1/x^2} \cdot x^{-m/n} = \lim_{t \to \infty} \frac{t^{m/n}}{e^{t^2}} = \lim_{t \to \infty} \frac{m}{n} \frac{t^{m/n-1}}{2te^{t^2}} = \lim_{t \to \infty} \frac{m}{n} \frac{t^{m/n-2}}{2e^{t^2}} = \ldots = 0.$$

Here, we used the substitution $x = \frac{1}{t}$ and then applied L'Hospital's rule sucessively until the power of t in the numerator becomes negative or zero. Based on this result, we readily conclude that

$$\lim_{x \to 0} h(x) = \lim_{x \to 0} \left(c^2 - 4ce^{-1/x^2} \cdot x^{-m/n} + 3e^{-2/x^2} \cdot x^{-2m/n}\right) = c^2 > \frac{c^2}{2}.$$

It means, that there exists a neighborhood $(-\delta, \delta)$ of $x = 0$ such that for $\forall x \in (-\delta, \delta)$ we have $h(x) > \frac{c^2}{2}$. Notice that $x^{2m/n} > 0$ for $\forall x \neq 0$. Therefore, $g(x) = x^{2m/n} \cdot h(x) > 0$ for $\forall x \in (-\delta, 0) \bigcup (0, \delta)$. Since $g(0) = 0$, this means that $x = 0$ is the strict local minimum of $g(x)$. Hence, $f(x, y)$ satisfies the statement conditions. At the same time, in any neighborhood of the origin,

$f(x,y)$ assumes both positive and negative values: $f(0,y) = y^2 > 0 = f(0,0)$, while $f\left(x, 2e^{-1/x^2}\right) = -e^{-2/x^2} < 0 = f(0,0)$. Therefore, it does not have local extrema at the origin.

Remark. Example 12 in section 7.3 and the last two Examples 21 and 22 represent a chain of refined, but still insufficient, conditions for a local maximum/minimum in terms of extremum properties of the function restrictions.

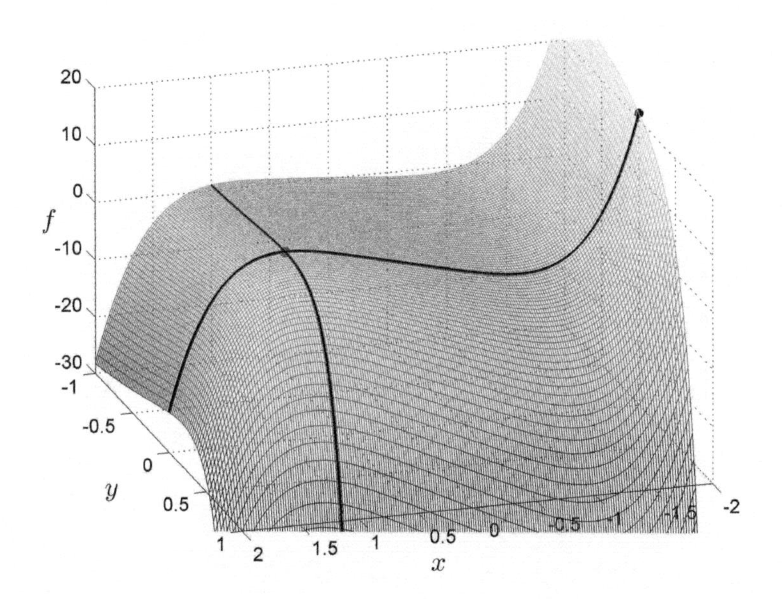

FIGURE 8.3.12: Example 23

Example 23. "If $f(x,y)$ is differentiable on $\mathbb{R}^2$ and has only one critical point c, which is a local maximum, then this point is a global maximum of $f(x,y)$."

Solution.

Let us consider the infinitely differentiable on $\mathbb{R}^2$ function $f(x,y) = 5xe^y - x^5 - e^{5y}$. The only critical point is found by solving the system of necessary conditions: $\begin{cases} f_x = 5e^y - 5x^4 = 0 \\ f_y = 5xe^y - 5e^{5y} = 0 \end{cases}$, which has the only solution $x = 1$, $y = 0$. Therefore, the only critical point is $c = (1,0)$. The second-order derivatives $f_{xx} = -20x^3$, $f_{xy} = 5e^y$, $f_{yy} = 5xe^y - 25e^{5y}$ give the positive determinant of the Hessian matrix at the point $c = (1,0)$:

$$\det \begin{pmatrix} f_{xx} & f_{xy} \\ f_{xy} & f_{yy} \end{pmatrix}(c) = \det \begin{pmatrix} -20 & 5 \\ 5 & -20 \end{pmatrix} = 375 > 0,$$

which means that the critical point is a local extremum. Since $f_{xx}(1,0) = -20 < 0$, one concludes that $c = (1,0)$ is a local maximum of $f(x,y)$. However, the point $c = (1,0)$ is not a global maximum, because $f(1,0) = 3$, while $f(-2,0) = 21$.

The function $f(x,y) = 3xe^y - x^3 - e^{3y}$ provides a similar counterexample.

Remark. Similar statement for one-variable functions is true.

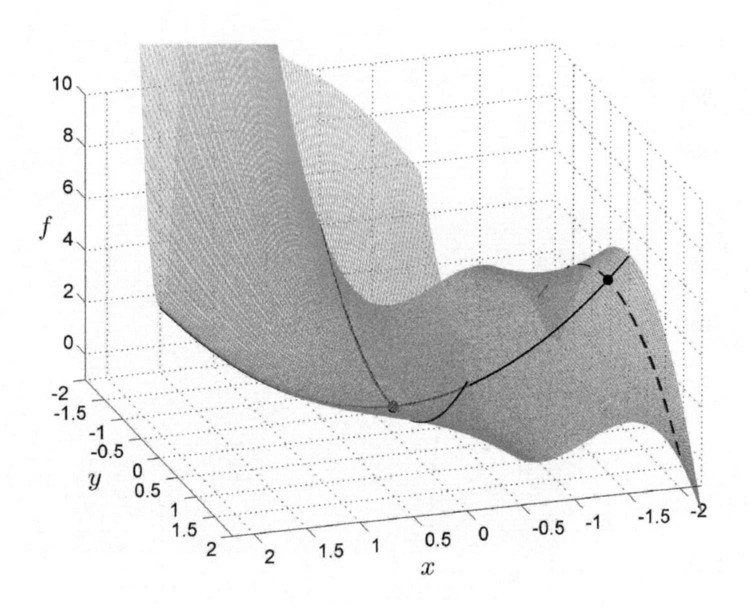

FIGURE 8.3.13: Example 24

Example 24. "If $f(x,y)$ is differentiable on $\mathbb{R}^2$, has only one local minimum and is unbounded below, then $f(x,y)$ achieves a local maximum somewhere."

Solution.

Let us consider the infinitely differentiable on $\mathbb{R}^2$ function $f(x,y) = x^2 + y^2(1+x)^3$. The system of necessary conditions for critical points $\begin{cases} f_x = 2x + 3y^2(1+x)^2 = 0 \\ f_y = 2y(1+x)^3 = 0 \end{cases}$ has the only solution $x = 0$, $y = 0$. The second-order derivatives

$$f_{xx} = 2 + 6y^2(1+x),\ f_{xy} = 6y(1+x)^2,\ f_{yy} = 2(1+x)^3$$

give the positive determinant of the Hessian matrix at the critical point $c =$

$(0,0)$:

$$\det\begin{pmatrix} f_{xx} & f_{xy} \\ f_{xy} & f_{yy} \end{pmatrix}(c) = \det\begin{pmatrix} 2 & 0 \\ 0 & 2 \end{pmatrix} = 4 > 0,$$

which means that the critical point is a local extremum. Since $f_{xx}(0,0) = 2 > 0$, it follows that $c = (0,0)$ is a local minimum. Besides, $f(x,y)$ is unbounded below since for $x = -2$ one has $f(-2,y) = 4 - y^2 \underset{y\to\infty}{\to} -\infty$. (It is also unbounded above, since for $x = 0$ one has $f(0,y) = y^2 \underset{y\to\infty}{\to} +\infty$.) However, $f(x,y)$ has no local maximum, because $c = (0,0)$ is the only critical point.

The function $f(x,y) = -3xe^y + x^3 + e^{3y}$ provides another counterexample.

Remark. Similar statement for one-variable functions is true.

Remark to Examples 20-24. The same statements are false for any type of a local extremum, either minimum or maximum.

Exercises

1. Let $f(x,y)$ be differentiable at (a,b). Show by example that the assertion that "in this case the remainder satisfies the inequality $|\gamma| = |\Delta f - df| \le C\rho^2$, where C is a non-negative constant and $\rho^2 = \Delta x^2 + \Delta y^2$" is not true.

2. If a function $f(x,y)$ has partial derivatives in a neighborhood of (a,b) it does not guarantee that $f(x,y)$ is differentiable or (at least) continuous at (a,b). Provide an example to this statement. Formulate this result in the form of a counterexample. Compare with Examples 1 and 2 in section 8.3.

3. If a function $f(x,y)$ has bounded partial derivatives in a neighborhood of (a,b) it does guarantee the continuity of $f(x,y)$ in the same neighborhood, but it is not yet sufficient for differentiability even at (a,b). Provide an example to the last part of this statement. Formulate in the form of a counterexample. Compare with Examples 3 and 4 in section 8.3.

4. Give an example of a function $f(x,y)$ that has continuous mixed derivative f_{yx}, but does not have the first derivative f_x. Formulate as a counterexample.

5. Give an example of a function $f(x,y)$ that is not differentiable at a point (a,b), but $|f(x,y)|$ is differentiable. What about a function differentiable (a,b) with the absolute value $|f(x,y)|$ not differentiable? Formulate these results in the form of counterexamples.

6. Show that the differentiability of $h(x,y) = f(x,y) \cdot g(x,y)$ at (a,b) does not imply the differentiability of $f(x,y)$ and $g(x,y)$. Compare with Examples 3 and 4 in section 8.2.

7. Show that each of the functions $f(x,y) = D(x-y)$, $g(x,y) = D(x+y)$ and $h(u,v) = D(u+v)$, where $D(t)$ is Dirichlet's function, is not differentiable anywhere, but the composed function $h(f(x,y), g(x,y))$ is differentiable everywhere.

8. Construct a counterexample to the following statement: "if there exists a tangent plane to the graph of $f(x,y)$ at (a,b), then $f(x,y)$ is differentiable at (a,b)".

9. Construct a counterexample to the following statement: "if $f(x,y)$ is differentiable and unbounded on a set S, then at least one of the partial derivatives $f_x(x,y)$ or $f_y(x,y)$ is unbounded on S".

10. Construct a counterexample to the following statement: "if $f(x,y)$ is differentiable and both partial derivatives are unbounded on a bounded set S, then $f(x,y)$ is unbounded on S". (Hint: use $f(x,y) = \begin{cases} (x^2+y^2)\cos\frac{1}{x^2+y^2}, & x^2+y^2 \ne 0 \\ 0, & x^2+y^2 = 0 \end{cases}$ on $S = x^2+y^2 \le 1$.)

11. Give a counterexample to the statement: "if $f(x,y)$ is continuous in a neighborhood of (a,b) and a tangent plane does not exist at (a,b), then $f(x,y)$ does not have a local extremum at (a,b)".

12. Show that the function $f(x,y) = |xy|$ is another counterexample to Example 15 in section 8.3.

13. Construct a differentiable at (a,b) function that does not possess partial derivatives in any other point. Formulate as a counterexample. (Hint: $f(x,y) = (x^2+y^2)\cdot D(x+y)$.)

14. Give an example of a function not differentiable at (a,b), but possessing a tangent plane at (a,b).

15. Construct a function that is not differentiable anywhere, but has a tangent plane at some points. (Hint: $f(x,y) = \sqrt[3]{x^2}\cdot D(x^2+y^2) + \sqrt[5]{x^2}\cdot(1 - D(x^2+y^2))$.)

16. Show that $f(x,y) = \sin x + y^2$ is a counterexample to the following statement similar to that in Example 20, section 8.3: "if $f(x,y)$ is differentiable on $\mathbb{R}^2$ and has two or more local minima, then it has also a local maxima". Use the functions in Example 20 to obtain other counterexamples to the last statement.

17. Show that $f(x,y) = (x^3-y)(y-3x^3)$ is a counterexample to the following statement similar to that in Example 21, section 8.3: "if $f(x,y)$ is differentiable on $\mathbb{R}^2$ and its restriction to an arbitrary straight line passing through the origin has a strict local maximum at the origin, then $f(x,y)$ has a local maximum at the origin". Use the functions in Example 21 to obtain other counterexamples to the last statement.

18. Provide a counterexample to the following statement: "if $d^2f(a,b) \geq 0$ at a critical point (a,b), then $f(x,y)$ has a local minimum at (a,b)". (Hint: try $f(x,y) = x^2+y^3$ at the origin.)

19. Show that $f(x,y) = \begin{cases} xy\frac{x^4-2y^4}{x^4+y^4}, & x^2+y^2 \neq 0 \\ 0, & x^2+y^2 = 0 \end{cases}$ is another counterexample to Example 11, section 8.3.

20. Give a counterexample to the following statement: "if $f(x,y)$ is continuously differentiable on $\mathbb{R}^2$ and $f_x(x,y) \underset{(x,y)\to\infty}{\to} 0$, $f_y(x,y) \underset{(x,y)\to\infty}{\to} 0$, then $f(x,y) \underset{(x,y)\to\infty}{\to} const$". (Hint: consider $f(x,y) = \cos\sqrt[3]{x^2+y^2}$.)

21. Give a counterexample to the following statement: "if $f(x,y)$ is continuously differentiable on $\mathbb{R}^2$ and $f(x,y) \underset{(x,y)\to\infty}{\to} 0$, then $f_x(x,y) \underset{(x,y)\to\infty}{\to} 0$ and $f_y(x,y) \underset{(x,y)\to\infty}{\to} 0$". (Hint: try $f(x,y) = \begin{cases} \frac{1}{x^2+y^2}\sin\left(x^2+y^2\right)^3, & x^2+y^2 \neq 0 \\ 0, & x^2+y^2 = 0 \end{cases}$.)

22. Show that $f(x,y) = \begin{cases} |x^3|, & y \geq 0 \\ x^3, & y < 0 \end{cases}$ considered on $D = \{y \in [0,1], x \in [-y,1]\} \cup \{y \in [-1,0], x \in [y,1]\}$ is another counterexample to Example 17 in section 8.3.

23. Is the continuity of f_{xx} and f_{yy} is sufficient for the twice differentiability of $f(x,y)$? If not, provide a counterexample. (Hint: consider $f(x,y) = xy\ln(-\ln(x^2+y^2))$ in a neighborhood of the origin.)

24. The implicit function theorem (one of its versions) states that if $F(x,y)$

is continuously differentiable in a neighborhood of a point (a, b) at which $F(a, b) = 0$ and $F_y(a, b) \neq 0$, then near a the equation $F(x, y) = 0$ determines the only function $y = f(x)$ such that $b = f(a)$, and $f(x)$ is continuously differentiable with the derivative defined by the formula $f'(x) = -\frac{F_x(x,y)}{F_y(x,y)}$. Show that the continuity of $F(x, y)$ in a neighborhood of (a, b) is not a necessary condition, but its violation may invalidate the results of the above theorem. (Hint: the equation $F(x, y) = 0$ considered in a neighborhood of the origin defines the only function if $F(x, y) = \begin{cases} x + y, \ x \neq y \\ -x, \ x = y \end{cases}$, but it defines two functions if $F(x, y) = \begin{cases} x + y, \ x \neq y \\ 0, \ x = y \end{cases}$.)

25. Show that the condition $F(a, b) = 0$ cannot be dropped in the formulation of the implicit function theorem.

26. Show that the condition $F_y(a, b) \neq 0$ in the implicit function theorem is not a necessary condition, but its violation may invalidate the results of the theorem. (Hint: the equation $F(x, y) = 0$ considered in a neighborhood of the origin defines the only function if $F(x, y) = x^3 - y^3$, but it does not define any function if $F(x, y) = x - y^2$.)

Chapter 9

Integrability

9.1 Elements of theory

Double (Riemann) Integral

1. Intergrability over a rectangle. Let $f(x, y)$ be defined on the rectangle $R = [a, b] \times [c, d] \subset \mathbb{R}^2$. A partition P of R into a set of elementary rectangles

$$R_{ij} = [x_{i-1}, x_i] \times [y_{j-1}, y_j], \; i = 1, \ldots, n, \; j = 1, \ldots, m$$

is made by using the partitions

$$P_x = \{x_i : \; a = x_0 < x_1 < \ldots < x_{n-1} < x_n = b\}$$

and

$$P_y = \{y_j : \; c = y_0 < y_1 < \ldots < y_{m-1} < y_m = d\}$$

of the intervals $[a, b]$ and $[c, d]$, respectively. The corresponding increments are denoted $\Delta x_i = x_i - x_{i-1}$ and $\Delta y_j = y_j - y_{j-1}$, the length of a diagonal of R_{ij} is denoted by $\Delta_{ij} = \sqrt{\Delta x_i^2 + \Delta y_j^2}$ and the partition diameter by $\Delta = \max\limits_{i,j} \Delta_{ij}$. The *(double) Riemann sum* of $f(x, y)$ corresponding to the specific partition P and the specific choice of points $(\xi_i, \eta_j) \in R_{ij}$ is $S(f) = \sum_{i=1}^{n} \sum_{j=1}^{m} f(\xi_i, \eta_j) \cdot \Delta x_i \Delta y_j$. If there exists a finite limit of the Riemann sums of $f(x, y)$, as Δ approaches 0, which does not depend on a choice of partitions and points (ξ_i, η_j), then this limit is called the *double (Riemann) integral* of $f(x, y)$ over R, is denoted as follows: $\lim\limits_{\Delta \to 0} S(f) = \iint_R f dA$ or $\lim\limits_{\Delta \to 0} S(f) = \iint_R f dx dy$, and the function $f(x, y)$ is called *(Riemann) integrable* over R.

2. Integrability over a general set. Let $f(x, y)$ be defined on a bounded closed set $D \subset \mathbb{R}^2$ whose boundary ∂D has zero area. (The last means that for any $\varepsilon > 0$ the boundary ∂D can be covered by a set of rectangles with the total area less than ε. Those familiar with the measure theory will immediately recognize the definition of the plane Lebesgue measure zero. In particular, any rectifiable curve has zero area.) Consider an arbitrary rectangle $R = [a, b] \times [c, d]$ containing the set D and define the following function on R: $F(x, y) = \begin{cases} f(x, y), & (x, y) \in D \\ 0, & otherwise \end{cases}$. If $F(x, y)$ is integrable over R, then $f(x, y)$ is *integrable over* D and $\iint_D f dA = \iint_R F dA$.

Remark. Hereinafter in this chapter we will consider integrability over sets D that represent the closures of bounded connected open sets with zero area boundary.

Theorem. If $f(x, y)$ is integrable over D, then it is bounded on D.

Theorem. If $f(x, y)$ is continuous on D, then it is integrable over D.

Theorem. If $f(x, y)$ is bounded on D and is continuous on D except at a finite number of smooth curves of zero area, then $f(x, y)$ is integrable over D.

Theorem. If $f(x, y)$ is bounded on D and is continuous on D except at a countable set of curves of zero area, then $f(x, y)$ is integrable over D.

Lebesgue's Criterion. $f(x, y)$ is integrable over D if, and only if, $f(x, y)$ is bounded on D and the set of its discontinuity points has plane (Lebesgue) measure zero.

Comparative properties

1) If $f(x, y)$ and $g(x, y)$ are integrable over D and $f(x, y) \le g(x, y)$ on D, then

$$\iint_D f dA \le \iint_D g dA.$$

2) If $f(x, y)$ is integrable on D, then

$$f_{\inf} \cdot A(D) \le \iint_D f dA \le f_{\sup} \cdot A(D),$$

where $f_{\inf} = \inf_{(x,y) \in D} f(x, y)$, $f_{\sup} = \sup_{(x,y) \in D} f(x, y)$ and $A(D)$ is the area of D. In particular, if $f(x, y)$ is continuous on D, then

$$f_{\min} \cdot A(D) \le \iint_D f dA \le f_{\max} \cdot A(D),$$

where $f_{\min} = \min_{(x,y) \in D} f(x, y)$, $f_{\max} = \max_{(x,y) \in D} f(x, y)$.

Arithmetic (algebraic) properties

1) If $f(x, y)$ and $g(x, y)$ are integrable over D, then $f(x, y) + g(x, y)$ is also

integrable over D and

$$\iint_D (f+g)\,dA = \iint_D f dA + \iint_D g dA.$$

2) If $f(x,y)$ is integrable over D and c is an arbitrary constant, then $cf(x,y)$ is also integrable over D and

$$\iint_D cf dA = c\iint_D f dA.$$

3) If $f(x,y)$ is integrable over D, then $|f(x,y)|$ is also integrable over D and

$$\left|\iint_D f dA\right| \le \iint_D |f|\,dA.$$

4) If $f(x,y)$ and $g(x,y)$ are integrable over D, then $f(x,y)\,g(x,y)$ is also integrable over D.
5) $f(x,y)$ is integrable over D, if and only if, $f(x,y)$ is integrable over D_1 and D_2, where $D_1 \cup D_2 = D$ and $D_1 \cap D_2$ is a zero area curve. Furthermore,

$$\iint_{D_1} f dA + \iint_{D_2} f dA = \iint_D f dA.$$

The Mean Value Theorem for double integral. If $f(x,y)$ is integrable on D, then there exists a constant γ, $f_{\text{inf}} \le \gamma \le f_{\text{sup}}$, such that $\iint_D f dA = \gamma \cdot A(D)$. In particular, if $f(x,y)$ is continuous on a connected set D, then there exists a point $(a,b) \in D$ such that $\iint_D f dA = f(a,b)\cdot A(D)$.

Iterated Integrals

1. Rectangles. Let $f(x,y)$ be defined on the rectangle $R = [a,b]\times[c,d] \subset \mathbb{R}^2$ and for each $x \in [a,b]$ there exists the Riemann integral $\int_c^d f(x,y)\,dy = g(x)$. If the Riemann integral $\int_a^b g(x)\,dx$ also exists, it is called the *iterated integral* of $f(x,y)$ and denoted as follows: $\int_a^b g(x)\,dx \equiv \int_a^b dx \int_c^d f(x,y)\,dy$. In the same way can be defined the second iterated integral $\int_c^d dy \int_a^b f(x,y)\,dx$.

2. General domains. Let $f(x,y)$ be defined on a bounded closed set $D \subset \mathbb{R}^2$ whose boundary ∂D has zero area. Additionally, suppose that each line parallel to y-axis crosses ∂D at two points at most, let say $y_1(x)$ and $y_2(x)$, $y_1(x) \le y_2(x)$, or along the entire segment $[y_1(x), y_2(x)] \subset \partial D$. In this case the set D can be represented as $D = [a,b] \times [y_1(x), y_2(x)]$ and it is frequently called the type I region. If for each $x \in [a,b]$ there exists the Riemann integral $\int_{y_1(x)}^{y_2(x)} f(x,y)\,dy = g(x)$ and the Riemann integral $\int_a^b g(x)\,dx$ also exists, the latter is called the *iterated integral* of $f(x,y)$ and denoted as follows

$\int_a^b dx \int_{y_1(x)}^{y_2(x)} f(x,y)\,dy$. In a similar way can be defined the second iterated integral $\int_c^d dy \int_{x_1(y)}^{x_2(y)} f(x,y)\,dx$ on the type II region $D = [x_1(y), x_2(y)] \times [c,d]$.

Fubini's Theorem. Let $f(x,y)$ be defined on a type I region D. If $f(x,y)$ is Riemann integrable over D and for each $x \in [a,b]$ there exists the Riemann integral $\int_{y_1(x)}^{y_2(x)} f(x,y)\,dy$, then the iterated integral $\int_a^b dx \int_{y_1(x)}^{y_2(x)} f(x,y)\,dy$ exists and is equal to the double integral:

$$\iint_D f(x,y)\,dA = \int_a^b dx \int_{y_1(x)}^{y_2(x)} f(x,y)\,dy.$$

A similar result is true for type II regions.

In the special case when $f(x,y)$ is continuous on D, the existence of double and both iterated integrals is guaranteed and the three integrals give the same value.

Change of variables. Let $f(x,y)$ be defined on a bounded closed connected set D. Consider two functions $x = \varphi(u,v)$, $y = \psi(u,v)$ defined on $(u,v) \in \tilde{D}$ and such that this transformation (the change of variables) is one-to-one correspondence between $\tilde{D}$ and D. If both functions φ and ψ have continuous partial derivatives on $\tilde{D}$ and the Jacobian $\frac{\partial(\varphi,\psi)}{\partial(u,v)} \equiv \det\begin{pmatrix} \varphi_u & \varphi_v \\ \psi_u & \psi_v \end{pmatrix} \neq 0$ on $\tilde{D}$, then for $f(x,y)$ integrable over D the following formula holds:

$$\iint_D f(x,y)\,dxdy = \iint_{\tilde{D}} f(u,v) \left| \frac{\partial(\varphi,\psi)}{\partial(u,v)} \right| dudv.$$

Line Integrals

Line integral. Let C be a plane rectifiable simple curve given in the parametric form $x = \varphi(t)$, $y = \psi(t)$, $t \in [a,b]$, where functions $\varphi(t)$, $\psi(t)$ are continuous on $[a,b]$, and let $f(x,y)$ be defined on C. (Recall that a curve is called simple if it does not have self-intersections, except probably for the initial and final points.) Consider an arbitrary partition

$$P = \{t_i : a = t_0 < t_1 < \ldots < t_{n-1} < t_n = b\}$$

of $[a,b]$ with the diameter $\Delta = \max_i \Delta t_i$, $\Delta t_i = t_i - t_{i-1}$. A partition P corresponds to a certain partition of C by the points $(x_i, y_i) = (\varphi(t_i), \psi(t_i))$, $i = 1, \ldots, n$, and an increment Δt_i is associated with the following increments of coordinates and length of the line segment joining two successive points (x_{i-1}, y_{i-1}) and (x_i, y_i):

$$\Delta x_i = x_i - x_{i-1},\ \Delta y_i = y_i - y_{i-1},\ \Delta l_i = \sqrt{(x_i - x_{i-1})^2 + (y_i - y_{i-1})^2}.$$

Choose an arbitrary point $\tau_i \in [t_{i-1}, t_i]$ in each partition interval, which corresponds to the point $(x(\tau_i), y(\tau_i))$ on the arc of C between (x_{i-1}, y_{i-1}) and (x_i, y_i), and form the three integral sums:

$$S_l(f) = \sum_{i=1}^{n} f(x(\tau_i), y(\tau_i)) \cdot \Delta l_i,$$

$$S_x(f) = \sum_{i=1}^{n} f(x(\tau_i), y(\tau_i)) \cdot \Delta x_i,\ S_y(f) = \sum_{i=1}^{n} f(x(\tau_i), y(\tau_i)) \cdot \Delta y_i.$$

If finite limits of these sums, as the partition diameter Δ approaches 0, exist and do not depend on a choice of partitions and points τ_i, then they are called the *line integral* of $f(x, y)$ along C and are denoted as follows:

$$\lim_{\Delta \to 0} S_l(f) = \int_C f(x, y)\, dl,$$

$$\lim_{\Delta \to 0} S_x(f) = \int_C f(x, y)\, dx,\ \ \lim_{\Delta \to 0} S_y(f) = \int_C f(x, y)\, dy.$$

In order to distinguish one line integral from another, the first one is called the *line integral with respect to arc length*, the second - *with respect to* x, and the third - *with respect to* y. Frequently, the first integral is called the *line integral of the first kind*, and two others - the *line integrals of the second kind.* The sum of two second kind integrals

$$\int_C f(x, y)\, dx + \int_C g(x, y)\, dy \equiv \int_C f(x, y)\, dx + g(x, y)\, dy$$

is called a general line integral of the second kind.

Remark 1. The value of the line integral of the first kind does not depend on the parametrization of C.

Remark 2. The absolute value of the line integrals of the second kind does not depend on the parametrization of C.

Remark 3. If C is a simple closed curve representing the boundary of a bounded set D, then the positive orientation of C is such that D is always on the left under the traversal of C.

Theorem. If C is a smooth curve without singular points (that is, in the curve definition $x = \varphi(t)$, $y = \psi(t)$ are continuously differentiable on $[a, b]$, and $(\varphi'(t))^2 + (\psi'(t))^2 \neq 0$, $\forall t \in [a, b]$), and if $f(x, y)$ is continuous on C, then the line integrals of $f(x,y)$ along C exist and can be calculated by the following formulas:

$$\int_C f(x,y)\,dl = \int_a^b f(\varphi(t),\psi(t))\sqrt{(\varphi'(t))^2+(\psi'(t))^2}dt,$$

$$\int_C f(x,y)\,dx = \int_a^b f(\varphi(t),\psi(t))\,\varphi'(t)\,dt,$$

$$\int_C f(x,y)\,dy = \int_a^b f(\varphi(t),\psi(t))\,\psi'(t)\,dt,$$

where the integrals in the right-hand sides are the Riemann integrals.

Remark. The theorem is true also for a piecewise smooth curve, that is, for a continuous curve represented as a union of a finite number of smooth curves.

The line integrals have the same comparative and arithmetic properties as the double (and "ordinary") Riemann integral.

Path Independence. The line integral $\int_C fdx+gdy$ is *path independent* in a set D if $\int_{C_1} fdx+gdy = \int_{C_2} fdx+gdy$ for any two smooth curves (paths) in D that have the same initial and final points.

Theorem. If $f(x,y)$ and $g(x,y)$ are continuous and have continuous partial derivatives f_y and g_x on an open simply-connected set D, then the following four conditions are equivalent:
1) the line integral $\int_C fdx+gdy$ is path independent in a set D
2) the line integral $\int_C fdx+gdy = 0$ for every closed path in D (that is, for every path whose initial and final points coincide)
3) the expression $fdx+gdy$ represents the differential of a function $\eta(x,y)$ on D
4) $f_y = g_x$ on D.

Remark 1. The conditions 1) and 2) are equivalent for any line integrable (for example, continuous) on D functions f and g.

Remark 2. The conditions 1) and 3) are equivalent for any continuous functions $f(x,y)$ and $g(x,y)$ on an open connected set D.

Remark 3. The condition 4) follows from 1) (or 2), or 3)) for any open connected set D.

Remark 4. The condition 3) can be rewritten in the following form: there exists a function $\eta(x,y)$ such that $\nabla\eta = (f,g)$ in D. In this case the vector field $F=(f,g)$ is called conservative (or potential) on D and the function η is called a potential function of F.

Green's formulas. If $f(x,y)$ and $g(x,y)$ have continuous partial derivatives on a closed bounded set D with a smooth boundary ∂D, then the following formulas hold:

$$\iint_D f_y dA = -\int_{\partial D} f dx, \iint_D g_x dA = \int_{\partial D} g dy,$$

$$\iint_D (g_x - f_y)\, dA = \int_{\partial D} f dx + g dy, \iint_D (f_x + g_y)\, dA = \int_{\partial D} -g dx + f dy,$$

where all the line integrals have the positive orientation, that is, a single positive traversal of ∂D.

Integration by parts. If $f(x)$ and $g(x)$ have continuous partial derivatives on a closed bounded set D with a smooth boundary ∂D, then the following formulas hold:

$$\iint_D f_y g dA = -\iint_D f g_y dA - \int_{\partial D} f g dx,$$

$$\iint_D f_x g dA = -\iint_D f g_x dA + \int_{\partial D} f g dy,$$

where all the line integrals have the positive orientation.

9.2 One-dimensional links

General Remark. Like in sections 7.2 and 8.2, the examples in this section have direct connections with some examples for one-variable functions given in Part I and, correspondingly, the majority of the provided below counterexamples are adapted to $\mathbb{R}^2$ versions of the functions introduced in chapter 5. Each example in this section has the reference to the corresponding one-dimensional counterpart.

Example 1. "If $f(x,y)$ is bounded on D, then there exists the double integral of $f(x,y)$ over D."

(compare with Example 1, section 5.3)

Solution.

The two-dimensional version of Dirichlet's function $D(x,y) = \begin{cases} 1, & \sqrt{x^2+y^2} \in \mathbb{Q} \\ 0, & \sqrt{x^2+y^2} \in \mathbb{I} \end{cases}$ is bounded but not integrable over any bounded closed set D. Let us consider for simplicity a rectangle $R = [a,b] \times [c,d]$. For an arbitrary partition of R, each elementary rectangle $R_{ij} = [x_{i-1}, x_i] \times [y_{j-1}, y_j]$ contains the points with rational and irrational distance from the origin. Choosing rational distance points in the definition of Riemann sums, one obtains $S_{rat}(f) = (b-a)(d-c)$, while for irrational distance points one gets $S_{irr}(f) = 0$. Therefore, the limits for these sums are different:

$$\lim_{\Delta \to 0} S_{rat}(f) = (b-a)(d-c) \neq 0 = \lim_{\Delta \to 0} S_{irr}(f),$$

which means that the double integral does not exist.

Remark. The graph of this function is similar to that of the two-dimensional modified Dirichlet's function shown in Example 1, section 7.2.

Example 2. "If $f(x,y)$ is integrable over D, then $f(x,y)$ is continuous on D."

(compare with Example 2, section 5.3)

Solution.

The function $f(x,y) = \operatorname{sgn} x$ is discontinuous on $D = [0,1] \times [0,1]$, since

$$\lim_{(x,y_0)\to(0,y_0)} f(x,y) = \lim_{x\to 0} \operatorname{sgn} x = 1 \neq 0 = f(0,y_0).$$

On the other hand, for an arbitrary choice of partition and points $(\xi_i, \eta_j) \in R_{ij}$, the following evaluation of the Riemann sums for $f(x,y)$ is true:

$$1 - \Delta x_1 = \sum_{i=2}^{n}\sum_{j=1}^{m} 1 \cdot \Delta x_i \Delta y_j \leq S(f)$$

$$= \sum_{i=1}^{n}\sum_{j=1}^{m} f(\xi_i, \eta_j) \cdot \Delta x_i \Delta y_j \leq \sum_{i=1}^{n}\sum_{j=1}^{m} 1 \cdot \Delta x_i \Delta y_j = 1.$$

Since $\Delta x_1 \to 0$ when $\Delta \to 0$, it follows that the limit of the Riemann sums, as Δ approaches 0, exists and does not depend on the choice of partition and points (ξ_i, η_j): $\lim\limits_{\Delta\to 0} S(f) = \iint_D \operatorname{sgn} x dA = 1$, that is, $f(x,y)$ is integrable over D.

Example 3. "If $|f(x,y)|$ is integrable over D, then $f(x,y)$ is also integrable over D."

(compare with Example 5, section 5.3)

Solution.

The two-dimensional modified Dirichlet's function $\tilde{D}(x,y) = \begin{cases} 1, & \sqrt{x^2+y^2} \in \mathbb{Q} \\ -1, & \sqrt{x^2+y^2} \in \mathbb{I} \end{cases}$ is not integrable over any set D, in particular, it is not integrable over a rectangle $R = [a,b] \times [c,d]$ (it can be shown like in Example 1), but $\left|\tilde{D}(x,y)\right| = 1$ is integrable over any set D.

Remark 1. The Remarks to Example 5 of section 5.3 are applicable here with corresponding adjustments to the case of $\mathbb{R}^2$.

Remark 2. The graph of this function is shown in Example 1, section 7.2.

Example 4. "If $f(x,y)$ is integrable over D and $\iint_D f^2 dA = 0$, then $f(x,y) = 0$ on D."

(compare with Example 9, section 5.3)

Solution.

According to the arithmetic rules for integrals, the function $f(x,y) = 1 - \operatorname{sgn} x$ is integrable on $D = [0,1] \times [0,1]$ as a difference of two integrable functions (see Example 2 for integrability of $\operatorname{sgn} x$) and

$$\iint_D f^2 dA = \iint_D 1 - \operatorname{sgn} x dA = \iint_D 1 dA - \iint_D \operatorname{sgn} x \, dA = 1 - 1 = 0.$$

However, $f(x,y) \neq 0$ on D since $f(0,y) = 1$.

Remark. The Remarks to Example 9 of section 5.3 are applicable here with corresponding adjustments to the case of $\mathbb{R}^2$.

Example 5. "If $f(x,y) + g(x,y)$ is integrable over D, then both $f(x,y)$ and $g(x,y)$ are integrable over D."

(compare with Example 11, section 5.3)

Solution.

$f(x,y) = D(x,y)$ and $g(x,y) = 1 - D(x,y)$ is a simple counterexample for an arbitrary set D.

Remark. Similar statements for other arithmetic operations are also false.

Example 6. "If double integral and both iterated integrals of $f(x,y)$ exist and are equal, then these integrals can be calculated using antiderivtives of $f(x,y)$ with respect to x (when y is fixed) or with respect to y (when x is fixed)."

(compare with Example 14, section 5.3)

Solution.

The function $f(x,y) = \operatorname{sgn}(xy)$ is integrable on $[0,1] \times [0,1]$ and its iterative integrals also exist (to show this one can use the arguments similar to those applied in Example 2). However, the first iterated integral $\int_0^1 dx \int_0^1 \operatorname{sgn}(xy) dy$ requires first the evaluation of $\int_0^1 \operatorname{sgn}(x_0 y) dy$ for every fixed $x_0 \in [0,1]$, that is possible to make obtaining the result $\int_0^1 \operatorname{sgn}(x_0 y) dy = \operatorname{sgn} x_0$, but not through antiderivatives because there is no antiderivative for $\operatorname{sgn}(x_0 y)$ for any $x_0 \neq 0$. Additionally, for the same reason, the subsequent calculation of $\int_0^1 \operatorname{sgn} x dx$ is also not possible through antiderivative. Due to the function symmetry, the same considerations are true for the second iterative integral.

Example 7. "If $f(x,y)$ is integrable over D, then there exists a point $(a,b) \in D$ such that $\iint_D f dA = f(a,b) \cdot A(D)$."

(compare with Example 15, section 5.3)

Solution.

The Heaviside function $H(x,y) = \begin{cases} 0, x < 0 \\ 1, x \geq 0 \end{cases}$ is integrable over $D =$

$[-1,1] \times [-1,1]$, since it is a bounded function on D and it is continuous on D with the only exception being the simple line $x = 0$, $y \in [-1,1]$. Due to the arithmetic rules for double integrals

$$\iint_D H dA = \iint_{D_1} 0 dA + \iint_{D_2} 1 dA = 0 + 2 = 2,$$

where $D_1 = [-1,0] \times [-1,1]$ and $D_2 = [0,1] \times [-1,1]$. At the same time , there is no point $(a,b) \in D$ such that $H(a,b) = \frac{\iint_D H dA}{A(D)} = \frac{1}{2}$.

Remark. The additional condition of the continuity of $f(x,y)$ on D will turn this false statement in the Mean Value Theorem for double integrals.

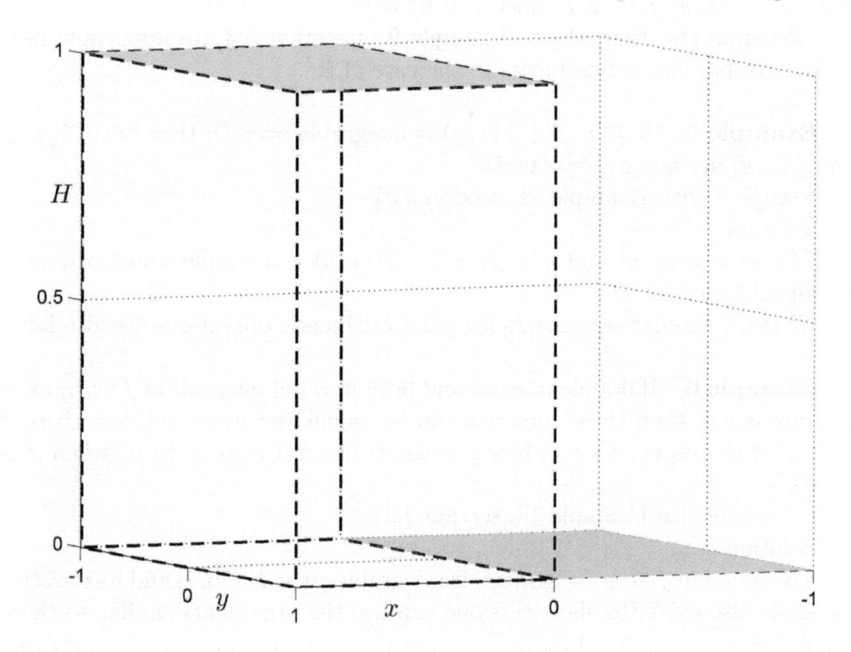

FIGURE 9.2.1: Example 7

Example 8. "If $f(x,y)$ is integrable (or even continuous) over D, then the value of the integral does not change if one alters the values of $f(x,y)$ at a countable number of points."

(compare with Example 18, section 5.3)

Solution.

The functions $f(x,y) \equiv 0$ and

$$\hat{D}(x,y) = D(x) \cdot D(y) = \begin{cases} 1, & x \in \mathbb{Q},\, y \in \mathbb{Q} \\ 0, & \textit{otherwise} \end{cases}$$

(here $D(x)$ and $D(y)$ are one-dimensional Dirichlet's functions) differ only at points with rational coordinates, that is on a countable set, however $f(x,y)$ is Riemann integrable and $\hat{D}(x,y)$ is not integrable over any set D (it can be shown in the same way as in Example 1 of this section).

Remark. The statement becomes true if both functions are integrable.

Example 9. "If $f(x,y)$ is continuous on D except at only one point, then $f(x,y)$ is integrable over D."

(compare with Example 19, section 5.3)

Solution.

The function $f(x,y) = \cot\left(x^2 + y^2\right)$ has the only discontinuity point $(0,0)$ on $D = [0,1] \times [0,1]$, but $f(x,y)$ is not integrable over D because it approaches infinity as (x,y) approaches $(0,0)$.

Remark. The statement becomes true if $f(x,y)$ is additionally bounded on D.

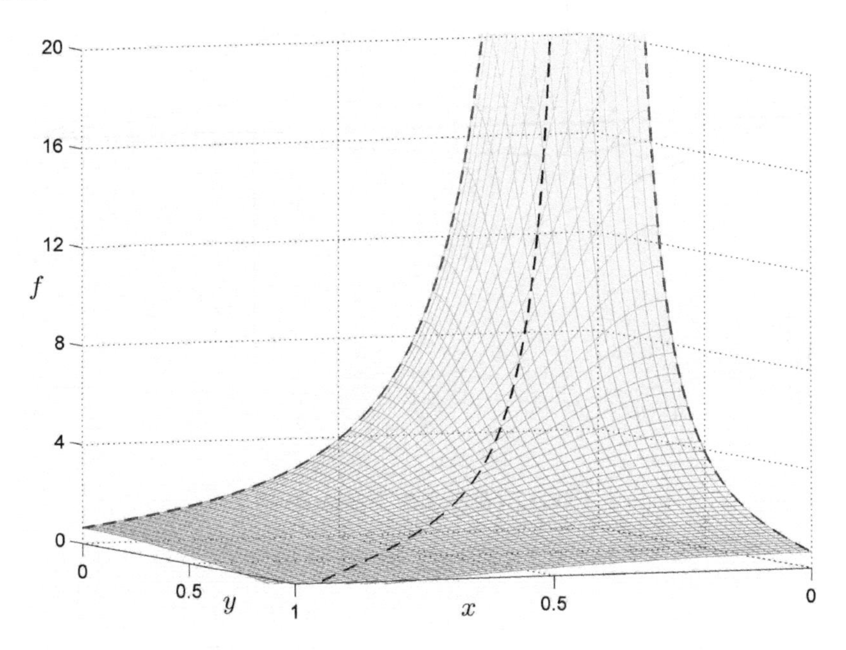

FIGURE 9.2.2: Example 9

Example 10. "If $f(x,y)$ is continuous on D, then the volume of the figure between the graph of $f(x,y)$ and the base D in xy-plane (and the corresponding lateral walls) can be found by the formula $V = \iint_D f dA$."

(compare with Example 4, section 5.5)

Solution.

A simple counterexample is $f(x,y) = x + y$ on $[-1,1] \times [-1,1]$. In this case the volume of the figure is the sum of the volumes of two equal tetrahedrons:

$V = V_1 + V_2 = \frac{4}{3} + \frac{4}{3} = \frac{8}{3}$, but the integral formula gives

$$\iint_D f dA = \int_{-1}^{1} dx \int_{-1}^{1} (x+y)\, dy = \int_{-1}^{1} \left(xy + \frac{y^2}{2} \right) \Big|_{-1}^{1} dx = \int_{-1}^{1} 2x dx = 0.$$

Remark. The condition of the non-negativity of $f(x,y)$ is missing in the statement conditions, or without the non-negativity the correct formula is $V = \iint_D |f|\, dA$.

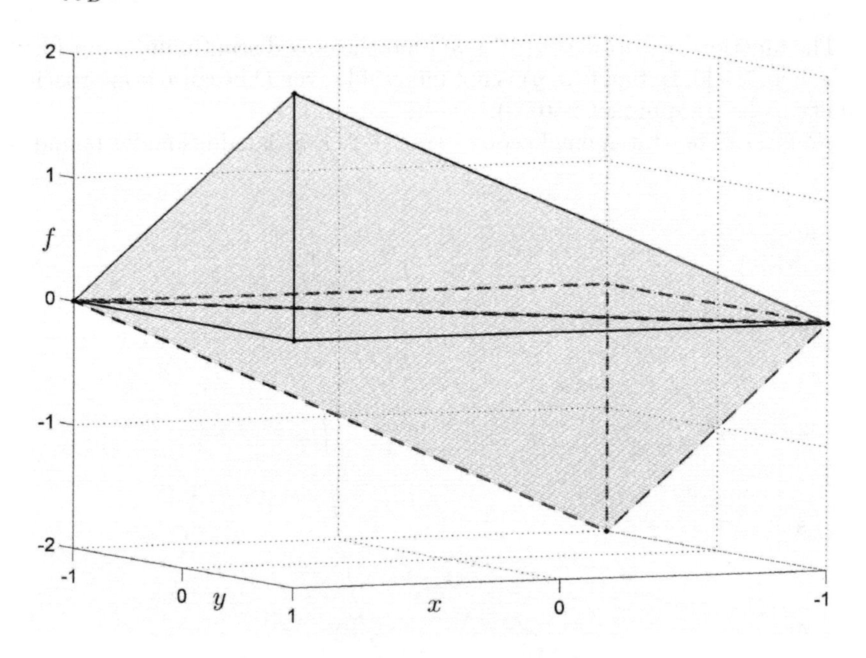

FIGURE 9.2.3: Example 10

9.3 Multidimensional essentials

Example 1. "Denote $r_{ij} = \Delta x_i \cdot \Delta y_j$ the area of an elementary rectangle R_{ij} in partition of a rectangle R and $r = \max_{i,j} r_{ij}$. Then the definition of the double integral over a rectangle R can be given as follows: $\iint_R f dA = \lim_{r \to 0} S(f)$, where the limit does not depend on a choice of partitions and points (ξ_i, η_j)."

Solution.

Let us consider the function $f(x,y) = x$ on the rectangle $R = [0,1] \times [0,1]$ and choose the partition into a set of elementary rectangles

$$R_{1j} = [0,1] \times [y_{j-1}, y_j]\,,\ j = 1, \ldots, m,$$

that is, keep the entire interval in x-axis. Then $\lim\limits_{r \to 0} S(f)$ depends on the choice of points (ξ_i, η_j): if one use the points $(0, \eta_j)$, then $f(0, \eta_j) = 0$ and

$$S(f) = \sum_{i=1}^{1} \sum_{j=1}^{m} f(0, \eta_j) \cdot \Delta x_i \Delta y_j = 0,$$

while for $(1, \eta_j)$ one has $f(1, \eta_j) = 1$ and

$$S(f) = \sum_{i=1}^{1} \sum_{j=1}^{m} f(1, \eta_j) \cdot \Delta x_i \Delta y_j = 1.$$

Therefore, the double integral of $f(x,y) = x$ does not exist according to the proposed "definition" with the parameter r. However, the existence of the double integral can be immediately confirmed using the properties of integrals ($f(x,y) = x$ is continuous on $R = [0,1] \times [0,1]$) or directly by the definition. In the last case, one can employ the following evaluations:

$$S(f) = \sum_{i=1}^{n} \sum_{j=1}^{m} f(\xi_i, \eta_j) \cdot \Delta x_i \Delta y_j \le \sum_{i=1}^{n} \sum_{j=1}^{m} x_i \cdot \Delta x_i \Delta y_j$$

$$= \sum_{j=1}^{m} \Delta y_j \cdot \sum_{i=1}^{n} x_i \cdot \Delta x_i = \sum_{i=1}^{n} x_i \cdot (x_i - x_{i-1})$$

$$= \frac{1}{2} \sum_{i=1}^{n} (x_i + x_{i-1})(x_i - x_{i-1}) + \frac{1}{2} \sum_{i=1}^{n} (x_i - x_{i-1})(x_i - x_{i-1})$$

$$= \frac{1}{2} \sum_{i=1}^{n} (x_i^2 - x_{i-1}^2) + \frac{1}{2} \sum_{i=1}^{n} \Delta x_i^2 \le \frac{1}{2}(x_n^2 - x_0^2) + \frac{1}{2} \max_i \Delta x_i \cdot \sum_{i=1}^{n} \Delta x_i$$

$$\le \frac{1}{2}(x_n^2 - x_0^2) + \frac{1}{2}\Delta = \frac{1}{2} + \frac{1}{2}\Delta$$

and

$$S(f) = \sum_{i=1}^{n} \sum_{j=1}^{m} f(\xi_i, \eta_j) \cdot \Delta x_i \Delta y_j \ge \sum_{i=1}^{n} \sum_{j=1}^{m} x_{i-1} \cdot \Delta x_i \Delta y_j$$

$$= \sum_{j=1}^{m} \Delta y_j \cdot \sum_{i=1}^{n} x_{i-1} \cdot \Delta x_i = \sum_{i=1}^{n} x_{i-1} \cdot (x_i - x_{i-1})$$

$$= \frac{1}{2} \sum_{i=1}^{n} (x_i + x_{i-1})(x_i - x_{i-1}) - \frac{1}{2} \sum_{i=1}^{n} (x_i - x_{i-1})(x_i - x_{i-1})$$

$$\geq \frac{1}{2}(x_n^2 - x_0^2) - \frac{1}{2}\max_i \Delta x_i \cdot \sum_{i=1}^{n} \Delta x_i \geq \frac{1}{2} - \frac{1}{2}\Delta.$$

Thus, for every partition and an arbitrary choice of the points (ξ_i, η_j) it holds

$$\frac{1}{2} - \frac{1}{2}\Delta \leq S(f) \leq \frac{1}{2} + \frac{1}{2}\Delta$$

and, consequently,

$$\lim_{\Delta \to 0} S(f) = \iint_R f dA = \frac{1}{2}.$$

Remark. The problem with the use of the maximum area r of elementary rectangles as a parameter for the limit of the Riemann sums is that approaching r to 0 does not necessarily require a refinement of partitions in both directions. For instance, in the above counterexample, when r approaches 0 for the chosen set $R_{1j} = [0,1] \times [y_{j-1}, y_j]$, $j = 1, \ldots, m$, the refinement occurs only in y-direction, while in x-direction always is used the entire interval $[0,1]$. It should be noted that the same situation can not occur with the partition diameter Δ. Indeed, for the partitions provided in the counterexample the diameter cannot approach 0, because $\Delta > 1$ for each of the above partitions (Δ approaches 1 when r approaches 0). In a general case, according to its definition $\Delta = \max_{i,j} \Delta_{ij}$, $\Delta_{ij} = \sqrt{\Delta x_i^2 + \Delta y_j^2}$, the condition that Δ approaches 0 means that the maximum diagonal of the elementary rectangles should approach 0, which in its turn implies that the maximum length of each side of elementary rectangles should approach 0, that is, the refinement is compulsorily in both directions.

Example 2. "If $f(x,y)$ is discontinuous on a countable set of curves in a set D, then $f(x,y)$ is not integrable on D."

Solution.

Let $D = R = [0,1] \times [0,1]$ and

$$E = \cup_{k=1}^{+\infty} E_k,\ E_k = \{(x,y) : x = \frac{1}{2^k}, y \in [0,1]\},\ \forall k \in \mathbb{N}.$$

Consider the following function on D: $f(x,y) = \begin{cases} 1, & (x,y) \in E \\ 0, & (x,y) \in D \backslash E \end{cases}$ that is discontinuous at every point of E. To show that $f(x,y)$ is integrable over D let us estimate the integral Riemann sums. Since $f(x,y) \geq 0$ on D, an integral sum is non-negative for an arbitrary partition P of R into a set of elementary rectangles

$$R_{ij} = [x_{i-1}, x_i] \times [y_{j-1}, y_j],\ i = 1, \ldots, n,\ j = 1, \ldots, m$$

and for an arbitrary choice of the points $(\xi_i, \eta_j) \in R_{ij}$:

$$S(f) = \sum_{i=1}^{n} \sum_{j=1}^{m} f(\xi_i, \eta_j) \cdot \Delta x_i \Delta y_j \geq 0.$$

On the other hand, choosing the points (ξ_i, η_j) to be in E each time when R_{ij} contains at least one point of E, we get upper estimation for any integral sum:

$$S(f) = \sum_{i=1}^{n} \sum_{j=1}^{m} f(\xi_i, \eta_j) \cdot \Delta x_i \Delta y_j$$

$$\leq S_1 + S_2 = \sum_{i,j} 0 \cdot \Delta x_i \Delta y_j + \sum_{i,j} 1 \cdot \Delta x_i \Delta y_j,$$

where the first sum corresponds to all R_{ij} that have no common point with E and the second - to the rectangles containing points of E. Notice that $S_1 = 0$ and each term in S_2 equals to the area of the corresponding rectangle R_{ij}. Therefore, we should now evaluate the area of all the rectangles containing the points of E.

Choose an arbitrary $\epsilon > 0$ and find a corresponding K such that $\frac{1}{2^K} < \frac{\epsilon}{4} \leq \frac{1}{2^{K-1}}$ (such K exists, because $\frac{1}{2^k} \underset{k\to\infty}{\to} 0$). Then $\frac{1}{2^k} < \frac{\epsilon}{4}$ for $\forall k \geq K$, which means that all segments $E_k, k \geq K$ lie in the strip $0 < x < \frac{\epsilon}{4}, 0 \leq y \leq 1$ and only $K-1$ segments $E_k, k = 1, \ldots, K-1$ lie out of this strip. Consider now an arbitrary partition P with the diameter $\Delta < \frac{\epsilon}{4K}$ and choose the smallest group of the leftmost rectangles $R_{ij}, i = 1, \ldots, I$, which contain all the segments $E_k, k \geq K$. Evidently, all these rectangles, except probably for those in the rightmost column, belong to the the strip $[0, \frac{\epsilon}{4}] \times [0,1]$. However, the area of the last column is easily estimated in terms of ϵ: $1 \cdot \Delta x_I < 1 \cdot \Delta < \frac{\epsilon}{4K}$. Therefore, the overall area of these rectangles can be evaluated as follows:

$$A_1 < \frac{\epsilon}{4K} + \frac{\epsilon}{4} \leq \frac{\epsilon}{2}.$$

It remains to estimate the area of those R_{ij}, which contain all the points of $E_k, k = 1, \ldots, K-1$. Notice that each of these E_k lie inside a strip $[x_{i_k-1}, x_{i_k}] \times [0,1]$ of the width $\Delta x_{i_k} = x_{i_k} - x_{i_k-1}$ or belongs to a border $x = x_{i_k}, y \in [0,1]$ between two subsequent strips. Therefore, the overall area of these rectangles satisfies the following evaluation:

$$A_2 \leq \sum_{k=1}^{K-1} (\Delta x_{i_k} + \Delta x_{i_k+1}) < 2\Delta \cdot (K-1) < \frac{\epsilon}{2}.$$

Summarizing, for an arbitrary partition with the diameter $\Delta < \frac{\epsilon}{4K}$, the overall area of the elementary rectangles R_{ij}, which contain all the segments $E_k, \forall k \in \mathbb{N}$, has the following evaluation:

$$A = A_1 + A_2 < \frac{\epsilon}{2} + \frac{\epsilon}{2} = \epsilon.$$

Since $S_2 \leq A < \epsilon$, it follows from the above estimates of the integral sums that $0 \leq S(f) \leq S_2 < \epsilon$. The last inequality implies that $\lim_{\Delta\to 0} S(f) = 0$, that is, $f(x,y)$ is integrable over R and $\iint_R f dA = 0$.

Remark 1. Since the constructed set of the curves

$$E_k = \{(x, y) : x = \frac{1}{2^k}, y \in [0, 1]\}, \ \forall k \in \mathbb{N}$$

has the plane measure zero and the function $f(x, y)$ is bounded on D, the existence of the double integral follows from Lebesgue's Criterion of integrability. However, we think that an explicit evaluation of the Riemann sums is instructive to prove integrability in this case.

Remark 2. The above function $f(x, y)$ differs from 0 on a countable set of segments, which is a larger set then a countable set of points, like that used in Example 8, section 9.2. Nevertheless, the function in Example 8 is not integrable while the above function gives the double integral equal to 0. Therefore, one can conclude that not only the "quantity" of discontinuity points is important for integrability, but also their distribution. In Example 8 the discontinuities were introduced in such a way that every point in D became a discontinuity point, while in the last Example the set of discontinuity points is not dense in D.

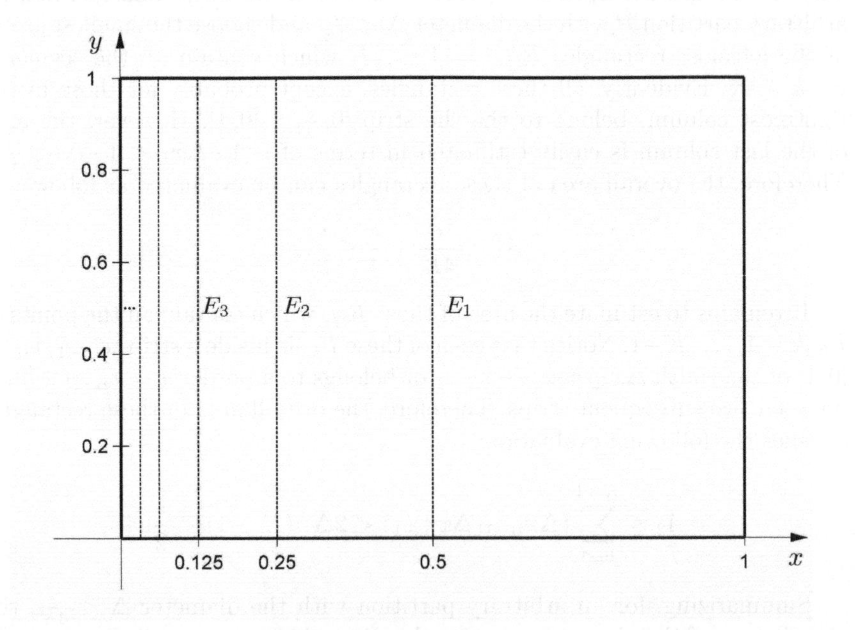

FIGURE 9.3.1: Example 2

Example 3. "If $f(x, y)$ is discontinuous at each point of a bounded set S, then both iterated integrals over S do not exist."

Solution.

Consider the function $f(x, y)$ defined on $S = [0, 1] \times [0, 1]$ as follows:

$f(x,y) = \begin{cases} 0, \ x \in \mathbb{Q} \\ y - \frac{1}{2}, \ x \in \mathbb{I}, \ y \neq \frac{1}{2} \\ -1, \ x \in \mathbb{I}, \ y = \frac{1}{2} \end{cases}$. This function is discontinuous at every point of S: at every point (x_0, y_0), $x_0 \in \mathbb{Q}$, $y_0 \neq \frac{1}{2}$ the function has two different partial limits

$$\lim_{\substack{(x,y)\to(x_0,y_0) \\ x=x_0}} f(x,y) = 0, \quad \lim_{\substack{(x,y)\to(x_0,y_0) \\ x\in\mathbb{I},\ y=y_0}} f(x,y) = y_0 - \frac{1}{2} \neq 0;$$

at the points (x_0, y_0), $x_0 \in \mathbb{Q}$, $y_0 = \frac{1}{2}$ two different partial limits are

$$\lim_{\substack{(x,y)\to(x_0,y_0) \\ x=x_0}} f(x,y) = 0, \quad \lim_{\substack{(x,y)\to(x_0,y_0) \\ x\in\mathbb{I},\ y=y_0}} f(x,y) = -1 \neq 0;$$

for each (x_0, y_0), $x_0 \in \mathbb{I}$, $y_0 \neq \frac{1}{2}$ two different partial limits are

$$\lim_{\substack{(x,y)\to(x_0,y_0) \\ x\in\mathbb{Q},\ y=y_0}} f(x,y) = 0, \quad \lim_{\substack{(x,y)\to(x_0,y_0) \\ x=x_0}} f(x,y) = y_0 - \frac{1}{2} \neq 0;$$

finally, for (x_0, y_0), $x_0 \in \mathbb{I}$, $y_0 = \frac{1}{2}$ the different partial limits are

$$\lim_{\substack{(x,y)\to(x_0,y_0) \\ x\in\mathbb{Q},\ y=1/2}} f(x,y) = 0, \quad \lim_{\substack{(x,y)\to(x_0,y_0) \\ x\in\mathbb{I},\ y=1/2}} f(x,y) = -1 \neq 0;$$

Nevertheless, the iterated integral $\int_0^1 dx \int_0^1 f(x,y)\,dy$ exists. In fact, for any fixed $x_{rac} \in \mathbb{Q}$ one has $\int_0^1 f(x_{rac}, y)\,dy = \int_0^1 0 dy = 0$. Further, for any fixed $x_{irr} \in \mathbb{I}$ the function $f(x_{irr}, y)$ is continuous with respect to y on $[0,1]$ except at the only point $y = \frac{1}{2}$ where $f(x_{irr}, y)$ has a removable discontinuity, that is, $f(x_{irr}, y)$ is Riemann integrable on $[0,1]$ and $\int_0^1 f(x,y)\,dy = \int_0^1 y - \frac{1}{2} dy = 0$. Therefore, $g(x) = \int_0^1 f(x,y)\,dy = 0$ for every fixed $x \in [0,1]$. Consequently, the iterated integral is defined and $\int_0^1 dx \int_0^1 f(x,y)\,dy = \int_0^1 g(x)\,dx = 0$.

Remark 1. Notice that the second iterated integral $\int_0^1 dy \int_0^1 f(x,y)\,dx$ does not exist, since $\int_0^1 f(x,y)\,dx$ does not exist: for any fixed y, the function $f(x,y)$ represents a Dirichlet's like function, which is not Riemann integrable.

Remark 2. Evidently, $f(x,y)$ is not integrable in the sense of double integral, because the number of its discontinuities is not of the measure 0.

Example 4. "If a region D is divided into regions D_1 and D_2 by a curve of an infinite length, then the arithmetic property 5) in section 9.1 does not hold."

Solution.

Consider the rectangle $D = [-1,1] \times [-2,2]$ divided into two regions by the saw-like curve defined as follows: $h(x) = \begin{cases} g(x), \ x \in [0,1], \\ -g(-x), \ x \in [-1,0] \end{cases}$ is the function composed of the left "negative" saw (the second sentence) and the right "positive" saw (the first sentence) $g(x) = \begin{cases} g_n, \ x \in [\frac{1}{2^n}, \frac{1}{2^{n-1}}], \ \forall n \in \mathbb{N} \\ 0, \ otherwise \end{cases}$, containing an infinite number of teeth in the form

$$g_n(x) = \begin{cases} 2^{n+1}x - 2, \ x \in [\frac{1}{2^n}, \frac{3}{2^{n+1}}] \\ -2^{n+1}x + 4, \ x \in [\frac{3}{2^{n+1}}, \frac{1}{2^{n-1}}] \end{cases}, \ \forall n \in \mathbb{N}.$$

Notice, that each point of the segment $[-1,1]$ on y-axis is the limit point of the curve $h(x)$.

Consider first the "positive" saw $g(x)$ located in the rectangle $[0,1]\times[-2,2]$. Since each tooth $g_n(x)$ has the length greater than 2 and the number of these teeth is infinite, the curve $g(x)$ has an infinite length. Since the "negative" saw $-g(-x)$ is symmetric to $g(x)$ with respect to the origin, its length is also infinite. Nevertheless, the plane (Lebesgue) measure of the curve $h(x)$ is zero. In fact, one can show that this curve can be covered by a finite set of rectangles whose overall area is less than a given positive number. To this end, consider an arbitrary $\epsilon > 0$ and choose the number N such that $\frac{1}{2^N} < \frac{\epsilon}{24} \le \frac{1}{2^{N-1}}$ (such number always exists, because $\frac{1}{2^N} \underset{N\to\infty}{\to} 0$). Construct now the rectangle $R_\epsilon = [-\frac{\epsilon}{24}, \frac{\epsilon}{24}] \times [-\frac{3}{2}, \frac{3}{2}]$ of the area $A_\epsilon = \frac{2\epsilon}{24} \cdot 3 = \frac{\epsilon}{4}$, which contains all the teeth of the saw $h(x)$ with $n \ge N$ (both of the left negative and right positive parts) and also all the points of the segment $[-1,1]$ on y-axis. The remaining $N-1$ teeth of the "positive" saw $g(x)$ represent a rectifiable curve and, consequently, have the plain measure zero, that is, this part of saw can be covered by a finite set of rectangles with the overall area less than $\frac{\epsilon}{4}$. The same is true for the remaining $N-1$ teeth of the "negative" saw $-g(-x)$. Therefore, all the saw $h(x)$ can be covered by a finite number of rectangles with the overall area $A < \frac{\epsilon}{4} + \frac{\epsilon}{4} + \frac{\epsilon}{4} < \epsilon$, which means that the curve $h(x)$ has zero area. Since the applied splitting of D is such that $D_1 \cup D_2 = D$ and $D_1 \cap D_2$ is the zero area curve $h(x)$, all the conditions of the arithmetic property 5) in section 9.1 are satisfied and for any integrable on D function $f(x,y)$ it holds

$$\iint_{D_1} f dA + \iint_{D_2} f dA = \iint_D f dA.$$

For instance, if $f(x,y) \equiv 1$, then one obtains the area of D at the right-hand side of the last formula

$$\iint_D f dA = \int_{-1}^{1} dx \int_{-2}^{2} dy = 8$$

and the areas of D_1 and D_2 at the left-hand side

$$\iint_{D_1} f dA + \iint_{D_2} f dA = 4 + 4 = 8$$

(due to the symmetry of D_1 and D_2 and the symmetry of $f(x,y)$ each area is just a half of the area of D).

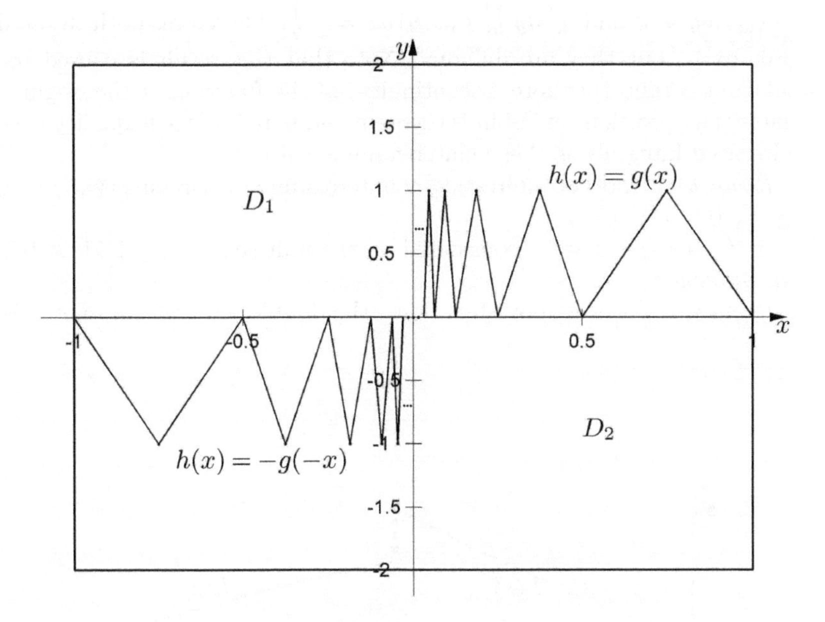

FIGURE 9.3.2: Example 4

Example 5. "If both iterated integrals exist, then they are equal."
Solution.

Let us consider the function $f(x,y) = \begin{cases} \frac{x-y}{(x+y)^3}, & (x,y) \neq (0,0) \\ 0, & (x,y) = (0,0) \end{cases}$ defined on the unit square $S = [0,1] \times [0,1]$. For any $x \neq 0$ we have

$$I_1(x) = \int_0^1 \frac{x-y}{(x+y)^3} dy = \int_0^1 \frac{-x-y}{(x+y)^3} + \frac{2x}{(x+y)^3} dy = \left(\frac{1}{x+y} - \frac{x}{(x+y)^2} \right) \Bigg|_0^1$$

$$= \frac{1}{x+1} - \frac{1}{x} - x \left(\frac{1}{(x+1)^2} - \frac{1}{x^2} \right) = \frac{1}{(x+1)^2}.$$

Since the last function is continuous (and consequently integrable) on $[0,1]$ and the value of the definite integral does not depend on the function definition in a single point, the iterated integral $\int_0^1 dx \int_0^1 f(x,y)\, dy$ exists and is equal

$$\int_0^1 dx \int_0^1 f(x,y)\, dy = \int_0^1 \frac{1}{(x+1)^2} dx = -\frac{1}{x+1} \Bigg|_0^1 = \frac{1}{2}.$$

Due to the antisymmetry of the function with respect to the interchange of the variables, for the second iterated integral we obtain $I_2(y) = \int_0^1 \frac{x-y}{(x+y)^3} dx = -\frac{1}{(y+1)^2}$, $\forall y \neq 0$ and $\int_0^1 dy \int_0^1 f(x,y)\, dx = -\frac{1}{2}$. Therefore, both iterated integrals exist, but they are different. Note that this result is caused by unboundedness (and therefore discontinuity) of the function at the origin. This violates the conditions of Fubini's theorem and may lead to inequality between the iterated integrals as this counterexample shows.

Remark 1. Another interesting counterexample is the function $f(x,y) = \begin{cases} y^{-2}, \ 0 < x < y < 1 \\ -x^{-2}, \ 0 < y < x < 1 \\ 0, \ otherwise \end{cases}$ considered on the unit square $S = [0,1] \times [0,1]$.

Remark 2. For the above functions, the double integral does not exist.

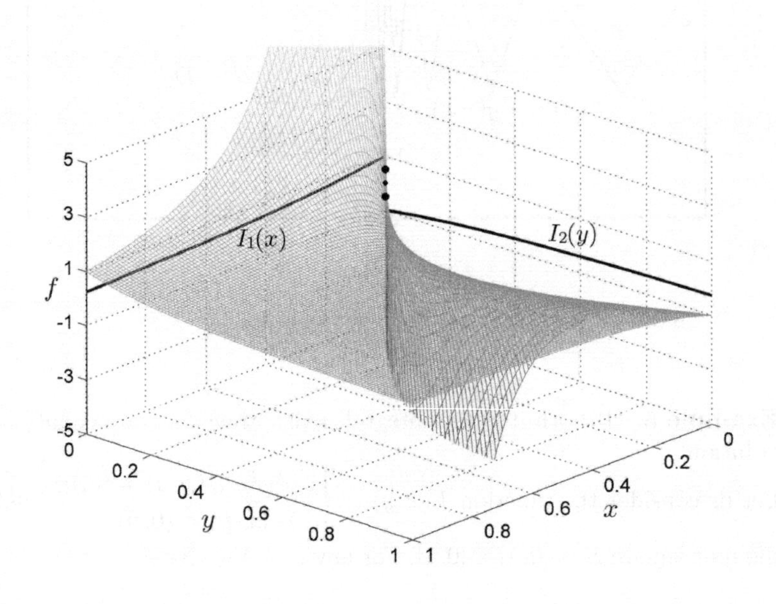

FIGURE 9.3.3: Example 5

Example 6. "If $\int_a^b dx \int_c^d f(x,y)\, dy = \int_c^d dy \int_a^b f(x,y)\, dx$, then $f(x,y)$ is continuous on $[a,b] \times [c,d]$."

Solution.

The function

$$f(x,y) = \begin{cases} y^{-1}, \ (x,y) \in D_1 : \ 0 < x < y < 1 \\ x^{-1}, \ (x,y) \in D_2 : \ 0 < y < x < 1 \\ 0, \ otherwise \end{cases}$$

defined on the unit square $S = [0,1] \times [0,1]$ has discontinuities on the square

boundary and also on the line $y = x$. In fact, on the boundary points (x_0, y_0) we have the following cases:
1) if $y_0 = 0$, $0 < x_0 \le 1$, then $(x, y) \in D_2$ and

$$\lim_{(x,y)\to(x_0,0)} f(x,y) = \lim_{x\to x_0} \frac{1}{x} = \frac{1}{x_0} \ne 0 = f(x_0, 0);$$

2) if $x_0 = 0$, $0 < y_0 \le 1$, then $(x, y) \in D_1$ and

$$\lim_{(x,y)\to(0,y_0)} f(x,y) = \lim_{y\to y_0} \frac{1}{y} = \frac{1}{y_0} \ne 0 = f(0, y_0);$$

3) if $y_0 = 1$, $0 < x_0 < 1$, then $(x, y) \in D_1$ and

$$\lim_{(x,y)\to(x_0,1)} f(x,y) = \lim_{y\to 1} \frac{1}{y} = 1 \ne 0 = f(x_0, 1);$$

4) if $x_0 = 1$, $0 < y_0 < 1$, then $(x, y) \in D_2$ and

$$\lim_{(x,y)\to(1,y_0)} f(x,y) = \lim_{x\to 1} \frac{1}{x} = 1 \ne 0 = f(1, y_0);$$

5) if $(x_0, y_0) = (0, 0)$, then the partial limit for $(x, y) \in D_1$ is

$$\lim_{(x,y)\to(0,0)} f(x,y) = \lim_{y\to 0} \frac{1}{y} = \infty;$$

6) if $(x_0, y_0) = (1, 1)$, then the partial limit for $(x, y) \in D_1$ is

$$\lim_{(x,y)\to(1,1)} f(x,y) = \lim_{y\to 1} \frac{1}{y} = 1 \ne 0 = f(1, 1).$$

Hence, all the boundary points are the discontinuity points. Besides, for the points (x_0, x_0), $0 < x_0 < 1$, the partial limit when $(x, y) \in D_1$ is

$$\lim_{(x,y)\to(x_0,x_0)} f(x,y) = \lim_{y\to x_0} \frac{1}{y} = \frac{1}{x_0} \ne 0 = f(x_0, x_0).$$

Therefore, all the points on the line $y = x$ are also discontinuity ones.

Nevertheless, the iterated integrals exist and are equal:

$$\int_0^1 dx \int_0^1 f(x,y)\, dy = \int_0^1 \left(\int_0^x \frac{1}{x} dy + \int_x^1 \frac{1}{y} dy \right) dx$$

$$= \int_0^1 \left(\frac{y}{x}\Big|_0^x + \ln y|_x^1 \right) dx = \int_0^1 (1 - \ln x)\, dx = (2x - x\ln x)|_0^1 = 2;$$

$$\int_0^1 dy \int_0^1 f(x,y)\, dx = \int_0^1 \left(\int_0^y \frac{1}{y} dx + \int_y^1 \frac{1}{x} dx \right) dy$$

$$= \int_0^1 \left(\frac{x}{y}\Big|_0^y + \ln x\Big|_y^1 \right) dy = \int_0^1 (1 - \ln y)\, dy = (2y - y\ln y)|_0^1 = 2.$$

Remark 1. This shows that the conditions in Fubini's theorem are sufficient and even if they are not satisfied, the iterated integrals may exist and be equal.

Remark 2. The double integral in the above counterexample does not exist, because the function is unbounded on S.

Remark 3. A similar counterexample can be given with the function

$$g(x,y) = \begin{cases} y^{-1}, & 0 < x < y < 1 \\ -x^{-1}, & 0 < y < x < 1 \\ 0, & otherwise \end{cases}$$

defined on the unit square $S = [0,1] \times [0,1]$. The principal difference between $f(x,y)$ and $g(x,y)$ is that along the line $y = x$ the former has removable discontinuity (just redefine the function as $f(x,x) = \frac{1}{x}$ for $0 < x < 1$), while the latter has a "wilder" discontinuity of the jump type.

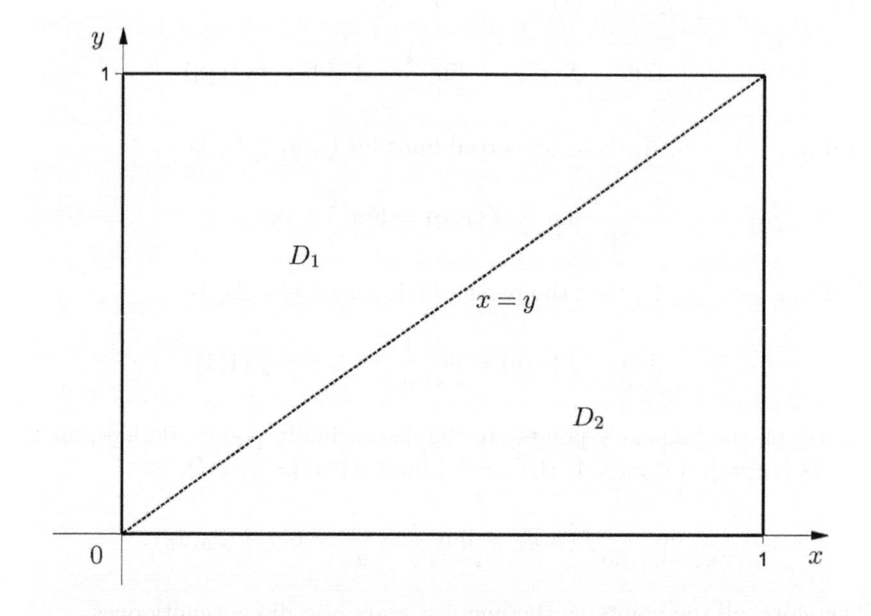

FIGURE 9.3.4: Example 6, division of the domain S

Example 7. "If both iterated integrals exist and have the same value, then the double integral also exists."

Solution.

Let us consider the function $f(x,y)$ defined on the unit square $S = [0,1] \times [0,1]$ as follows: $f(x,y) = \begin{cases} 1, & (x,y) \in T \\ 0, & (x,y) \notin T \end{cases}$, where T consists of

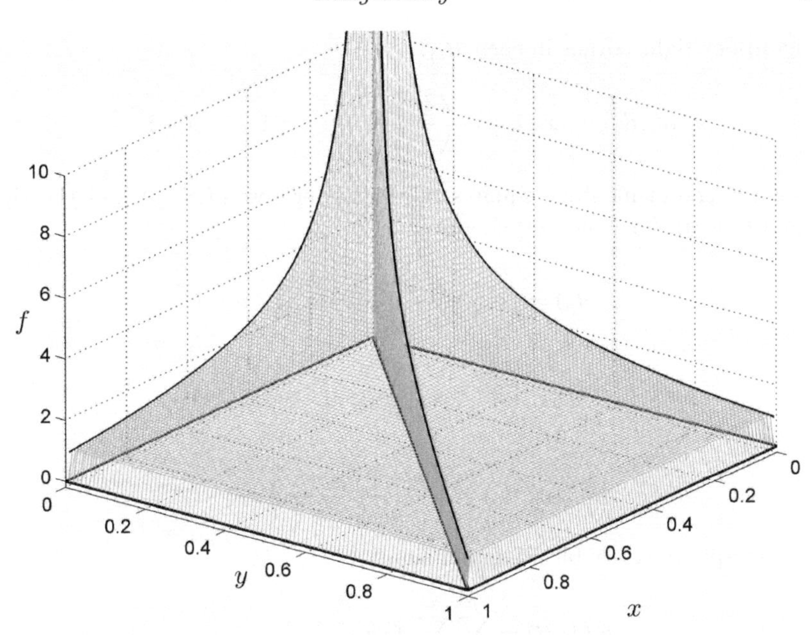

FIGURE 9.3.5: Example 6, graph of the function f

the points of S such that $(x, y) = \left(\frac{m}{p}, \frac{n}{p}\right)$, $m, n, p \in \mathbb{N}$, where both fractions are in lowest terms (note that the denominator is the same). First, let us show that the iterated integral $\int_0^1 dy \int_0^1 f(x, y)\, dx$ exists. In fact, if we choose any irrational y_0, then $f(x, y_0) = 0$ and $\int_0^1 f(x, y_0)\, dx = 0$. On the other hand, for a rational $y_0 = \frac{n}{p}$ (with the fraction in lowest terms), $f(x, y_0)$ is different from zero only in the finite number of the points, because for a given natural p there are at most p numbers of the form $x = \frac{m}{p}$, $m \in \mathbb{N}$, in $[0, 1]$. Therefore, again $\int_0^1 f(x, y_0)\, dx = 0$. Since the inner integral is zero for any y_0, the iterated integral exists and $\int_0^1 dy \int_0^1 f(x, y)\, dx = 0$. Due to the symmetry of the function and domain with respect to interchange of variables, the same result is true for the second iterated integral: $\int_0^1 dx \int_0^1 f(x, y)\, dy = 0$.

Now let us prove that the double integral $\iint_S f(x, y)\, dA$ does not exist. To this end, let us consider the special uniform partitions obtained by division of each side of S in p equal parts, where p is a prime number. Note that the diameter of such partitions approaches 0 as p approaches infinity. Each elementary square

$$R_{ij} = [x_{i-1}, x_i] \times [y_{j-1}, y_j]\,, \ i, j = 1, \ldots, p$$

contains the points $(c_i, d_j) \notin T$ and also the points $(c_i, d_j) \in T$. The possibility of the former choice is evident, since in any interval $[x_{i-1}, x_i]$ (or $[y_{j-1}, y_j]$) there are irrational points. The latter choice can be made by using the points of the upper right corner in each R_{ij}:

$$(c_i, d_j) = (x_i, y_j) = \left(\frac{i}{p}, \frac{j}{p}\right), \; i, j = 1, \ldots, p-1,$$

except for the elementary squares of the upper row ($j = p$) and the right column ($i = p$), for which we choose

$$(c_i, d_p) = \left(\frac{i}{p}, \frac{p-1}{p}\right), \; i = 1, \ldots, p-1,$$

$$(c_p, d_j) = \left(\frac{p-1}{p}, \frac{j}{p}\right), \; j = 1, \ldots, p-1,$$

and

$$(c_p, d_p) = \left(\frac{p-1}{p}, \frac{p-1}{p}\right).$$

The corresponding double Riemann sums

$$S(f; P) = \sum_{j=1}^{p} \sum_{i=1}^{p} f(c_i, d_j) \cdot \Delta x_i \Delta y_j$$

are equal 0 in the case $(c_i, d_j) \notin T$ and their partial limit is 0, but these sums are equal 1 if $(c_i, d_j) \in T$ and the corresponding partial limit is 1. Since two partial limits of the Riemann sums are different, there is no limit of the double Riemann sums as the partition diameter approaches 0, which means that the double integral does not exist.

Remark. The statement of this example is a strengthened (but still false) version of the statement in Example 6.

Example 8. "If $f(x, y)$ is bounded and integrable on a bounded set S, then both iterated integrals over S exist and are equal to the double integral."

Solution.

Let us define $f(x, y)$ on the square $S = [0,1] \times [0,1]$ as follows: $f(x, y) = \begin{cases} \frac{1}{n}, \; y \in \mathbb{Q}, \; x = \frac{m}{n} \in \mathbb{Q} \\ 0, \; otherwise \end{cases}$, where $m, n \in \mathbb{N}$ and the fraction $\frac{m}{n}$ is in lowest terms (i.e., m and n are relatively prime integers). It can be shown that this function is discontinuous at every point in $(0,1] \times [0,1]$ with both rational coordinates and continuous at all remaining points. Indeed, in any neighborhood of a rational point (x_0, y_0), $x_0, y_0 \in \mathbb{Q}$, $x_0 \neq 0$, there are points with irrational coordinates, and the partial limit along these points $\lim\limits_{\substack{(x,y)\to(x_0,y_0) \\ x\in\mathbb{I} \; or \; y\in\mathbb{I}}} f(x, y) = 0$ is different from $f(x_0, y_0) = f\left(\frac{m}{n}, y_0\right) = \frac{1}{n}$. It means that $f(x, y)$ is discontinuous at every point with rational coordinates in $(0,1] \times [0,1]$. For an arbitrary point (x_0, y_0) such that $x_0 \in \mathbb{I} \cup \{0\}$ or $y_0 \in \mathbb{I}$ (or both), we split all the paths of approaching (x_0, y_0) into two groups. First, if a path consists of points (x, y) such that $x \in \mathbb{I} \cup \{0\}$ or $y \in \mathbb{I}$ (or both), then we readily obtain $\lim\limits_{\substack{(x,y)\to(x_0,y_0)\\ x\in\mathbb{I}\, or\, y\in\mathbb{I}}} f(x,y) = 0 = f(x_0, y_0)$. In the second group we place all paths containing only rational points with $x \neq 0$. For any path of the second group, since rational abscissas $x = \frac{m}{n}$ approach x_0, it implies that the denominator n should approach infinity, because if it would not be so, that is, if the denominator n would be bounded, say $n \leq n_0$, then it will be only finitely many numbers of the form $x = \frac{m}{n}$, $n \leq n_0$ in a neighborhood of the point x_0. However it will contradict to the well-known fact that in any neighborhood of an arbitrary real point there are infinitely many rational points. Since $n \to \infty$ when $x = \frac{m}{n}$ approach x_0, it follows that $\lim\limits_{\substack{(x,y)\to(x_0,y_0)\\ x\in\mathbb{Q}\, and\, y\in\mathbb{Q}}} f(x,y) = \lim\limits_{n\to\infty} \frac{1}{n} = 0 = f(x_0, y_0)$. The two considered groups of paths involve all the points in S, and therefore $\lim\limits_{(x,y)\to(x_0,y_0)} f(x,y) = 0 = f(x_0, y_0)$, i.e., $f(x,y)$ is continuous at every point (x_0, y_0), $x_0 \in \mathbb{I} \cup \{0\}$ or $y_0 \in \mathbb{I}$ (or both).

Now let us show that $f(x,y)$ is integrable on S (the boundedness of the function is evident). Recall that a countable union of countable sets is again a countable set. For any fixed rational y the number of points with rational abscissas is countable. Since the number of rational ordinates y is countable too, the set of points with both rational coordinates is countable (it is true both for S and $\mathbb{R}^2$). Therefore it has the two-dimensional Lebesgue measure zero and, according to the Lebesgue criterion, $f(x,y)$ is integrable on S. Further, we can find the value of the integral by noting that all the lower Riemann sums are equal to 0:

$$\underline{S}(f) = \sum_{j=1}^{n}\sum_{i=1}^{m} \inf_{R_{ij}} f(x,y) \cdot \Delta x_i \Delta y_j = \sum_{i=1}^{n} 0 \cdot \Delta x_i \Delta y_j = 0$$

(since any elementary rectangle of an arbitrary partition contains the point with irrational coordinates). Since the Riemann integral exists, its value is equal to the lower Riemann integral, hence $\iint_S f(x,y)\, dA = 0$.

For any fixed $y \in \mathbb{I}$ we have $\int_0^1 f(x,y)\, dx = \int_0^1 0 dx = 0$, and for any fixed $y \in \mathbb{Q}$ the given function is the one-dimensional Riemann function, which is known to be integrable (see Remark 3 to Example 2 in section 5.3) and $\int_0^1 f(x,y)\, dx = 0$. So the first iterated integral exists and equals zero:

$$\int_0^1 dy \int_0^1 f(x,y)\, dx = \int_0^1 0 dy = 0.$$

However, the second iterated integral does not exist, because for any fixed $x \in \mathbb{Q}\backslash\{0\}$ the integral $\int_0^1 f(x,y)\, dy$ does not exist. In fact, for $x = \frac{m}{n}$ (with m, n being fixed natural numbers) we have

$$\varphi(y) = f\left(\frac{m}{n}, y\right) = \begin{cases} 1/n\,, & y \in \mathbb{Q}, \\ 0\,, & otherwise \end{cases} = \frac{1}{n} D(y)\,,$$

where $D(y)$ is Dirichlet's function, which is known to be non-integrable in the sense of the Riemann integral (see Example 1 in section 5.3). Therefore, $\varphi(y)$ is also non-integrable and the second iterated integral is not defined.

Remark. A straightforward generalization of this counterexample gives the function whose double integral exist, but both iterated integrals does not: $g(x,y) = \begin{cases} \frac{1}{n}+\frac{1}{q}, & y=\frac{p}{q}\in\mathbb{Q},\ x=\frac{m}{n}\in\mathbb{Q} \\ 0, & otherwise \end{cases}$, defined on the square $S=[0,1]\times[0,1]$, $m,n,p,q\in\mathbb{N}$ and the fractions $\frac{m}{n}$ and $\frac{p}{q}$ are in lowest terms. As above, it can be shown that this function is discontinuous at every point in $(0,1]\times(0,1]$ with both rational coordinates and continuous at all remaining points. Again, just like for $f(x,y)$, the number of discontinuity points is countable and consequently the double integral of $g(x,y)$ exists and $\iint_S g(x,y)\,dA = 0$. The difference occurs when we study the first iterated integral: for any fixed $y=\frac{p}{q}\in\mathbb{Q}$ the function $\psi(x) = g\left(x,\frac{p}{q}\right) = \begin{cases} \frac{1}{n}+\frac{1}{q}, & x=\frac{m}{n}\in\mathbb{Q} \\ 0, & otherwise \end{cases}$ is discontinuous at each point of the interval $[0,1]$, because $\frac{1}{q}$ is a fixed positive number. Therefore, the integral $\int_0^1 g(x,y)\,dx$ does not exist and the first iterated integral is not defined. Similarly, for any fixed $x=\frac{m}{n}\in\mathbb{Q}$, the integral $\int_0^1 g(x,y)\,dy$ does not exists, since the function $\varphi(y) = g\left(\frac{m}{n},y\right) = \begin{cases} \frac{1}{n}+\frac{1}{q}, & y=\frac{p}{q}\in\mathbb{Q} \\ 0, & otherwise \end{cases}$ is discontinuous at each point of the interval $[0,1]$. Consequently, the second iterated integral is also not defined.

Example 9. "If $f(x,y)$ is continuous on $\mathbb{R}^2$, then the double integral can be calculated exactly using any of the iterated integrals."

Solution.

The function $f(x,y) = y\left(e^{x^2y}+e^{-x^2y}\right)$ is continuous on $\mathbb{R}^2$ according to the arithmetic and composition rules. Since the conditions of Fubini's theorem are satisfied for any bounded set S in $\mathbb{R}^2$, the double integral and both iterated integrals exist and are equal. However, if we choose the square $S=[-a,a]\times[-b,b]$, then it is very simple to calculate the double integral using one of the iterated integrals

$$\iint_S f(x,y)\,dA = \int_{-a}^{a} dx \int_{-b}^{b} y\left(e^{x^2y}+e^{-x^2y}\right)dy = \int_{-a}^{a} 0dx = 0$$

(since $f(x,y)$ is odd with respect to y), but it cannot be done by using another iterated integral $\int_{-b}^{b} dy \int_{-a}^{a} y\left(e^{x^2y}+e^{-x^2y}\right)dx$, because the original function does not have any helpful symmetry with respect to x and does not have an antiderivative in the class of elementary functions. The last integral still can be calculated by expending the function in power series, but it will result (after integration) in a numerical series whose sum usually cannot be found exactly.

Example 10. "If a transformation $x = \varphi(u,v)$, $y = \psi(u,v)$ between sets $\tilde{D}$ and D is not a one-to-one mapping and/or its Jacobian vanishes at some points $(u,v) \in \tilde{D}$, then the corresponding change of variables in the double integral is not applicable."

Solution.

For simplicity we assume that the function $f(x,y) \equiv 1$, that is, the double integral $\iint_D f dxdy$ in the Cartesian coordinates (x,y) represents the area of the figure in D. Let us consider the transformation of the rectangle $\tilde{D} = [0,1] \times [0,2\pi]$ onto the disk $D = \bar{B}_1(0,0) = \{(x,y) : \ x^2+y^2 \le 1\}$ defined by the formulas $x = u\cos v$, $y = u\sin v$, representing the relation between the Cartesian and polar coordinates. The Jacobian of this transformation $\det\begin{pmatrix} x_u & x_v \\ y_u & y_v \end{pmatrix} = u$ vanishes at the points where $u=0$. It can also be noticed that the entire line segment $u = 0$, $v \in [0,2\pi]$ in $\tilde{D}$ corresponds to the only point $(x,y) = (0,0)$ in D, and that two different segments $u \in [0,1]$, $v = 0$ and $u \in [0,1]$, $v = 2\pi$ in $\tilde{D}$ are transformed onto the same segment $x \in [0,1]$, $y = 0$. Therefore, this correspondence is not one-to-one. However, the result on the change of variables is still valid and can be justified by the following reasoning. Consider a slight smaller rectangle $\tilde{D}_\delta = [\delta_u, 1] \times [\delta_v, 2\pi]$, δ_u, $\delta_v > 0$, which is transformed under the same mapping to the disk D with two peaces removed - small disk $B_\delta(0,0) = \{(x,y) : \ x^2+y^2 < \delta_u^2\}$ and small sector $C_\delta(0,0) = \{(x,y) : \ x^2+y^2 \le 1, \ 0 < y < \tan\delta_v \cdot x\}$, that is $D_\delta = D \backslash (B_\delta(0,0) \cup C_\delta(0,0))$. Under this transformation, both problems of vanishing of the Jacobian and non-uniqueness correspondence between the points are eliminated, and all the conditions of the theorem on change of variables are satisfied, that makes possible the use of the formula:

$$\iint_{D_\delta} 1dxdy = \iint_{\tilde{D}_\delta} \left| \frac{\partial(\varphi,\psi)}{\partial(u,v)} \right| dudv = \iint_{\tilde{D}_\delta} |u|\, dudv$$

$$= \int_{\delta_v}^{2\pi} dv \int_{\delta_u}^{1} udu = (2\pi - \delta_v)\frac{1-\delta_u^2}{2}.$$

Since this result is true for any small δ_u, $\delta_v > 0$, applying the limit, as δ_u, δ_v approaches 0, to the both sides of the formula, one obtains:

$$\iint_D 1dxdy = \iint_{\tilde{D}} \left| \frac{\partial(\varphi,\psi)}{\partial(u,v)} \right| dudv = \int_0^{2\pi} dv \int_0^1 udu = \pi.$$

Remark 1. This counterexample shows that the conditions of the theorem on the change of variables are sufficient, but not necessary. In practice, we frequently need to apply the theorem on change of variables in the cases when the conditions of non-zero Jacobian and one-to-one correspondence are violated at some points. The use of the popular polar coordinates (like in the counterexample above) is just one of typical examples of such situation. In general, if the two required conditions are not satisfied at a finite number of points or on a finite number of simple rectifiable curves, then the theorem on the change of variables is still applicable, with justification similar to the presented above.

Remark 2. The next example shows that when one of the conditions is violated on a large part of a set, then the change of variables may be unjustifiable.

Example 11. "If $f(x,y) \equiv 1$ and a transformation $x = \varphi(u,v)$, $y = \psi(u,v)$ between sets $\tilde{D}$ and D has continuous partial derivatives on $\tilde{D}$ and the Jacobian is positive on $\tilde{D}$, except possibly at the only point, then the result of the theorem on the change of variables is valid."

Solution.

Let us consider the transformation $x = u^2 - v^2$, $y = 2uv$ defined and continuously differentiable on $\tilde{D} = [-1,1] \times [-1,1]$ with the image D being the set enclosed by the parabolas $x = 1 - \frac{y^2}{4}$ and $x = \frac{y^2}{4} - 1$. The corresponding Jacobian $\det \begin{pmatrix} x_u & x_v \\ y_u & y_v \end{pmatrix} = 4\left(u^2 + v^2\right)$ is positive at every point, but at the origin. The last fact alone does not impede the use of the change of variables (see justification in the previous Example 10), and applying formally the change formula we obtain

$$\iint_D 1dxdy = \iint_{\tilde{D}} \left| \frac{\partial(\varphi,\psi)}{\partial(u,v)} \right| dudv = \int_{-1}^{1} dv \int_{-1}^{1} 4\left(u^2 + v^2\right) du = \frac{32}{3}.$$

However, the alternative evaluation of the area of figure D through the Riemann integral gives a different result:

$$\int_{-2}^{2} \left(1 - \frac{y^2}{4}\right) - \left(\frac{y^2}{4} - 1\right) dy = \frac{16}{3}.$$

The problem with the change of variables in this case is that the used transformation is not one-to-one mapping between $\tilde{D}$ and D. It happens that a right "half" of $\tilde{D}$, $\tilde{D}_r = [0,1] \times [-1,1]$ already covers all the image D, and the correspondence between $\tilde{D}_r$ and D is one-to-one with the only exception on the boundary points $u = 0$, $v \in [-1,1]$, which cover twice the segment $y = 0$, $x \in [-1,0]$ in D. The same is true for the left "half" $\tilde{D}_l = [-1,0] \times [-1,1]$. In this way, the transformation from $\tilde{D}$ to D covers the image twice, violating the condition of one-to-one correspondence on the entire domain. This is the cause of the wrong use of the change of variables above. The correct mode is to select one of the "halves" of $\tilde{D}$, say $\tilde{D}_r$, and then to apply (justifiably) the change of variables, obtaining the correct result:

$$\iint_D 1dxdy = \iint_{\tilde{D}_r} \left| \frac{\partial(\varphi,\psi)}{\partial(u,v)} \right| dudv = \int_{-1}^{1} dv \int_{0}^{1} 4\left(u^2 + v^2\right) du = \frac{16}{3}$$

(the use of $\tilde{D}_l$ gives the same result and this is why the wrong result obtained for $\tilde{D}$ is twice as much as the correct one).

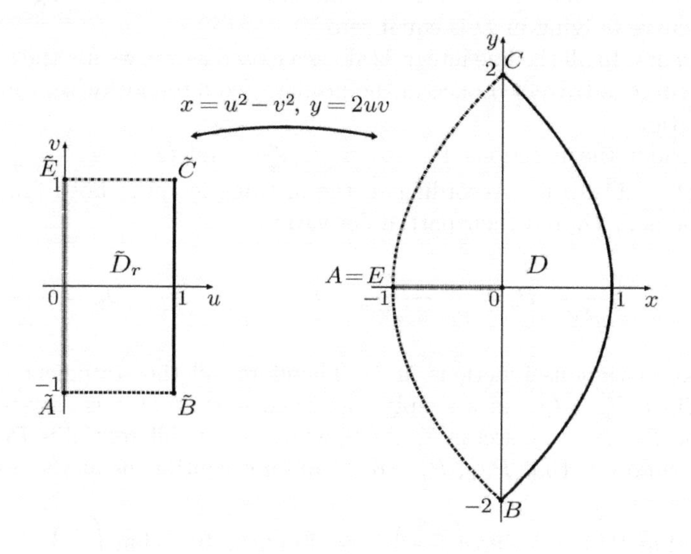

FIGURE 9.3.6: Example 11, right rectangle

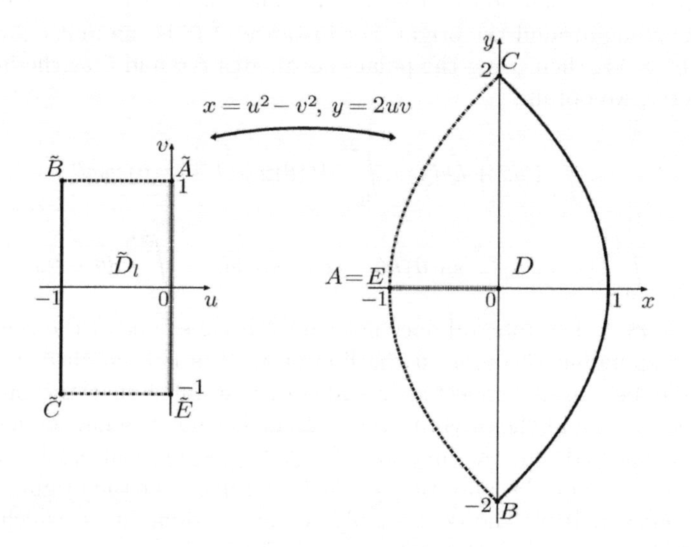

FIGURE 9.3.7: Example 11, left rectangle

Example 12. "If $P(x,y)$ and $Q(x,y)$ are continuously differentiable, and $P_y = Q_x$ in a domain D, then the line integral $\int_C Pdx+Qdy$ along any smooth closed curve C lying in D is equal zero."

Remark. In all the line integrals along a closed curve, we use the convention that a curve is traversed once in the positive (counterclockwise) direction.

Solution.

Consider the functions $P(x,y) = -\frac{y}{x^2+y^2}$ and $Q(x,y) = \frac{x}{x^2+y^2}$ on domain $D = \mathbb{R}^2 \backslash (0,0)$. According to the arithmetic rules, both functions are continuous in D, and their partial derivatives

$$P_x = \frac{2xy}{(x^2+y^2)^2},\ P_y = \frac{y^2-x^2}{(x^2+y^2)^2},\ Q_x = \frac{y^2-x^2}{(x^2+y^2)^2},\ Q_y = -\frac{2xy}{(x^2+y^2)^2}$$

are also continuous functions in D. Therefore, all the statement conditions hold. Since $P_y = Q_x$, in a simply connected part of D it is easy to find the function $F(x,y) = -\arctan\frac{x}{y}$, $y \neq 0$, whose exact differential is $Pdx + Qdy$. Notice, however, that P,Q, P_y and Q_x are not continuous at the origin:

$$\lim_{y\to 0} P(0,y) = \lim_{y\to 0}\left(-\frac{1}{y}\right) = \infty,\ \lim_{x\to 0} Q(x,0) = \lim_{x\to 0}\left(\frac{1}{x}\right) = \infty,$$

$$\lim_{y\to 0} P_y(0,y) = \lim_{y\to 0} Q_x(0,y) = \lim_{y\to 0}\frac{1}{y^2} = \infty,$$

and this causes violation of the integral formula in the statement for any curve C going around the origin. For instance, if C is the unit circumference $\{x^2+y^2=1\}$, then using the polar coordinates r,θ and θ as the integration parameter, we obtain:

$$\int_C Pdx + Qdy = \int_0^{2\pi} P(\theta)\, x_\theta d\theta + Q(\theta)\, y_\theta d\theta$$

$$= \int_0^{2\pi} (-\sin\theta)(-\sin\theta)\, d\theta + \cos\theta\cos\theta d\theta = \int_0^{2\pi} d\theta = 2\pi \neq 0.$$

Remark 1. The integral does not vanish because one of the conditions of the corresponding theorem on the line integral is not satisfied: the domain D should be simply-connected in order to guarantee that the line integral vanishes. If we add the origin to the domain D, then the new domain will be simply connected, but the functions P, Q, P_y and Q_x will not be continuous in the new domain, because they are not continuous at the origin.

Remark 2. If we choose a simply connected domain D, which does not contain the origin, then the statement will be true.

Remark 3. Notice, that the condition of continuity of P, Q, P_y and Q_x is only sufficient in the referred theorem, but not necessary as shown in the next example.

Example 13. "If at least one of the functions $P, Q, P_y = Q_x$ has a discontinuity at a point (x_0, y_0) of a simply connected domain D, then there exists a smooth closed curve C lying in D and not passing through the point (x_0, y_0) such that $\int_C Pdx + Qdy \neq 0$."

Solution.

Set $D = \mathbb{R}^2$ and consider the functions $P(x,y) = \frac{2xy}{(x^2+y^2)^2}$ and $Q(x,y) = \frac{y^2-x^2}{(x^2+y^2)^2}$, which are continuous in $\mathbb{R}^2 \backslash (0,0)$ together with their partial derivatives $P_y = Q_x = \frac{2x^3-6xy^2}{(x^2+y^2)^3}$ (according to the arithmetic rules). All these functions have a discontinuity at the origin:

$$\lim_{x \to 0} P(x,x) = \lim_{y \to 0} \frac{1}{2x^2} = \infty, \lim_{x \to 0} Q(x,0) = \lim_{x \to 0} \frac{-1}{x^2} = \infty,$$

$$\lim_{x \to 0} P_y(x,0) = \lim_{x \to 0} Q_x(x,0) = \lim_{x \to 0} \frac{2}{x^3} = \infty.$$

Now let us choose a smooth closed curve C that does not pass through the origin. If the point $(0,0)$ lies outside the domain enclosed by C, it is always possible to construct an auxiliary simply connected domain, which contains C and does not contain $(0,0)$. Therefore, for such domain the conditions of the theorem are satisfied and $\int_C Pdx + Qdy = 0$. If the point $(0,0)$ lies inside the domain $\tilde{D}$ enclosed by C, then the situation is more complex. Let us introduce a small circumference $C_a = \{x^2 + y^2 = a^2\}$ inside $\tilde{D}$ and note that $\int_C Pdx + Qdy = \int_{C_a} Pdx + Qdy$, since $Pdx + Qdy$ is an exact differential in $\mathbb{R}^2 \backslash (0,0)$ (i.e. $Pdx + Qdy = df$). Calculating the last integral in the polar coordinates r, θ, with θ being the integration parameter, we obtain:

$$\int_{C_a} Pdx + Qdy = \int_0^{2\pi} P(\theta)\, x_\theta d\theta + Q(\theta)\, y_\theta d\theta$$

$$= \int_0^{2\pi} \frac{2a^2 \cos\theta \sin\theta}{a^4} a \cdot (-\sin\theta) + \frac{a^2 \left(\sin^2\theta - \cos^2\theta\right)}{a^4} a \cdot \cos\theta d\theta$$

$$= -\frac{1}{a} \int_0^{2\pi} \cos\theta d\theta = 0.$$

Therefore, the line integral along any smooth closed curve C lying in D and not passing through the origin is zero.

Example 14. "If $f(x,y)$ and $g(x,y)$ have the sets of discontinuity points dense in D, then Green's formula $\iint_D (g_x - f_y)\, dA = \int_{\partial D} fdx + gdy$ does not hold."

Solution.

Consider $f(x,y) = 2xy + R(x)$ and $g(x,y) = x^2 - y^2 + R(y)$ on $D = [0,1] \times [0,1]$ (here $R(x)$ and $R(y)$ are the one-dimensional Riemann functions). In

Example 4, section 3.3 it was shown that $R(x)$ is continuous at every irrational point and discontinuous at every rational point. Therefore, the function $f(x,y)$ is discontinuous at each point of the segments $y \in [0,1], \forall x \in \mathbb{Q} \cap [0,1]$ and the function $g(x,y)$ is discontinuous at each point of the segments $x \in [0,1], \forall y \in \mathbb{Q} \cap [0,1]$. Let us denote the sets of discontinuity points of the functions $f(x,y)$ and $g(x,y)$ by E_f and E_g, respectively. Evidently, both sets are dense in D, that is, $\bar{E}_f = D$ and $\bar{E}_g = D$. Nevertheless, Green's formula

$$\iint_D (g_x - f_y)\, dA = \int_{\partial D} f dx + g dy$$

is satisfied. To show this, let us evaluate the left- and right-hand sides of the formula. First, $g_x = 2x = f_y$ for $\forall (x,y) \in D$, and consequently $\iint_D (g_x - f_y)\, dA = 0$. The line integral along ∂D can be divided in the four line integrals along the sides of the square D:

$$\int_{\partial D} f dx + g dy = \int_{AB} f dx + g dy + \int_{BC} f dx + g dy + \int_{CK} f dx + g dy + \int_{KA} f dx + g dy,$$

where

$$AB = \{(x,y) : y = 0,\ 0 \le x \le 1\},\ BC = \{(x,y) : x = 1,\ 0 \le y \le 1\},$$

$$KC = \{(x,y) : y = 1,\ 0 \le x \le 1\},\ AK = \{(x,y) : x = 0,\ 0 \le y \le 1\}.$$

Calculating each line integral separately, we obtain:

$$\int_{AB} f dx + g dy = \int_0^1 f(x,0) dx = \int_0^1 2x \cdot 0 + R(x) dx = 0,$$

$$\int_{CK} f dx + g dy = -\int_0^1 f(x,1) dx = -\int_0^1 2x \cdot 1 + R(x) dx = -\left. x^2 \right|_0^1 = -1,$$

$$\int_{BC} f dx + g dy = \int_0^1 g(1,y) dy = \int_0^1 1 - y^2 + R(y) dy = \left.\left(y - \frac{y^3}{3} \right)\right|_0^1 = \frac{2}{3},$$

$$\int_{KA} f dx + g dy = -\int_0^1 g(0,y) dy = -\int_0^1 -y^2 + R(y) dy = \left. \frac{y^3}{3} \right|_0^1 = \frac{1}{3}.$$

(Integrability of the Riemann function and the fact that its definite integral is zero was shown in Remark 3 to Example 2, section 5.3.) Joining all the calculations, we get

$$\int_{\partial D} f dx + g dy = 0 - 1 + \frac{2}{3} + \frac{1}{3} = 0,$$

that is, Green's formula holds in this case.

Exercises

1. Show by example that the following statement is false: "if $f(x,y)$ is bounded on a rectangle $R=[a,b]\times[c,d]$, then the double integral $\iint_R f dA$ can be defined by partitioning R in elementary rectangles R_{ij} of the same size and choosing the points (ξ_i,η_j) in the Riemann sum to be the centerpoints of R_{ij}". What if $f(x,y)$ is integrable on R?

2. Give an example that shows that the condition of independence of the limit on the choice of points (ξ_i,η_j) is important in the definition of the double integral. (Hint: use the two-dimensional version of Dirichlet's function $\hat{D}(x,y)=\begin{cases}1, & x,y\in\mathbb{Q}\\ 0, & otherwise\end{cases}$ and apply two different rules for the choice of the points (ξ_i,η_j): choose the points with rational coordinates and then with irrational coordinates in each R_{ij}.)

3. Show that the partition diameter Δ in the double integral definition cannot be substituted by any combination of the numbers of intervals n and m in x- and y-direction. In particular, construct a counterexample to the following false definition of the double integral of $f(x,y)$ over a rectangle R: "if there exists a finite limit of the Riemann sums, as n and m approach ∞, which does not depend on a choice of points (ξ_i,η_j), then this limit is called the double integral of $f(x,y)$ over R". (Hint: divide each side $[a,b]$ and $[c,d]$ of the rectangle $R=[a,b]\times[c,d]$ in two equal parts and then apply refinement only two one of the obtained halves of each interval.)

Show that the limits $\lim\limits_{\Delta\to 0} S(f)$ and $\lim\limits_{m,n\to\infty} S(f)$ are equal if one uses the uniform partitions in x- and y-directions.

4. Show by example that the converse to the comparative property 1) in section 9.1 is not true, that is, the following statement is false: "if $f(x,y)$ and $g(x,y)$ are integrable over D and $\iint_D f dA \le \iint_D g dA$, then $f(x,y)\le g(x,y)$ on D".

5. Show that $\hat{D}(x,y)=\begin{cases}1, & x,y\in\mathbb{Q}\\ 0, & otherwise\end{cases}$ is another counterexample to the statement in Example 1, section 9.2.

6. Show that $\tilde{D}(x,y)=\begin{cases}1, & x,y\in\mathbb{Q}\\ -1, & otherwise\end{cases}$ is another counterexample to the statement in Example 3, section 9.2.

7. Provide a counterexample to the following statement: "if $f(x,y)$ is discontinuous at a countable set of points in a set D, then it is not integrable over D". Compare this result with Example 2, section 9.3. (Hint: consider $D=[0,1]\times[0,1]$, $E=\cup_{n=1}^{+\infty}e_n$, $e_n=(1/n,1/n), \forall n\in\mathbb{N}$, $f(x,y)=\begin{cases}1, & (x,y)\in E\\ 0, & (x,y)\in D\backslash E\end{cases}$)

8. Show that another counterexample to the statement in Example 4, sec-

tion 9.3. can be composed as follows: the region $D = [-1,1] \times [-2,2]$ and the dividing curve is $h(x) = \begin{cases} \sin\frac{\pi}{x}, \ x \in [-1,0) \cup (0,1] \\ 0, \ x = 0 \end{cases}$. Use the reasoning of Example 4 to show that $h(x)$ is a zero area curve.

9. Show that $f(x,y) = \begin{cases} \frac{1}{y^2}, \ 0 < x < y < 1 \\ -\frac{1}{x^2}, \ 0 < y < x < 1 \\ 0, \ otherwise \end{cases}$ considered on $[0,1] \times [0,1]$ is another counterexample for Example 5 in section 9.3.

10. Show that $f(x,y) = D(x) \cdot R(y)$, where $D(x)$ and $R(y)$ are one-dimensional Dirichlet's and Riemann's functions, is another counterexample for Example 8 in section 9.3.

11. Verify that the transformation $x = \frac{u^2-v^2}{(u^2+v^2)^2}$, $y = -\frac{2uv}{(u^2+v^2)^2}$ has continuous partial derivatives and positive Jacobian on the set $\tilde{D}$ located between the circles $u^2 + v^2 = \frac{1}{2}$ and $u^2 + v^2 = 1$ and bounded by the segments $\frac{1}{2} \le u \le 1, v = 0$ and $-1 \le v \le -\frac{1}{2}, u = 0$ (this is three-quarters of the ring without the part located in $u > 0, v < 0$), but the theorem on the change of variables does not hold even for $f(x,y) \equiv 1$. Find out what is the problem with this transformation and formulate the result as a counterexample.

12. Provide a counterexample to the statement: "if a transformation $x = \varphi(u,v)$, $y = \psi(u,v)$ between sets $\tilde{D}$ and D has zero Jacobian on a subset $\tilde{D}_1 \subset \tilde{D}$ of non-zero area, then the corresponding change of variables in the double integral is not applicable". Compare this result with Example 10. (Hint: consider $f(x,y) \equiv 1$ and the transformation $x = \varphi(u,v) = \begin{cases} 0, \ u^2+v^2 < 1 \\ (u^2+v^2-1)^2, \ u^2+v^2 \ge 1 \end{cases}$, $y = \psi(u,v) = u \cdot v$ on the set $\tilde{D}$ enclosed by the rays $u = v, u = -v$ for $u > 0$ and by the arc of the circumference $u^2 + v^2 = 2$.)

13. Show that $P(x,y) = -\frac{y}{(5x+2y)^2+(3x+2y)^2}$ and $Q(x,y) = \frac{x}{(5x+2y)^2+(3x+2y)^2}$ considered on $D = \mathbb{R}^2$ give another counterexample for Example 12, section 9.3.

14. Prove that Green's formula does not hold for the functions of Example 12, section 9.3, $P(x,y) = -\frac{y}{x^2+y^2}$ and $Q(x,y) = \frac{x}{x^2+y^2}$ considered on the disk $D = x^2 + y^2 \le 1$ or any domain containing the origin. Prove also that Green's formula is satisfied for the functions of Example 13, section 9.3, $P(x,y) = \frac{2xy}{(x^2+y^2)^2}$ and $Q(x,y) = \frac{y^2-x^2}{(x^2+y^2)^2}$ on the same unit disk or any domain containing the origin. Notice that in both cases the conditions of Green's formula are satisfied at all points except only for the origin. Compare results and formulate them as counterexamples.

15. Show that $P(x,y) = \frac{x}{x^2+y^2}$ and $Q(x,y) = \frac{y}{x^2+y^2}$ considered on $D = \mathbb{R}^2$ give another counterexample for Example 13, section 9.3.

16. Provide a counterexample to the following statement: "if $f(x,y)$ has the set of discontinuity points dense in D, then Green's formula $\iint_D f_y dA = -\int_{\partial D} f dx$ does not hold". Compare this result with Example 14 in section 9.3. (Hint: consider $f(x,y) = x^2 + y^2 + R(x)$, where $R(x)$ is the Riemann

function, on the set D bounded by the curves $y = x^2 + 1$, $y = -x^2$, $x = 0$ and $x = 1$.)

17. Provide a counterexample to the following statement: "if $g(x, y)$ has the set of discontinuity points dense in D, then Green's formula $\iint_D g_x dA = \int_{\partial D} g dy$ does not hold". Compare this result with Example 14 in section 9.3 and with the previous exercise. (Hint: consider $g(x, y) = 3xy^2 + R(y)$, where $R(y)$ is the Riemann function, on the set D bounded by the curves $x = y^2$, $x = y^2 + 2$, $y = 0$ and $y = 1$.)

Bibliography

[1] S. Abbott. *Understanding Analysis.* Springer, New York, NY, 2002.

[2] P. Biler and A. Witkowski. *Problems in Mathematical Analysis.* Marcel Dekker, New York, NY, 1990.

[3] B. Demidovich. *Problems in Mathematical Analysis.* Mir Pub., Moscow, 1989.

[4] G.M. Fichtengolz. *Differential- und Integralrechnung, Vol.1-3.* V.E.B. Deutscher Verlag Wiss., Berlin, 1968.

[5] B.R. Gelbaum and J.M.H. Olmsted. *Counterexamples in Analysis.* Dover Pub., Mineola, NY, 2003.

[6] V.A. Ilyin and E.G. Poznyak. *Fundamentals of Mathematical Analysis, Vol.1,2.* Mir Pub., Moscow, 1982.

[7] W.J. Kaczor and M.T. Nowak. *Problems in Mathematical Analysis, Vol.1-3.* AMS, Providence, RI, 2001.

[8] B.M. Makarov, M.G. Goluzina, A.A. Lodkin, and A.N. Podkorytov. *Selected Problems in Real Analysis.* AMS, Providence, RI, 1992.

[9] G. Polya and G. Szego. *Problems and Theorems in Analysis, Vol.1-2.* Springer, Berlin, 1998.

[10] T.L.T. Radulesku, V.D. Radulesku, and T. Andreescu. *Problems in Real Analysis: Advanced Calculus on the Real Axis.* Springer, New York, NY, 2009.

[11] W. Rudin. *Principles of Mathematical Analysis.* McGraw-Hill, New York, NY, 1976.

[12] M. Spivak. *Calculus.* Publish or Perish, Houston, TX, 2008.

[13] J. Stewart. *Calculus: Early Transcendentals.* Brooks/Cole, Belmont, CA, 2012.

[14] V.A. Zorich. *Mathematical Analysis I, II.* Springer, Berlin, 2004.

Symbol Description

$=$	equality, definition, notation
$\to, \Rightarrow$	logical implication, tendency, correspondence
$\{\ldots\}$	collection of elements
$\in$ ($\notin$)	belongs (does not belong) to
$\subset$ ($\not\subset$)	is contained (is not contained) in
$\forall$	any, every, each, for all
$\exists$ ($\nexists$)	existence (non-existence)
$\mathbb{N}$	set of natural numbers
$\mathbb{Z}$	set of integers
$\mathbb{Q}$	set of rational numbers
$\mathbb{R}$	set of real numbers, one-dimensional real space
$\mathbb{R}^n$	n-dimensional real space
$\varnothing$	empty set
$\bar{D}$	closure of set D
∂D	boundary of set D
$\cup$	union of sets
$\cap$	intersection of sets
$\setminus$	difference of sets
$\times$	Cartesian product of sets
$d(x, y)$, $\lvert x - y\rvert$	distance between two points
$S_{c,r}$	sphere with radius r and centerpoint c
$B_{c,r}$	open ball with radius r and centerpoint c
$\bar{B}_{c,r}$	closed ball with radius r and centerpoint c
(a, b)	open interval
$[a, b]$	closed interval
$(a, b]$, $[a, b)$	half-open (half-closed) intervals
sup	supremum, least upper bound
inf	infimum, greatest lower bound
max	maximum
min	minimum
$f(x)$	function
$f\colon X \to Y$	function, correspondence, mapping from X to Y
$f^{-1}(x)$	inverse function
$D(x)$	Dirichlet's function
$R(x)$	Riemann function
$\lim\limits_{x\to a} f$	limit of $f(x)$ as x approaches a
$f \underset{x\to a}{\to}$	limit of f as x approaches a
$\lim\limits_{x\to a_+} f$	right-hand limit of f as x approaches a
$f \underset{x\to a_+}{\to}$	right-hand limit of f as x approaches a
$\lim\limits_{x\to a_-} f$	left-hand limit of f as x approaches a
$f \underset{x\to a_-}{\to}$	left-hand limit of f as x approaches a
Δx	increment of variable x
Δf	increment of function $f(x)$
f', $\frac{df}{dx}$	derivative of f
f'_+	right-hand derivative of f
f'_-	left-hand derivative of f
f'', $\frac{d^2f}{dx^2}$	second derivative of f
f^n, $\frac{d^nf}{dx^n}$	n-th derivative of f
df	differential of f
d^2f	second order differential of f
d^mf	m-th order differential of f

$\int f dx$ integral, indefinite integral

$\int_a^b f dx$ definite (Riemann) integral, improper integral of the second kind

$S(f; P, c)$,$S(f)$ Riemann sum

Δ partition diameter

$\int_a^{+\infty} f dx$,$\int_{-\infty}^b f dx$ improper integral of the first kind

$\int_{-\infty}^{+\infty} f dx$ improper integral of the first kind

$\{a_n\}_{n=1}^{\infty}$,a_n numerical sequence

$\lim\limits_{n\to\infty} a_n$ limit of sequence a_n

$\sum a_n$ sum of sequence a_n, numerical series

$\sum_{n=1}^{+\infty} a_n$ numerical series

$\sum_{n=0}^{\infty} c_n(x-a)^n$ power series

$\lim\limits_{(x,y)\to(a,b)} f$ limit of function of two variables

$f \underset{(x,y)\to(a,b)}{\to}$ limit of function of two variables

$\lim\limits_{y\to b}\lim\limits_{x\to a} f$ iterated limit of function of two variables

$\lim\limits_{x\to a}\lim\limits_{y\to b} f$ iterated limit of function of two variables

f_x,$\partial_x f$ partial derivative in x

$\frac{\partial f}{\partial x}$ partial derivative in x

f_{xx},$\partial_{xx} f$ second-order partial derivative in x, x

$\frac{\partial^2 f}{\partial x^2}$ second-order partial derivative in x, x

f_{x^m},$f_{x\ldots x}$ m-th order partial derivative in $x, \ldots, x$

$\frac{\partial^m f}{\partial x^m}$ m-th order partial derivative in $x, \ldots, x$

∇f,grad$f$ gradient

$D_v f$ directional derivative

$H(f)$ Hessian matrix

$\iint_D f dA$ double integral over D

$\iint_D f dxdy$ double integral over D

$\int_a^b dx \int_c^d f dy$ iterated integral

$\frac{\partial(\varphi,\psi)}{\partial(u,v)}$ Jacobian

$\det\begin{pmatrix} \varphi_u & \varphi_v \\ \psi_u & \psi_v \end{pmatrix}$ Jacobian

$\int_C f dl$ line integral in arc length

$\int_C f dx$ line integral in x

$\int_C f dy$ line integral in y

Index